Ecological Studies, Vol. 119

Analysis and Synthesis

Edited by

M.M. Caldwell, Logan, USA
G. Heldmaier, Marburg, Germany
O.L. Lange, Würzburg, Germany
H.A. Mooney, Stanford, USA
E.-D. Schulze, Bayreuth, Germany
U. Sommer, Kiel, Germany

Ecological Studies

Volumes published since 1989 are listed at the end of this book.

Springer

New York
Berlin
Heidelberg
Barcelona
Budapest
Hong Kong
London
Milan
Paris
Santa Clara
Singapore
Tokyo

Alexander M. Milner   Mark W. Oswood

Editors

Freshwaters of Alaska
Ecological Syntheses

With 152 Illustrations

 Springer

Alexander M. Milner
Institute of Arctic Biology
University of Alaska Fairbanks
Fairbanks, AK 99775-7220, USA
and
Environment and Natural Resources
  Institute
University of Alaska Fairbanks
707 A Street
Anchorage, AK 99501, USA

Mark W. Oswood
Department of Biology and Wildlife
  and Institute of Arctic Biology
University of Alaska Fairbanks
Fairbanks, AK 99775-6100, USA

Cover illustration by Maureen Milner.

QH
105
A4
F74
1997

Library of Congress Cataloging in Publication Data
Freshwaters of Alaska : ecological syntheses / Alexander M. Milner and
    Mark W. Oswood, editors.
        p.   cm.—(Ecological studies ; v.)
    Includes bibliographical references and index.
    ISBN 0-387-94379-X (hc ; alk. paper)
    1. Freshwater ecology—Alaska.   I. Milner, A. M. (Alexander M.),
    1952–   .   II. Oswood, Mark W.   III. Series.
    QH105.A4F74   1995
    574.5′2632′09798—DC20   95-37682

Printed on acid-free paper.

Production coordinated by Chernow Editorial Services, Inc., and managed by Terry Kornak; manu-
facturing supervised by Jacqui Ashri.
Typeset by Best-set Typesetter Ltd., Hong Kong.
Printed and bound by Maple-Vail, York, PA.
Printed in the United States of America.

9 8 7 6 5 4 3 2 1

ISBN 0-387-94379-X Springer-Verlag New York Berlin Heidelberg   SPIN 10098500

*To Bert Hawkes and Roland Bailey, who
sparked my original interest in freshwaters*
AMM

*To my father, who built a laboratory in the
basement for a small boy*
MWO

Preface

The study of Alaska's freshwaters is in its childhood, mostly beyond the tentative explorations of infancy but well short of a perspective afforded by old age. Therefore, an attempt to synoptically review the limnology of Alaska would be premature; instead, we focus on major research themes and long-term studies of Alaskan freshwaters. The principal aim of this book is to provide reviews and syntheses of much of the recent research on freshwater ecosystems in Alaska.

Cold-dominance has shaped both the natural ecosystems and human ecology of Alaska in ways that have been selected for particular kinds of limnological studies and insights. Severe climate limits resident population sizes of both humans and many other animals. Alaska's sparse human population and large area make for a poorly developed road system. This has had at least two limnological consequences. First, the synoptic studies of freshwater organisms (e.g., such as those carried out for many decades by state natural history surveys) have not been done in Alaska because most areas are accessible only by aircraft. Second, the logistics of limnological research in Alaska have led to development of remote research stations and an eclectic range of transportation (cruise ships to dog sleds; see Chapter 2).

Alaska's economy is largely based on the export of raw materials to regions of higher population densities. Low population sizes in Alaska mean that human energy needs are low. Consequently, all of Alaska's major rivers are free flowing, in sharp contrast to the pervasive hydroelectric development of rivers in the continental U.S. and most other developed regions. The extreme seasonality of Alaska (short warm summers–long cold winters) has combined with spectacular and largely pristine scenery to produce a huge summer migration of tourists, an important and growing element of the Alaskan economy. Thus a major role of freshwater ecologists in managing Alaska's ecosystems has been and will likely continue to be understanding and mitigating the consequences of extractive resource use, e.g., timber harvest (Chapter 9) and mining (Chapter 10), as well as hydrocarbon (oil, gas, coal) extraction. In surprising parallel with tropical rainforests, future management of Alaska's landscapes will likely balance resource uses with the potential for large-scale changes in landscape function and appearance (e.g., forestry and mining) against resource uses requiring functionally and visually intact watersheds (e.g., fisheries and tourism). Such management will require a predictive understanding of Alaskan freshwater ecosystems.

The diversity of high latitude freshwater ecosystems are constrained by a constellation of factors manifested in large seasonal variations in physical factors and biological processes and in low annual system productivity (Chapter 13). Just as studies of organisms in harsh environments have provided insights into ecophysiological adaptations (e.g., water conservation in desert organisms), studies of ecosystem patterns and processes in high latitude freshwater ecosystems can provide exceptionally clear examples of thermal, nutrient, and trophic limitations. The nutrient limitations of Alaskan systems have produced opportunities to examine the consequences of nutrient limitations both experimentally (e.g., via nutrient additions to an arctic stream, as described in Chapter 4) and naturally (marine-derived nutrients from salmon carcasses (Chapters 7 and 8). The relatively low taxonomic diversity of high latitude freshwaters makes studies of food webs and trophic dynamics (e.g., in arctic Toolik Lake, in Chapter 3, and subarctic Smith Lake, in Chapter 5) easier than in more species-rich systems at lower latitudes.

The interaction of time and space is exceptionally strong at high latitudes. The tremendous seasonal amplitude in solar insolation and temperature causes some freshwater habitats to become uninhabitable to larger organisms in winter, forcing migrations of fish (Chapter 11) and waterfowl (Chapter 6). Finally, retreating glaciers create new streams (Chapter 12), a rare example of primary succession in running waters. Physical and biological development in these new streams occurs in both time and space (downstream of the retreating glacier) so that successionally (but not cosmologically) time equals space.

Human history is a story of dispersal and colonization and of frontiers retreating from people. Alaska is one of the last frontiers, and, like all last things, is vulnerable. We hope that this book contributes in a small way to an appreciation, both scientific and aesthetic, of Alaska's freshwaters. Alaska provides one of earth's last great opportunities to demonstrate wise and informed stewardship of the land.

Alexander M. Milner
Mark W. Oswood

Acknowledgments

During most of the gestation period of this book, our home institutions provided a safe haven and support for the respiratory activities of publishing; telephone and mailing expenses; and, most of all, access to a photocopy machine. Ms. Tina Picolo and Ms. Carol Button helped develop a bibliographic database of publications in Alaskan limnology. Thanks to Ian Phillips for assembling the basic elements of the index. This book was a long time aborning and the editorial staff at Springer-Verlag showed the patience of a sequoia in waiting for the all of the parts of this book to gel. We are grateful to the chapter reviewers for taking time out from their busy schedules to add their constructive comments: M.G. Butler, J. Cedarholm, R.A. Cunjack, C.E. Cushing, J. Griffth, K. Karle, J. LaPerriere, D. Lynch, D. Mann, T. McMahon, W. Meehan, R. Post, F.A. Reid, J. Stockner, J.V. Ward, T. Waters, R.G. Wetzel, and M.J. Winterbourn. AMM would like to thank Dale Brown, author of the chapter "A Region Reborn" in a volume of the Time Life Series World's Wild Places entitled *Alaska*, which inspired a visit to Glacier Bay National Park in 1976. As Henry Thoreau said, how many people can date a change in their life from reading a book? Maybe this book will also inspire a few potential limnologists to head northwards and dip their collecting nets in Alaska's freshwaters.

Contents

Contributors

Vera Alexander   Institute of Marine Science, School of Fisheries and Ocean Sciences, University of Alaska Fairbanks, Fairbanks, AK 99775-7220, USA

Michele Bahr   The Ecosystems Center, Marine Biological Laboratory, Woods Hole, MA 02543, USA

William B. Bowden   Department of Natural Resources, University of New Hampshire, Durham, NC 03824, USA

Linda A. Deegan   The Ecosystems Center, Marine Biological Laboratory, Woods Hole, MA 02543, USA

Jim A. Edmundson   Alaska Department of Fish and Game, Soldotna, AK 99669, USA

John J. Goering   Institute of Marine Science, School of Fisheries and Ocean Sciences, University of Alaska Fairbanks, Fairbanks, AK 99775, USA

Binhe Gu   Institute of Marine Science, School of Fisheries and Ocean Sciences, University of Alaska Fairbanks, Fairbanks, AK 99775-7220, USA

Anne E. Hershey   Department of Biology, University of Minne-
                           sota, Duluth, MN 55812, USA

John E. Hobbie   The Ecosystems Center, Marine Biological
                           Laboratory, Woods Hole, MA 02543, USA

John G. Irons III   Institute of Northern Forestry, USDA Forest
                           Service, Fairbanks, AK 99775, USA

George W. Kipphut   Department of Geosciences, Murray State
                           University, Murray, KY 42071, USA

Thomas C. Kline Jr.   Prince William Sound Science Center,
                           Cordova, AK 99574, USA

George W. Kling   Department of Biology, University of
                           Michigan, Ann Arbor, MI 48109, USA

Hedy Kling   Freshwater Institute, Winnipeg, Manitoba
                           R3T 2N6, Canada

Jeffrey P. Koenings   Alaska Department of Fish and Game, Juneau,
                           AK 99802, USA

Gary B. Kyle   Alaska Department of Fish and Game,
                           Soldotna, AK 99669, USA

Jacqueline D. LaPerriere   Alaska Cooperative Fish and Wildlife Re-
                           search Unit, National Biological Service, Uni-
                           versity of Alaska Fairbanks, Fairbanks, AK
                           99775-7020, USA

Maurice A. Lock   School of Biological Sciences, University of
                           Wales, Bangor, Gwynedd LL57 2UW, United
                           Kingdom

Michael McDonald   Department of Chemical Engineering, Uni-
                           versity of Minnesota, Duluth, MN 55812,
                           USA

Richard W. Merritt   Department of Entomology, Michigan State
                           University, East Lansing, MI 48824, USA

Michael C. Miller   Department of Biological Sciences, University
                           of Cincinnati, Cincinnati, OH 45221, USA

Alexander M. Milner — Institute of Arctic Biology, University of Alaska Fairbanks, Fairbanks, AK 99775-7220, USA; and School of Geography, University of Birmingham, Edgbaston, Birmingham B15 2TT, United Kingdom

Michael L. Murphy — National Marine Fisheries Service, Alaska Fisheries Science Center, National Marine Fisheries Service, NOAA, Auke Bay Laboratory, Juneau, AK 99801-8626, USA

W. John O'Brien — Department of Systematics and Ecology, Division of Biological Sciences, University of Kansas, Lawrence, KS 66045-2106, USA

Mark W. Oswood — Department of Biology and Wildlife and Institute of Arctic Biology, University of Alaska Fairbanks, Fairbanks, AK 99775-6100, USA

Bruce J. Peterson — The Ecosystems Center, Marine Biological Laboratory, Woods Hole, MA 02543, USA

Robert J. Piorkowski — Institute of Arctic Biology, University of Alaska Fairbanks, Fairbanks, AK 99775-7220, USA

James B. Reynolds — Alaska Cooperative Fish and Wildlife Research Unit, National Biological Service, University of Alaska Fairbanks, Fairbanks, AK 99775-7020, USA

Parke Rublee — Department of Biology, University of North Carolina, Greensboro, NC 27412, USA

Jeffrey A. Schuldt — Department of Forest Resources, University of Minnesota, St. Paul, MN 55108, USA

James S. Sedinger — Institute of Arctic Biology, University of Alaska Fairbanks, Fairbanks, AK 99775-7220, USA

J. Robie Vestal — Department of Biological Sciences, University of Cincinnati, Cincinnati, OH 45221, USA

1. The Alaskan Landscape: An Introduction for Limnologists

Alexander M. Milner, John G. Irons III, and Mark W. Oswood

Alaska is a land of vastness and contrast covering approximately 152 million ha. It contains many mountain ranges with peaks to 6106 m and extensive areas of lowlands. Climate is extremely variable across the state from the coastal rain forests of southeast Alaska through the boreal forest taiga of the interior to the treeless North Slope above the Arctic Circle. Mean monthly temperatures range from $-23°$ to $+17°C$ and annual precipitation ranges from 500 cm in southeast Alaska to 5 cm in the most northerly areas. Rivers, lakes, and wetlands dominate the landscape. There are more than 3 million lakes larger than 4.5 ha and 94 lakes larger than 2500 ha. The largest lake is Lake Iliamna covering 2.63×10^5 ha or 2622 km^2 (Kline, 1991). Ten rivers are longer than 500 km (see Fig. 1.1 for map of major rivers, lakes, and wetlands).

Alaska is the westernmost extension of the North American continent and stretches from about 130°W to 172°E and from 52° to 72°N (about 2200 km). Alaska's coastline is longer than the coastline of the conterminous United States. Most of the continental shelf under U.S. jurisdiction lies off the coast of Alaska.

In area, Alaska is the largest state in the United States but, with a population of only about 600,000, possesses the lowest population density of 0.39 persons/km^2. The U.S. average is 27.1 persons/km^2. Approximately half of the population of Alaska lives in the greater Anchorage area in the southcentral region, another 80,000 in the Fairbanks area, and 28,000 in

1

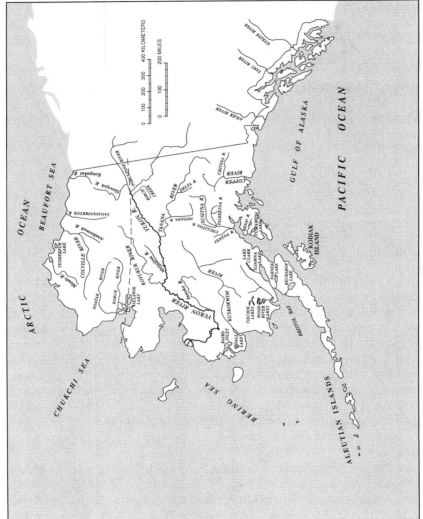

Figure 1.1. Major rivers and lakes of Alaska.

Juneau, the state capital, in southeast Alaska. Thus the population density in the rural part of the state is extremely low. For general reference, a map of Alaskan place names is provided in Fig. 1.2.

Alaska is composed of more than 50 terranes (rocks of different origins coalesced by motion of the earth's plates), many of which are separated by active faults. It was subduction of the Yakutat terrane along the eastern end of the Aleutian Trench fault that caused the 1964 Good Friday earthquake in southcentral Alaska. This earthquake was the largest recorded in North America (9.2 magnitude) and areas of southcentral Alaska were uplifted as much as 15 m, other areas subsided up to 2.25 m, and lateral displacements of 20 m were reported (Plafker, 1969). A tsunami wave generated by an underwater landslide destroyed significant areas of the coastal towns of Valdez, Seward, and Kodiak.

Alaska has been inhabited for more than 10,000 years by the native peoples of southeastern Alaska, the Tlingits, Haidas, and Tsimshians; by the Aleuts and Eskimos of the Aleutian Islands and western and arctic Alaska; and the Athapaskan groups of the Yukon and Tanana Basins and elsewhere. The first documented Europeans to arrive were the Russian explorers led by Vitus Bering's expedition of 1741. Traders soon followed. For the next 125 years, the Russians and their associates took possession, explored, and exploited the Aleutian Islands, the Pribilof Islands, and much of coastal southeast, southcentral, and western Alaska. Russian influence spread inland to upstream areas on the Yukon and Tanana Rivers, into the Copper River Basin, and to the southern Seward Peninsula. Alaska was part of the Russian Empire until 1867 when it was sold to the United States for $7.2 million. The Territory of Alaska was made the 49th state in 1959 by the U.S. Congress.

For thousands of years, rivers have played a major role as transportation corridors for native Alaskans, a tradition continued after European settlement. Today many residents, both native and nonnative, derive much of their livelihood from the use of aquatic resources, particularly the harvest of fish both commercially (Fig. 1.3) and for subsistence (Figs. 1.4, 1.5).

Topography

As a result of its tectonic history, Alaska has several major and numerous minor mountain ranges. As distinguished by Wahrhaftig (1965), Alaska contains four of the broadly defined physiographic divisions of North America: the Interior Plains; Rocky Mountain System; Intermontane Plateaus; and the Pacific Mountain System (Fig. 1.6). The Interior Plains are represented by the Arctic Coastal Plain, the Rocky Mountain System by the Arctic Foothills, the Brooks Range, and the southern foothills of the Brooks Range. The Intermontane Plateaus division is represented by low-

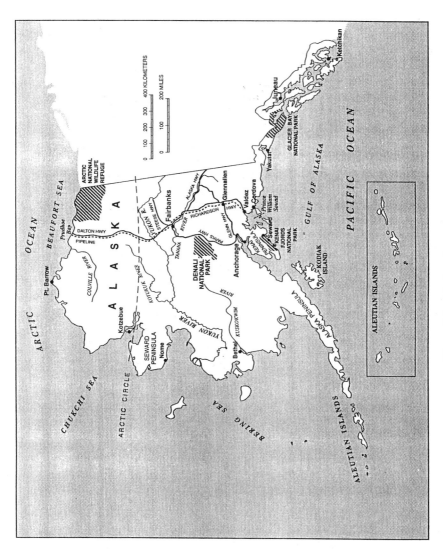

Figure 1.2. Alaska place names.

Figure 1.3. Fishing boats in Kodiak supported by Salmon runs in Alaskan rivers (photo Chris Arend, AEIDC, University of Alaska).

Figure 1.4. Subsistence fishing on the Copper River using a fish wheel (photo O. Eugene Coté, AEIDC, University of Alaska).

Figure 1.5. Dipnetting for salmon on the Copper River (photo M.W. Williams, AEIDC, University of Alaska).

lands along the Yukon including the Yukon Flats, the Yukon–Kuskokwim Lowland, the Kuskokwim, Tanana, and Northway Lowlands, and by highlands including the Yukon–Tanana Upland, the Kuskokwim Mountains, the Porcupine Plateau, and the Seward Pensinsula. The Pacific Mountain

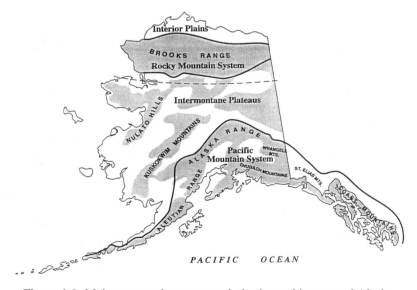

Figure 1.6. Major mountain ranges and physiographic areas of Alaska.

System, which includes the Alaska Range, the Chugach Mountains, the Wrangell Mountains and Saint–Elias Mountains, and the Coast Mountains, extends south to Baja California in Mexico.

Southeast Alaska includes the Coast Mountains and the land westward and encompasses an area of fjords and forests with many glaciers flowing down to tidewater. This area of Alaska, also known as the Alaska Panhandle, is approximately 600 km long by 200 km wide, much of which is made up of a large group of islands called the Alexander Archipelago. Many of the rivers in southeast Alaska are short, low order, and frequently steep. A number of the larger rivers rise in Canada and flow through the Coast Range, for example the Taku and the Stikine Rivers.

The Coast Mountains extend northward into southcentral Alaska to merge with the Saint Elias, Wrangell, and Chugach Mountain ranges (Fig. 1.6). The Wrangell Mountains are made up of shield and composite volcanoes, the majority of which have not been recently active except for Mt. Wrangell (Wahrhaftig, 1965). The highest peak is Mt. Blackburn at 5300 m. Although the Saint Elias Mountains are principally in Canada, they extend south into the panhandle and west into southcentral Alaska. Mt. Saint Elias itself is 5488 m and lies on the border between Alaska and Canada. Two large piedmont glaciers, the Bering (5625 km²) and the Malaspina (5120 km²), flow from these mountains and spread out onto forelands of the southcentral Alaskan coastline near Cordova and Yakutat, respectively. Unlike southeast Alaska, southcentral Alaska has substantial coastal lowlands where large rivers empty into the ocean across deltas. The Copper River and the Susitna River are two examples (Fig. 1.1). Similar to southeast Alaska, areas of the southcentral region contain a significant number of fjords including Prince William Sound and Kenai Fjords on the southeastern coast of the Kenai Peninsula.

The Alaska Range continues the Pacific Mountain complex northward, and then arches toward the southwest. This range contains the highest mountain in North America, Mt. McKinley at 6106 m, in Denali National Park (Fig. 1.2). However, fewer than 20 peaks exceed 3000 m and most average between 2000 and 3000 m. Numerous glaciers in this range feed some of the larger rivers in the state including the Nenana and the Delta Rivers flowing northward and the Chulitna and Susitna rivers flowing southward (Fig. 1.1).

The Alaska Range merges with the Aleutian Range that extends southwest along the Alaska Peninsula and along the Aleutian Island chain to the Commander Islands of Siberia (Fig. 1.6). The Aleutian Range (2560 km in length) contains 47 volcanoes active since the mid-1700s, and is part of what is frequently termed "the Pacific rim of fire." Only three of these volcanic peaks exceed 3000 m: Mt. Redoubt, Mt. Iliamna, and Mt. Spur. Mt. Redoubt was extremely active in the late 1980s when several notable eruptions caused debris flows within the Drift River basin and Mt. Spur became active for a period in 1992. Mt. Augustine, Mt. Veniaminof, and others have been active in the past decade. The Aleutian Chain is the longest archipelago of small volcanic islands in the world and consists of the crests of an arc of submarine volcanoes. Twenty volcanoes of the Aleutian Range have collapsed craters or calderas, some of which have become filled with fresh-water, such as the Aniakchak caldera.

Interior Alaska (Fig. 1.6), between the Brooks and Alaska Ranges, is a continuation of the Intermontane Plateaus of the North American Cordillera and contains a series of northeast to southwest trending uplands separated by broad lowland alluvial valleys. These lowland areas frequently contain numerous lakes and muskeg-filled basins caused largely by the thawing of permafrost and the meandering and flooding of large rivers as illustrated by the Yukon and Minto Flats. Only small sections of the interior uplands are believed to have been glaciated by alpine glaciers. The glacial outwash plains of the Pleistocene glaciers along the northern front of the Alaska Range provided source regions for the significant deposits of loess and silts that mantle the lowlands and the lower portions of the uplands of the interior. In many areas, the loess covers and conceals older layers of ice-rich permafrost. Also common are sand dunes resulting from winds blowing across glacial outwash plains. The Copper, Delta, and other rivers draining glaciers today are modern source regions for loess. The Yukon–Kuskokwim lowland, bounded on the north by the Yukon and on the south by the Kuskokwim rivers, consists of numerous thaw lakes, bogs, oxbow lakes, meander scars, delta lakes, and coastal lagoons. Its central areas contain small Tertiary volcanoes.

Above the Arctic Circle, the Brooks Range is a northward extension of the Rocky Mountain System (Fig. 1.6). Although typically less than 3000 m high, mountains in the Brooks Range are rugged and separate rivers flowing southward into the Yukon and its tributaries from those flowing northward into the Arctic Ocean. The entire region north of the Brooks Range is known as the arctic slope or North Slope (Fig. 1.1) and is underlain by continuous permafrost except under the largest lakes. The region can be divided into two distinct topographical areas: the Arctic foothills, an area of rolling hills and ridges, and to the north the arctic coastal plain, a vast flat area characterized by thousands of lakes and wetlands. This plain is a continuation of the interior plains of North America (Fig. 1.6; Wahrhaftig, 1965). The north slope of the Brooks Range is drained principally by the Colville River and its tributaries. Another river, the Sagavanirktok, essentially divides the Arctic coastal plain into two parts: a western, broader section, and an eastern narrower section (Fig. 1.1).

Glacial History

Alaska has experienced a number of repeated glaciations in her history, the most recent occurring during the Pleistocene epoch (2 million years to 10,000 years ago) and the Neoglacial stage of the Holocene epoch. These glaciations have had a great impact on the Alaskan landscape, even in unglaciated areas, by causing the formation of proglacial lakes, construction of out-wash terraces, loess deposition, and isostatic depression of coastal lowlands (Hamilton, 1994). Both ancient and modern glaciers have been most extensive in southern Alaska due to the proximity of moisture sources in the North Pacific Ocean and the Gulf of Alaska.

During the Wisconsin glaciation, the last of the Pleistocene stages (80,000 to 10,000 years ago), half of North America was covered by major ice sheets. At this time ice caps were present in the Brooks and Alaska Ranges, and the large Cordilleran glacier complex extended north and eastward into the adjacent Yukon Territory of Canada and south to Seattle. Nevertheless, the interior of Alaska and the arctic coastal plain remained free of ice (Fig. 1.7). Worldwide lowering of sea level exposed the shallow continental shelf beneath the Bering Sea, making Alaska more a part of Siberian Asia than North America.

The vast unglaciated plain that existed during glacial maxima is now termed Beringia, named after the Bering Sea that covered the area following the rise of sea levels after the ice retreat. Beringia provided a refugium for many Pleistocene mammalian megafauna and is thought to have provided a point of entry of humans into the Americas as they followed mammoths and other prey mammals across the land bridge. Although most evidence suggests a lack of tree species during the Pleistocene, it is thought

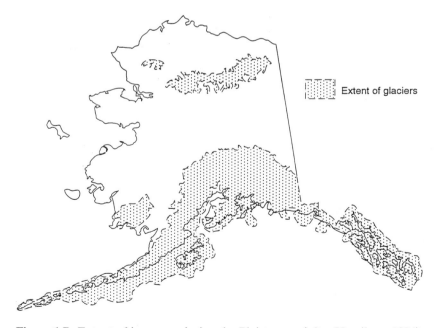

Figure 1.7. Extent of ice cover during the Pleistocene (after Hamilton, 1994).

that there may have been isolated refugia for some species, for example balsam poplar (Barnosky et al., 1987). The area was arid, with the amount of flowing water probably small, and thus likely did not serve as an important refugium for most aquatic species during the Pleistocene. Hence, a great majority of the taxa probably colonized Alaska's freshwaters, either from adjacent Siberia while the land bridge was extant, or from south of North American ice sheets following deglaciation some 14,000 to 10,000 years ago. Evidence from the caddisflies (Trichoptera) suggests that most species present in Alaska are either North American or Holarctic, suggesting that endemics did not evolve during the Pleistocene (Irons, 1988). However, Beringia was the only North American refuge for 10 freshwater fishes (McPhail and Lindsey, 1970), including one fish, the Alaska Blackfish (*Dallia pectoralis*), whose postglacial distribution is limited to Beringia (McPhail and Lindsey, 1970; Pielou, 1991).

Three thousand to 4000 years ago the climate again cooled. Coastal Alaska experienced the Neoglacial Ice Age when glaciers advanced and formed ice sheets that reached their maxima in the 18th century, although multiple advances and retreats occurred over this time period. The best documented examples of Neoglacial advance occurred during the last glaciation, the Little Ice Age, in Glacier Bay, southeast Alaska, and Kenai Fjords in southcentral Alaska (Calkin, 1988; Wiles and Calkin, 1994) (Fig. 1.2). Since that time glaciers have receded to approximately their preadvance size, although a number have shown minor readvances. Coastal land emerging from the Neoglacial ice is displaying uplift or isostatic rebound following removal of the great weight of the ice from the earth's crust. The greatest uplift rates have been measured at 4 cm/ year in Glacier Bay (Hicks and Shotnos, 1965 cited in Barnes, 1990), although a portion of this uplift may be due to tectonic forces (Barnes, 1990).

Of the glaciers in the United States, Alaska presently contains over 75% by number and 99% by surface area. Varying estimates of 65,800 km² (Brown, 1989) and 74,700 km² (Post and Meier, 1980) are given, although glaciers only cover 5% of the Alaskan land surface (Hamilton, 1994). Most of these glaciers are located in the Alaska Range, the Wrangell–Saint Elias Mountains, and the Coast Range, close to moisture sources in the Gulf of Alaska. Small scattered glaciers exist throughout the eastern Brooks Range. The effects of glaciers and their movement on the freshwater environment are reviewed in Chapter 12, including the potential influence of global climate change.

Minerals

Alaska is rich in minerals and has a long history of mining. Although the famous Gold Rush to Dawson in the Yukon Territory took place in the

1890s, placer gold mining has occurred in interior Alaska for well over a century, and both placer and lode gold mining for a longer period in southeast Alaska. Several types of mineral deposits form as a result of weathering and erosion, and gold placers are by far the best known type of these deposits in Alaska. Weathering frees the mineral from the parent rock, and because of its weight the deposit remains. Deposits may then be concentrated by stream action and most of the common gold deposits in Alaska are associated with stream placers. Although lode mining has been important in Alaska, placer mining has had the greatest effect on freshwaters (see Chapter 10).

Alaska contains a wide variety of other known mineral deposits including copper, rare earths, silver, antimony, tin, mercury, iron ore, and numerous other metals. About 50% of the United States' and 15% of the world's coal reserves are believed to occur in Alaska. The three main fields are located in northern Alaska, Cook Inlet–Susitna lowland, and the Nenana River basin. Presently the only working coal mine in Alaska is the Usibelli mine near Healy in the Nenana River basin.

Climate

Topographical features have a major influence on Alaska's climate that can be broadly categorized into the five principal zones (National Climate Center, 1982) shown in Fig. 1.8. These zones, with their principal precipita-

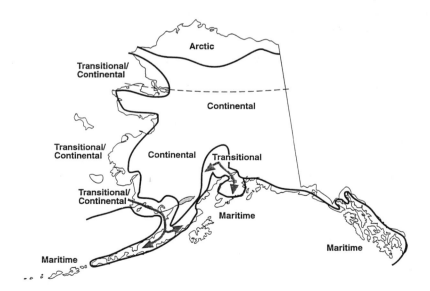

Figure 1.8. Climatic zones of Alaska.

tion and temperature characteristics, are as follows (data from University of Alaska, 1989);

1. *Maritime* zone: includes southeastern Alaska, the southcentral coast, and the southwestern islands. Coastal mountain ranges in this zone produce annual precipitation values of up to 500 cm for the southeastern panhandle, 380 cm for the southcentral coast, and 150 cm for the southwestern islands. Along the southcentral coast, snow makes up a large percentage of the total annual precipitation. Average temperatures typically range from −3°C in January to 13°C in July.

2. *Transitional–Continental* zone: includes the western portions of Bristol Bay and the west-central zones. Annual rainfall in this zone varies from 25 cm in the north to 65 cm in the south. Temperatures average around −23°C in January and 11°C in July. Winter temperatures are similar to the continental zone due to the Bering Sea being frozen, but are moderated in summer by the oceanic influence.

3. *Transitional* zone: between the maritime and continental zones in southcentral Alaska. This zone includes the northern part of the Cook Inlet area and the remainder of the region south of the Alaska range and west of the Talkeetna Mountains. Average annual rainfall varies from 48 cm in the Kenai area to an average of 31 cm in more northern areas. Average temperatures for the Kenai area in January are −11°C and in July are 12°C; northern parts of this zone, further from oceanic influence, have January averages of −13°C and July averages of 14°C.

4. *Continental* zone: encompasses the interior and the rest of the west-central areas and the Copper River drainage. Yearly average precipitation in this zone is around 30 cm. This area has wide extremes of temperature with average January values of −23°C and average July temperatures of 17°C. Summer temperatures in the Copper River drainage area of this zone are a little cooler due to proximity to the Gulf of Alaska.

5. *Arctic* zone: principally the area north of the Brooks Range. Annual precipitation is less than 15 cm with average temperatures ranging from −29°C in winter to 10°C in summer.

Vegetation Types

Viereck et al. (1992) divided the state into four major vegetation zones; coastal forest, boreal forest or taiga, lowland tundra, and upland tundra (Fig. 1.9).

The coastal forest is a northern temperate rainforest dominated by western hemlock (*Tsuga heterophylla*) and Sitka spruce (*Picea sitchensis*). Mountain hemlock (*Tsuga mertensiana*), Alaska cedar (*Chamaecyparis nootkatensis*), and a varied understory of herbs and shrubs (e.g., red alder, *Alnus rubra*) also characterize these forests. The tree line is at about 800 to 1000 m. Permafrost is not found in this region.

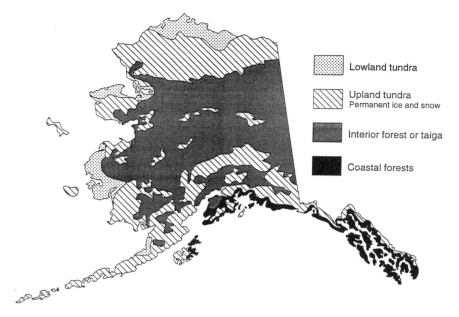

Lowland tundra

Upland tundra
Permanent ice and snow

Interior forest or taiga

Coastal forests

Figure 1.9. Major forest zones of Alaska (after Viereck et al., 1992).

The boreal forest occupies a vast area of the state, far in excess of that covered by the coastal forest, and principally covers the intermontane plateau between the Brooks Range and the Alaska Range (Van Cleve et al., 1983). It also extends south of the Alaska Range to the coastal forest, a latitudinal range of some 8° (Zasada, 1976). The interior boreal forest can be subdivided into moderate and sparse stands. The majority of the moderate growth stands lie within 50 km of the Yukon, Kuskokwim, and Tanana rivers. Boreal forest is composed chiefly of white spruce (*Picea glauca*) and black spruce (*P. mariana*). Other species include paper birch (*Betula papyrifera*), cottonwood (*Populus balsamifera and P. trichocarpa*), quaking aspen (*P. tremuloides*), and tamarack (*Larix laricina*). Black spruce is the dominant species on poorly drained sites, sites underlain by permafrost, and cool sites on the north facing slopes and covers 44% of the area below tree line (Van Cleve et al., 1983). Due to recurrent fire, trees in the boreal forest are seldom older than 200 years, except on river–floodplain islands or within moist drainages. Tree line is typically 600 to 800 m in the interior boreal forest.

In tundra areas, plants are short, and growth is restricted by the shallow, seasonally thawed soil above the permafrost, known as the active layer (ca. 30 to 50 cm in depth). Upland tundra is dominated by grasses and sedges with several species of tundra flowers, and dwarf willows and birches (ca. 1 m in height) along water courses. Upland tundra is common in the northern foothills of the Brooks Range, central Seward Peninsula, and the Bristol Bay region.

Lowland tundra is dominated by a wet sedge meadow of *Eriophorum angustifolium* and *Carex aquatilis* among numerous thaw lakes. Tundra tussock, dominated by *Eriophorum vaginatum*, occurs in mesic sites in both upland and lowland tundra and is common on the western North Slope, the northern part of the Seward Peninsula, and the Yukon–Kuskokwim deltas. Above the tree line at higher elevations, alpine tundra and barren rock and ice are found.

Typology of Alaskan Running Waters

Many schemes have been proposed for classifying running waters (reviewed by Gordon et al., 1992: 410–423). Unfortunately, these classification systems differ in purpose, in the variables used in classification (e.g., geomorphology, water chemistry, biota, wilderness status), and in scale (stream reach to physiographic regions). Although a standardized classification system for running waters is desirable, such a system is not yet available (Gordon et al., 1992).

At the landscape level, the longitudinal division of the the river continuum (Vannote et al., 1980) from source to mouth into headwater streams (orders 1 to 3), medium-sized rivers (orders 4 to 6), and large rivers (orders 7+) is widely used. These categories correspond principally with the zones of sediment production, transfer, and deposition determined by fluvial geomorphology (Gordon et al., 1992). Our typology of Alaskan running waters incorporates this division. We have superimposed local and regional modifiers (Fig. 1.10) on this continuum of river size. This scheme does not provide a simple classification of running waters because the number of possible combinations of descriptors and river sizes is huge. Rather, we see the descriptors as "ecological adjectives" describing the principal modifiers of the structure and function of lotic systems in Alaska.

Some of these descriptors are unique to Alaska (e.g., the hydrothermal regimes describing the seasonal patterns in discharge and water temperature characteristic of the major hydrologic/climatic regions of Alaska), while other descriptors (e.g., the modifications of streams by salmon and beavers or the sharp differences in streams derived from glacial, brownwater, or clear-water sources) are characteristic of northern ecosystems throughout the Holarctic region.

Large rivers (orders 7+) by definition drain very large catchments and so their chemical and physical characteristics are an amalgam of the often disparate characteristics of their tributaries. Further, the interface (riparian zone, atmosphere) of large rivers with their surroundings is small relative to the great volume of water flow, and so some descriptors (such as modifica-

Figure 1.10. Descriptors applicable to running waters of Alaska. Shaded boxes indicate descriptors not applicable or of negligible significance at a particular river size. See text for descriptors of hydrothermal regimes.

Descriptors	Stream Order		
	1 - 3	4 - 6	7 - 9
Water Regime			
Perennial			
Intermittent		███	███
Temporary		███	███
Water Salinity			
Fresh <0.5ppt			
Tidal 0.5-30ppt			
Water Source			
Glacial			
Clear-Water			
Brown-Water			███
Spring/Groundwater			
Thermal		███	███
Lake Outlet			
External Modifications			
Beaver			███
Salmon			
Human			
Large Woody Debris			
Hydrothermal Regime			
Southeast			
Southcentral			
Yukon			
Arctic			
Northwest			

tion by beavers or by woody debris derived from riparian vegetation) are likely to be much less applicable to large rivers than to streams and smaller rivers. Likewise, at the terminus of these large rivers, tidal influences and delta formation produce ecosystems distinct from wholly freshwater large rivers. Therefore, we discuss separately the characteristics of tidal rivers, large rivers, and small rivers and streams.

Tidal Rivers and Deltas

A number of Alaska's large rivers possess significant deltas as they empty into the ocean over broad flats. The deltas are frequently associated with tidal wetlands and are areas of very high habitat value, both for fisheries and wildfowl. The largest of the river deltas is associated with the Yukon River that rises in Canada and flows across the entire width of interior Alaska before emptying into the Bering Sea over a vast delta. Other large river deltas include the Kuskokwim, the Colville, the Copper, and the Susitna. Tidal wetlands of streams and rivers in southeast and southcentral Alaska are important feeding areas for deer in winter and bears in spring (Fig. 1.11).

Large Rivers

Alaska possesses several major river systems, most of which flow through lowland areas of Alaska between the mountain ranges. The most notable is the Yukon River (Fig. 1.12) that drains approximately 855,000 km^2, ranking 23rd among world rivers in drainage area (Czaya, 1981). The Yukon River arises in Canada and, because of the east–west orientation of the Brooks and Alaska mountain ranges, runs westward for 725 km across the inter-montane plateau of interior Alaska before emptying into the Bering Sea. The Yukon River, together with the Kuskokwim River, which is also a large lowland river, drain most of interior Alaska. Other large, lowland rivers include the Colville River that drains the north slopes of the Brooks Range and flows northward to enter the Beaufort Sea, the Copper River that rises

Figure 1.11. Tidal wetland on a small river system on Kodiak Island (photo William J. Wilson, AEIDC, University of Alaska.)

Figure 1.12. Junction of the Yukon and Tanana Rivers in interior Alaska with side sloughs and side channels (photo C.D. Evans, AEIDC, University of Alaska).

in the eastern Alaska Range and the Wrangell Mountains and flows south through the Chugach Range to enter the Gulf of Alaska near Cordova, and the Susitna River that rises on the south side of the Alaska Range and flows through the Susitna valley to enter Cook Inlet over the Susitna flats. The above rivers typically have vast delta formations near their mouths. In the Arctic, the Noatak and the Kobuk rivers, which flow westward on the south side of the Brooks Range and enter Kotzebue Sound, belong to this group. Some large rivers in southeast Alaska typically do not form large deltas at their mouths due to valley constraints; these include the Taku, the Stikine, the Alsek, and the Chilkat.

These large rivers typically possess a broad valley, containing numerous side channels and sloughs (Fig. 1.12), and the channel is meandering (Fig. 1.13) and frequently braided. Although many of these large rivers receive numerous clear-water tributaries, they are frequently dominated by glacial runoff and carry enormous silt loads (Fig. 1.14). The influence of riparian vegetation on these large rivers is minimal except perhaps where it borders side channels and sloughs. The rivers act as corridors for fish to migrate to clear-water tributaries, although side channels and sloughs can provide important rearing habitat for salmonids.

Benke (1990), in a perspective on America's vanishing streams, pointed out that few large rivers remain in the conterminous United States that are unregulated or uninfluenced by humans. Alaska is fortunate in that all her large rivers are still virtually pristine and unregulated. A hydroelectric dam near Whitehorse, Canada in the upper reaches of the Yukon River is the only dam influencing a large Alaskan river. Two dams on large rivers were proposed and planned but never constructed (Watana dam on the Susitna River and Rampart dam on the Yukon River).

Small Rivers and Streams

As noted above, large rivers take on the composite characteristics of many upstream tributaries (the lotic "melting pot") and their sheer size makes large rivers resistant to all but large scale perturbations. In contrast to the ecological similarity of large rivers, small rivers and streams show tremendous individuality, even within a small region. There are an uncountable number of streams in the coastal rain forest, interior taiga, and arctic tundra of Alaska. Milky streams fed by glacial meltwaters flow side by side with clear water streams. In small streams, a run of a few spawning salmon might

Figure 1.13. The meanders of the Nowitna River, a tributary of the Yukon River (photo C.D. Evans, AEIDC, University of Alaska).

Figure 1.14. Clearwater tributary of the turbid Yukon River (photo C.D. Evans, AEIDC, University of Alaska).

constitute a substantial subsidy of marine-derived nutrients, and beavers cause significant changes in local hydrology, nutrient cycling, and biota. Although large rivers in Alaska have escaped significant human impacts, much of the Alaskan economy is based upon extractive resources and so small rivers and streams have been perturbed by placer and other kinds of mining, logging, and urbanization. There are a number of hydroelectric dams on the lake outlets of small rivers and streams in southcentral and southeast Alaska.

Local and Regional Modifiers

Water Regime

Perennial streams flow year-round, intermittent streams flow only during certain times of the year, and temporary streams flow only during or after precipitation (Gordon et al., 1992). In arid regions, lack of surface flow is often the result of a protracted dry season. However, in northern Alaska and other high latitude regions, intermittent streams often flow only during the summer, with flow derived from snow melt and summer precipitation. Stream channels may be dewatered during the winter because surface runoff ceases and water in channels freezes (Craig, 1989). Streams derived from glacial meltwater show a similar flow regime across the state.

Salinity

Many streams and rivers in Alaska are affected by tides in their lower reaches, with associated periodic increases in salinity. These salinity changes are accompanied by shifts in biotic community structure. For example, in Porcupine Creek (on Etolin Island in southeast Alaska), Hansen (1980) found a sharp transition in benthic communities, with freshwater sites dominated by insects and an estuarine site (average salinity 3‰) dominated by crustaceans (mostly isopods). Our classification scheme fol-

Figure 1.15. Glacier-fed stream in Wrangell Saint Elias National Park (photo AEIDC, University of Alaska).

Figure 1.16. A turbid glacier-fed river with high suspended sediment load in Denali National Park (Photo AEIDC, University of Alaska).

lows the "Venice System" (as described in Cowardin et al., 1979) in which freshwaters have salinities of <0.5‰, and tidal waters have salinities of 0.5 to 30‰.

Water Source

Water source is a major variable determining the characteristics of Alaskan streams and rivers and the biotic communities present.

Glacier fed: water flow derived from glacial meltwater (Fig. 1.15), producing cold turbid water (generally $T_{max} \leq 10°C$; see page 26) with a seasonal discharge peak typically in mid to late summer. Turbidities exceed 30 NTU and suspended sediment concentrations are greater than $50 \, mg \, L^{-1}$ during the summer melt period (Fig. 1.16). Suspended sediment levels in glacier-fed rivers in Denali National Park have been reported to $1800 \, mg \, L^{-1}$ (Parm Edwards, unpublished data). Studies have indicated that even a few percent glacierization of the basin can modify the hydrograph from snowmelt/rainfall dominated rivers. For example, in the Tanana River in interior Alaska where 5% of the basin is glacierized, 47% of the summer runoff was derived from glacial runoff (Anderson, 1970). Channels are frequently braided (Fig. 1.17), except where a proglacial lake is present (Milner and Petts, 1994).

Clear water: water flow derived from precipitation and shallow ground water, showing seasonal discharge peaks from snow melt and/or precipita-

Figure 1.17. The Tolkat River, typical of many braided glacier-fed systems in Denali National Park (photo AEIDC, University of Alaska).

tion. Turbidities are generally less than 10 NTU and suspended sediment levels below $20\,mg\,L^{-1}$.

Brown water: water flow derived from peaty soils, producing stained water with high concentrations of dissolved organic compounds or DOC (e.g., approaching $30\,mg\,L^{-1}$ DOC in a tundra stream; Oswood et al., 1996). Brown-water streams are typically acidic in nature (pH < 6.0).

Spring/groundwater: water flow derived from deep aquifers producing near constant year-round flow and little seasonal variation in water temperature.

Thermal hot springs: water flow derived from deep aquifers producing near constant year-round flow and little seasonal variation in water temperature. Water temperatures are elevated and water is high in solute concentrations.

Lake outlet: water derived as outflow of lake, with thermal regime determined by lake surface water temperatures and generally with high concentrations of lake-derived particulate organic matter (especially plankton) and low sediment loads (lake acts as settling basin).

Biotic Modifications

Beaver activity: Beaver dams (Fig. 1.18) modify streams by impounding water, increasing sediment retention, altering the amount of spawning and rearing habitat, changing the processing and export of carbon and nutrients, and altering the riparian vegetation (Naiman et al., 1988). Where the impoundment is large, downstream thermal regimes may be altered.

Salmon spawning: Spawning streams with high densities of spawning salmon receive (via decay of carcasses, urine, and feces) a substantial subsidy of marine-derived nutrients (see Chapter 7 for review). Also, redd construction scours stream bed materials, and respiration of living spawners and microbial decay of carcasses may severely depress dissolved oxygen concentrations. Salmon eggs may also be a food source for other fish.

Human activity: The Alaskan economy depends heavily upon utilization of natural resources. However, the huge size of the state, combined with low population density and limited road access in most areas, have combined to limit environmental degradation of freshwater resources. Extensive timber harvest to date has largely been confined to the productive forests of southeast Alaska. Clear-cut logging has sometimes resulted in degradation of spawning and rearing habitat of stream salmonids and so resulted in a classic multiple-use resource conflict (reviewed in Chapter 9). The impacts of mining on running waters in Alaska have waxed and waned depending upon mining economics, the contingencies of two world wars, gradual depletion of readily accessible mineral deposits, and more recently, environmental regulations. Placer mining for gold (reviewed in Chapter 10)

Figure 1.18. Pond on Kodiak Island created by the action of beavers (photo AEIDC, University of Alaska).

has been important in the cultural and economic history of interior and northwest Alaska, with localized but sometimes catastrophic effects on streams and rivers. The Exxon Valdez oil spill focused world attention on the potentially catastrophic impacts of oil development, especially to marine ecosystems. In contrast, construction and operation of the trans-Alaska pipeline from Valdez to Prudhoe Bay has so far resulted in relatively minor impacts to freshwater systems and, in fact, has fostered long-term studies of arctic ecosystems (see Chapters 3 and 4) and extensive experimental examination of the effects of oil on freshwater ecosystems (reviewed by Alexander and VanCleve, 1983). Finally, urbanization impacts on freshwater systems in Alaska seem to be largely confined to Anchorage and Juneau, major urban centers of Alaska. Ongoing studies of the urban streams of Anchorage show the usual complex brew of urbanization effects, including inputs of toxic and organic substances and channelization.

Hydrothermal Regimes

Six major hydrologic regions, defined by major watersheds, exist within Alaska (Fig. 1.19). Alaska has the highest average water yield per unit area in the United States, accounting for about one-third of the total U.S. runoff. Rivers in different hydrologic regions have characteristic hydrographs and temperature regimes, largely driven by the climatic regime and topography of the region. The summary hydrographs presented in Fig. 1.20 are integrations of a number of hydrographs from rivers and

Figure 1.19. Hydrologic regions of Alaska.

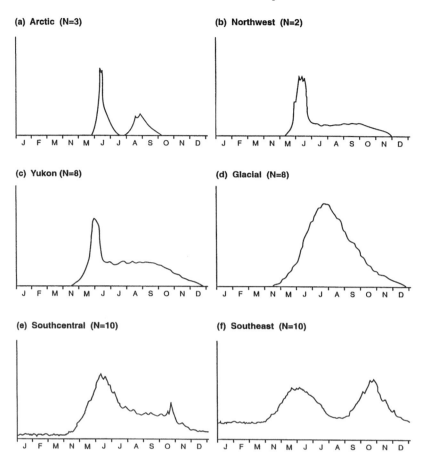

Figure 1.20. Composite representative hydrographs for rivers in different hydrologic regions and for glacier-fed rivers (N = number of rivers used for composite. Y axes show relative discharge). (data compiled from Chapman, 1982.)

streams within each region (no hydrography provided for the southwest hydrologic region). The hydrographs are usually derived from smaller systems (catchment < 1000 km²), except for the glacier-fed rivers, and typically are not fed by a lake. Temperature data are mostly from midsized rivers to reflect the climate influence of the hydrologic region; one first-order stream is included for comparison in the Yukon region. Annual degree days are in degrees centigrade (°C). The hydrograph and thermal regime of predominantly glacier-fed rivers are strikingly similar across the state, regardless of hydrologic region.

Glaciers account for 35% of the total runoff in Alaska (Mayo, 1986). Where glacial runoff contributes 50% or greater to stream flow, seasonality is strongly influenced. As indicated previously, this can occur with as little as 5% of the basin glacierized. Glacial rivers and streams typically have very low or no discharge in winter, then flows begin rising in early May with increased solar radiation to reach a summer peak at maximum glacier melt (Fig. 1.20d). Discharge then declines gradually until freeze-up in November and December. The timing of the peak and the onset of freeze-up vary according to watershed size, climatic zone, and extent of winter snowfall. Interestingly, the hydrograph of glacier-fed rivers in southeast Alaska does not reflect the characteristic October peak evident in nonglacial streams from heavy autumn rains. Marked diel fluctuations in flow in glacially influenced streams are also prevalent. Most arctic streams have negligible glacial influence. Maximum water temperature is generally less than 10°C close to the glacier and in streams and smaller rivers; but in larger slower rivers further from their glacial source, water temperature may rise substantially above 10°C. For example, daily temperatures to 16°C have been recorded for the Sustina River near its mouth at Susitna Flats (USGS, 1986). We suggest that the thermal regime of large glacier-fed rivers near their mouths will be similar to the thermal regime of a nonglacial river of the same size and hydrologic region.

Arctic streams vary according to permafrost characteristics and length of seasonal thaw. Some streams have no discharge from October through early May and then rapidly peak due to spring snowmelt in June. Low discharges then occur until late July or early August when convective thunderstorms increase discharge, but with significantly lower peaks than spring (Fig. 1.20a). In a beaded tundra stream, maximum monthly temperature was reached in July at about 13.5°C, but daily maxima may be as high as 21°C. Accumulated degree-days were about 1000 (Irons and Oswood, 1992). A nearby third-order river followed a similar thermal regime (Fig. 1.21).

Streams from the northwest hydrologic region differ from arctic streams in that substantial discharge is maintained during July to October due to less severe permafrost conditions in northwest Alaska than arctic Alaska (Fig. 1.20b). No satisfactory annual temperature data from this hydrologic region were available.

Nonglacial streams in the Yukon hydrologic region start to rise rapidly after breakup and flow peaks in late May during breakup snowmelt. Late summer storms cause high discharges that are often localized and brief (Kane et al., 1992). Discharges decrease in October (Fig. 1.20c). Streams are generally frozen for 6 to 6.5 months and maximum monthly temperature is generally reached in July, ranging typically from 8° to 10°C (Fig. 1.21). Accumulated degree days generally range from 300 in cold headwater streams (e.g., Little Poker Creek) to 900 in second-order streams (e.g., Monument Creek) to 1200 in fourth-order rivers (e.g., Little Chena River).

In the southeast and southcentral hydrologic regions, the proportion of the precipitation that falls as snow has an important influence on the flow of nonglacial streams. In southeast Alaska, a large proportion of the winter

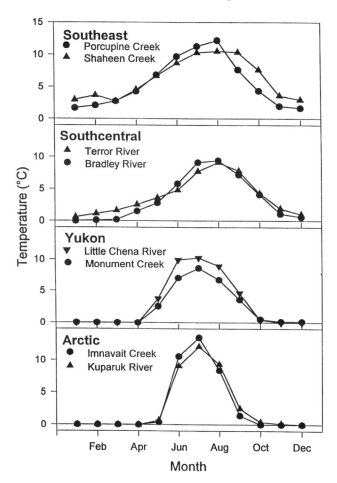

Figure 1.21. Representative temperature curves for rivers from four hydrologic regions. Data sources and length of record as follows; Porcupine Creek 5 years (Koski, 1982), Shaheen Creek 1 year (J. Thedinger personal communication), Terror River and Bradley River 1 year (USGS), Little Chena River 6 years (USGS), Monument Creek 2.5 years (Irons and Oswood, 1992), Little Poker Creek 1 year (Irons and Oswood, 1992), Imnavait Creek 4 years (Irons and Oswood, 1992), Kuparak River 2 years (C. Harvey, personal communication).

precipitation occurs as rain and consequently a smaller snowpack exists in watersheds; in southcentral Alaska a large percentage of winter precipitation falls as snow. The snow versus rain effect is clearly demonstrated by comparing the hydrographs of streams in southeast Alaska and southcentral Alaska (Fig. 1.20**e,f**). In southcentral Alaska, winter flows in nonglacial streams are low and the principal peak is during spring snowmelt, often with a second lower peak in the fall (Fig. 1.20e). In streams of this hydrologic region farther from maritime influence, the fall peak is typically smaller and occurs earlier, and streams do not return to base flow during the summer because of the prolonged snowmelt season. In contrast, hydrographs for nonglacial rivers in southeast Alaska have significant winter flow and two major peaks: one peak during spring snowmelt runoff and a second larger peak during the fall rainy season (Fig. 1.20f). Stream hydrographs from southcentral Alaska show significantly greater variation than those from southeast Alaska, which are remarkably consistent across streams.

The thermal regime of streams in southeast Alaska is influenced by proximity to the ocean. Nonglacial streams freeze only occasionally and monthly mean water temperature is always above 0°C (Fig. 1.21). Maximum monthly mean temperature of undisturbed streams is about 10° to 12°C (although daily maxima may be higher) and is reached in August or September. Accumulated annual degree-days are on the order of 2500. In contrast, streams in the southcentral hydrologic region may freeze in winter, particularly away from oceanic influence. Because this region crosses a number of climatic zones, a range of thermal regimes with more extremes of temperature exists away from the maritime influence. Here streams are generally frozen for about 5 to 5.5 months, while closer to the ocean, the period of freeze-up is reduced to 4 to 4.5 months or in some years rivers may not freeze (Fig. 1.21). Accumulated degree days are about 1400 to 1500 for streams near the coast.

There were no suitable data for hydrothermal regimes from the southwest hydrologic region of the state.

In summary, distinct hydrothermal regimes can be discerned for the different hydrologic regions of Alaska. The hydrographs differ in the amount of winter base flow, and the timing and relative magnitudes of spring snowmelt and late summer/autumn high flow peaks. Glacier-fed dominated systems have a similar hydrograph across the state. The thermal regimes differ primarily in length of the ice cover season rather than summer maxima. Indeed the highest monthly means in Fig. 1.21 are in Arctic streams, while the ice cover season increases from negligible in southeast Alaska to more than 7 months in the Arctic. In addition, the maximum monthly mean temperature tends to be later in the more southerly streams: in August or September in southeast and southcentral streams, compared to July in Yukon and Arctic streams. To conclude we suggest that:

1. The length of the ice cover season is determined by the regional climate, and differs more among hydrologic regions and their associated climates than among rivers of similar size and distance from source within a hydrologic region;

2. Summer thermal regimes (e.g., maximum summer temperature, accumulated degree-days) may differ more within a region than among regions; and

3. Proximity to an ice-free ocean buffers the thermal regime by ameloriating the extremes and delaying the summer maximum in the hydrologic regions of southeast and southcentral Alaska.

Freezing Cycle of Rivers and Streams

One of the principal features of rivers and streams in Alaska is the effects of low temperatures during the winter time on the hydrology and overall habitat characteristics. These effects, which include the formation of several different types of ice in winter (e.g., frazil ice, anchor ice, and aufeis), the length of the ice-free season, and minimum winter temperature reached in frozen substrates, have significant implications for the biotic communities living in these habitats and for their productivity (Oswood et al., 1991; Beltaos et al., 1993; Power et al., 1993; Scrimgeour et al., 1994). These effects are more magnified the greater the latitude. The role of ice in Alaskan running waters is reviewed in Chapter 13.

Typology of Alaskan Lakes

Lakes are a dominant feature of the Alaskan landscape, particularly in low-lying areas, and cover approximately 2.1 million ha of land surface. The efforts of limnologists have tended to cover specific areas of Alaska such that lakes in many regions of the state have not been well studied. The different types of lakes typically found in Alaska are summarized in Fig. 1.22. Their mode of origin has been described extensively by Hutchinson (1975) and so will not be duplicated here. Examples and typical localities of some of these lake types are provided. However, because themokarst processes are a unique part of the northern regions of the state, these processes and the associated lakes are outlined in detail.

Thermokarst

Permafrost is soil or rock material that has remained at or below 2°C for 2 or more years (Muller, 1947). Approximately 85% of Alaska is within the permafrost region that can be divided into two major zones, continuous and discontinuous (Fig. 1.23). The continuous zone makes up 20% of the permafrost (Benson et al., 1986) and is underlain by permafrost everywhere except for bodies of water large enough not to freeze to the bottom. The discontinuous zone has areas without permafrost (Ferrians, 1994). There are also sporadic areas that typically lack permafrost except in isolated areas. Above the permafrost the active layer freezes and thaws seasonally.

Climate is the major factor that determines the regional distribution of permafrost. Within the continuous permafrost zone of Alaska, Barrow

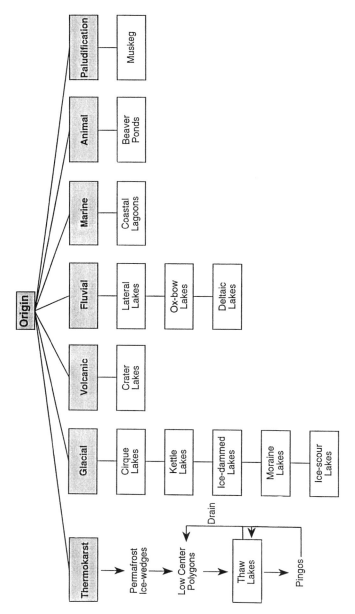

Figure 1.22. Summary of principal lake types in Alaska.

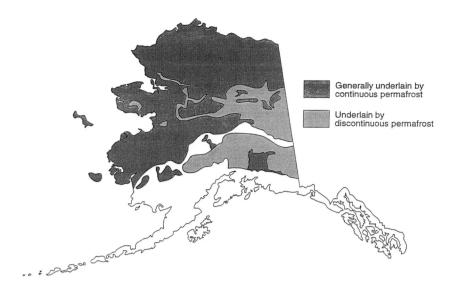

Generally underlain by
continuous permafrost

Underlain by
discontinuous permafrost

Figure 1.23. Extent of permafrost in Alaska (after Ferrians, 1969).

has the lowest mean annual air temperature of −12.2°C; Anchorage just outside the occurrence of sporadic permafrost has a mean annual air temperature of 1.7°C. The thickness of permafrost varies from 600 m on the arctic coastal plain to less than 1 m in the southern part of the permafrost region (Brown and Krieg, 1983). The thickness and areal distribution of permafrost are directly affected by surface features like bodies of water, topography, drainage, and vegetation that act as a heat source or heat sink or as insulation (Ferrians, 1994). In the arctic and subarctic regions of low rainfall and relatively level terrain, the ice-rich permafrost table prevents infiltration and outflow of water to the sea except through a small number of river valleys. Ice-free soils allow infiltration and lateral flow. Permafrost processes significantly influence the land forms (periglacial features) present and create a water logged surface characterized by different lake types.

Ice wedges are bodies of underground ice that originate from contraction cracking of frozen soils during extreme low winter temperatures. The eventual result is a series of intersecting cracks, polygonal in shape, into which blowing snow and meltwater accumulates during winter and spring. Because cracks extend into the permafrost, entering water turns to ice and expands. Through repeated winter cracking, ice wedges increase in width and may reach up to 10 m (Black, 1969; Billings and Peterson, 1980). These ice wedges in the underlying permafrost result in extensive areas of the tundra being characterized by polygonal networks of troughs and ridges. Two related types of polygons typically occur, depending on the state of the ice wedges: low-centered polygons form above active ice wedges, and are saucer-shaped depressions surrounded by a raised rim on either side of the ice wedge; and high-centered polygons form when the ice wedges under low-centered polygons melt, leaving depressions where there once were ridges. If the melting lowers the water table, high-centered polygons may have no surface water; however, if the water table remains high, the depressions become channels. If a topographic gradient exists, high-centered polygons may evolve into a beaded stream, where melting of ice wedges creates rounded pools connected by narrow channels (see Fig. 1.24).

Over 95% of the wet coastal tundra of the arctic coastal plain is made up of *thaw lakes* (Fig. 1.25) in one form or another that range in length from several hundred meters to several kilometers (Billings and Peterson, 1980). Thaw lakes form where the ground ice melts and the overlying soil collapses below the water table. These lakes are found in areas of both continuous and discontinuous permafrost. Lake depth is fairly uniform, typically less than 2 m, and the lakes commonly freeze to the bottom in winter (Sellman et al., 1975). Some thaw lakes are elliptical or rectangular in shape, and are typically oriented at right angles to the prevailing northeast–southwest winds. Many mechanisms for this phenomenon have been proposed, and are a subject of much controversy (Washburn, 1980). Livingstone (1954) and Carson and Hussey (1962) suggest that initially thaw lakes are not oriented but that the predominant winds rapidly enlarge the small lake and

Figure 1.24. Patterned ground and a beaded stream on the north slope (photo C.D. Evans, AEIDC, University of Alaska).

Figure 1.25. Thaw lakes on the arctic coastal plain (photo C.D. Evans, AEIDC, University of Alaska).

orient it in a north-northwest–south-southeast direction. As the shallow lake enlarges, sediments and peat from the surrounding tundra are washed into the lake that serve to insulate the ice wedges.

When ice wedges thaw along the edge of the lake allowing flow to enter a stream, mature thaw lakes may drain. Some of the ice wedges persist beneath the thaw lake and are important in reinitiating the polygonal, terrestrial part of the cycle and the eventual reformation of another thaw lake. The completion of this cycle probably requires 2000 to 3000 years (Billings and Peterson, 1980). Ice wedges will also form independently of this cycle.

In certain cases when thaw lakes drain, a large bulb of unfrozen soil (known by the Russian term *talik*) extends below the surrounding permafrost level. When the insulating effect of the lake is removed, this bulb begins to freeze inward from the top and the sides. Eventually the unfrozen talik is surrounded by frozen ground and pressure builds up in the confined water. As the expansion due to the freezing continues, a mound, known as a pingo (from an Eskimo word meaning hill), builds up on the surface (see Fig. 1.26.) Typical diameters range between 30 and 600 m with 85% of pingos less than 20 m in elevation, although large ones may reach 50 m (Pissart, 1988). These large, ice-cored hills are a common feature of the Alaskan north slope and Canada. As pingos grow, the insulating surface soil (primarily peat) begins to crack exposing the ice core that then begins to

Figure 1.26. Pingo (photo C.D. Evans, AEIDC, University of Alaska).

thaw. Eventually the pingo may crack and become a thaw lake that is progressively widened by thermokarst processes. Thus there occurs over a long time period a cycle of thaw lake, pingo, and back to thaw lake in many arctic regions. In interior Alaska, particularly in the eastern region, thaw lakes frequently form in depressions caused by the thawing of discontinuous permafrost even in those circumstances in which massive ground ice is not present. Such lakes may expand due to thermal erosion at a rate up to 20 cm per year (Hutchinson, 1975).

Glacial processes were very important in carving out river valleys and forming lakes. *Ice-scour lakes*, together with thaw lakes, are the principal lake types found in arctic Alaska. Many *kettle lakes* are found in southcentral and southeast Alaska. *Cirque lakes* are found predominantly in the Alaska Range and the coastal mountains of Alaska where the upper section of glaciated valleys have been shaped into "amphitheaters" by basal erosion through the freezing and thawing action of the ice. *Ice-dammed lakes* formed where active glaciers result in the damming of drainage from a valley, and is exemplified by Lake Malaspina in southcentral Alaska. These types of lakes may be subject to spectacular glacial outburst floods as reviewed by Stone (1963). Post and Mayo (1971) identified a total of 750 ice-dammed lakes in Alaska and adjacent parts of Canada. *Moraine lakes*, formed where terminal or recessional moraines have blocked the stream that replaces the retreating glacier, are particularly common along the southern flanks of the Brooks Range, the eastern flanks of the Ahklun Mountains near Togiak, the Alaska Peninsula and elsewhere in glaciated areas. (Fig. 1.27).

Many of the major rivers on the coastal plain and interior Alaska carry high silt loads due to glacial runoff in headwaters. Many rivers in this region meander extensively and as a consequence meanders may be cut off from the mainstem of the river creating *ox-bow lakes* (Fig. 1.28). If these lakes become filled with vegetation they are termed meander scars. Harding Lake is a lateral lake among several similar lakes in the Tanana River valley (Nakao et al., 1981) where the river has aggraded its course faster than a tributary river that then became obstructed by sediments deposited along the sides of the main valley. With the many large rivers in Alaska this has been a fairly common occurrence. In the large deltas of the major rivers (e.g., the Yukon, Kuskokwin, Copper) river flow becomes so reduced that sediment is deposited forming barriers that eventually isolate sections of the channel as shallow lakes. These *deltaic lakes* often receive saline water and hence are frequently brackish (Wetzel, 1983).

A number of the *crater lakes* lie in the Aleutian Range and southern part of the Alaska Range. The highly complex tectonic structure of Alaska has

Figure 1.27. Moraine dammed lake in the Brooks Range (photo AEIDC, University of Alaska).

also created numerous depressions that are filled with water and also areas that, due to tectonics, appear to be sinking and forming depressions that form sediment-filled lake basins. Numerous lakes, particularly in the Coast Ranges, have a dog-leg character and are termed *offset lakes* reflecting the lateral movements of fault systems.

In coastal areas, the accumulation of beach sands may close valleys or embayments and form *coastal lagoons*. These are frequently breached by the ocean and may drain. These lagoons are typically brackish in nature. Some of the best examples in Alaska occur in Cape Krusenstern National Monument in the northwest portion of the state.

In areas of southeast Alaska where the drainage is poor, the soils of the forest may become waterlogged and western hemlock/Sitka spruce trees are replaced by *Sphagnum* moss. This process is known as paludification. Where this phenomenon is particularly enhanced, acidic black-water ponds and small lakes are created (Fig. 1.29).

Cannon (1977) described a unique meteorite impact lake, Sithylemenkat Lake, in central Alaska, 250 km northwest of Fairbanks. This lake is ap-

Figure 1.28. Ox-bow lakes in the Yukon River floodplain (photo C.D. Evans, AEIDC, University of Alaska).

Figure 1.29. Muskeg pond in southeast Alaska (photo Alexander Milner).

proximately 500 m deep and is very high in nickel because impacting meteorites typically contain substantial amounts of this element.

Lake Descriptors

Similar to rivers, we can use a series of descriptors that are important in influencing the biotic diversity and productivity of lakes within Alaska, as summarized in Fig. 1.30.

Salinity

Some lakes in coastal Alaska have intrusions of saltwater, particularly deltaic lakes and coastal lagoons, that can increase salinity and cause shifts in biotic community structure.

Water Source

Glacial: Many lakes are influenced by meltwater from glaciers that increases turbidity, potentially reducing temperature and light penetration and thereby decreasing primary production (Koenings et al., 1986). Nevertheless these turbid lakes can still function as important nursery grounds for juvenile sockeye salmon, for example Kenai and Skilak lakes on the Kenai

Descriptors	Lake Size (ha)		
Water Salinity	<10	10-1000	>1000
Fresh <0.5ppt			
Tidal 0.5-30ppt			■
Water Source			
Glacial			
Clear-Water			
Brown-Water			■
Thermal			■
Mixing Patterns			
Holomictic			
Cold monomictic			■
Warm monomictic			
Dimictic			
Meromictic			
External Modifications			
Salmon			
Human			
Fertilization			■
Urbanization			■
Impoundment			■

Figure 1.30. Descriptors for lakes in Alaska.

River and Tustumena Lake on the Kasilof River (both rivers on the Kenai Peninsula, southcentral Alaska). These turbid lakes may become clear water during the winter at times when glacial runoff is absent.

Clear water: turbidities typically less than 10 NTU and in many oligotrophic lakes light penetration may exceed 10 m.

Brown water: water derived from peaty soils, producing stained water that also reduces light penetration and hence potential productivity. Water in these lakes may also be acidic.

Thermal: Ponds and small lakes may be influenced by thermal spring sources that significantly increase their temperature. Thermal lakes are located in interior and northwest Alaska and on a number of islands in southeast Alaska.

Mixing Patterns

Mixing patterns determine the stratification regimes of lakes and are dependent upon the morphometry of the lake basin including volume and depth, solar radiation regime, and wind patterns. Climate, whether oceanic or continental, has a major influence on stratification patterns. Stratification regimes can determine the thermal and chemical characteristics of the lake and its overall productivity. Of the *holomictic* (circulation throughout the entire water column) *lakes*, the principal circulation types in Alaska are *dimictic* (lakes circulate freely twice a year) and *cold monomictic* (water temperature never >4°C throughout the year and one circulation in summer). Some lakes may alternate from year to year between these patterns. For example Lake Schrader (Fig. 1.31) in the Brooks Range was dimictic and summer stratified in 1958 with the epilimnion at 10°C and the hypolimnion at 4°C. In 1959 it was cold monomictic with break-up occurring a month later than normal and the lake not warming above 4°C (Hobbie, 1961). There are no documented examples of *amictic lakes* (perennially ice covered) in Alaska, but there are a number of examples of *meromictic* (entire water column does not mix) *lakes*. Pingo Lake, in interior Alaska, has been shown to be permanently stratified with the absence of oxygen and presence of hydrogen sulfide in the deep waters (Likens and Johnson, 1966). These authors also described two other meromictic lakes in the Tanana River valley.

External Modifications

There are a number of biotic modifications, which may act as descriptors in our typology of Alaskan lakes, involving salmon and human influences.

Figure 1.31. Lake Schrader in the Brooks Range.

Salmon spawning: The presence of salmon spawners can provide a substantial input of marine-derived nutrients, particularly nitrogen and phosphorus, that may be limiting to lake productivity. This may occur through decay of carcasses from beach spawning in the lake or through decay of carcasses that spawned upstream with nutrients then being flushed downstream into the lake. Because many Alaskan lakes are oligotrophic, the input of these marine derived nutrients is important in maintaining salmon productivity and acts as a positive feedback mechanism. The reduction of the numbers of salmon spawners occurring from human harvest may potentially decrease on nutrient pools and lead to a subsequent reduction in salmon runs as discussed in more detail in Chapter 8.

Human

Fertilization: In a number of lakes in Alaska, fertilization with nutrients has been employed to improve salmon productivity as outlined in Chapter 8.

Urbanization: Similar to streams, a number of lakes in Anchorage and Juneau have been impacted by urban runoff that has resulted in algal blooms and sediment contamination from accumulated hydrocarbons and heavy metals.

Impoundment: A number of lakes with natural falls have been dammed for the creation of hydroelectric power, including Bradley Lake near Homer and Tyee Lake near Wrangell.

Summary

This chapter hopefully has provided the reader with an insight into the wide diversity of landscapes in Alaska from the southeast region of the state to the arctic slope. Across this landscape climate, topography, vegetation, and surface features vary enormously. Wetlands, lakes, rivers, and streams are a dominant feature. Across the hydrologic regions of Alaska, characteristic hydrothermal regimes of rivers and streams are of particular significance to the biotic communities, especially in relation to accumulated temperature units (degree days) and the length of the ice-free season. Ice is an increasingly important factor influencing biotic communities with increasing latitude. Virtually all of Alaska's streams and rivers are unregulated including all the major rivers: this is a unique feature compared with the remainder of North America. Water quality has been influenced in a number of geographical areas by specific disturbances; these include placer mining in interior Alaska, timber harvest in southeast Alaska, and urbanization in Anchorage and Juneau. Nevertheless most of the freshwater within the state is unimpaired.

There is no doubt that the unique features of Alaska's freshwaters offer limnologists an opportunity to answer some fundamental questions concerning the functioning of freshwater ecosystems. Some of these ideas are further expanded in chapter 13.

Acknowledgments. We sincerely thank Vera Alexander, Jackie LaPerrierre, Tom Kline, Gary Kyle, Bob Lynch, and Dan Mann for their constructive criticism of earlier versions of this manuscript. We are grateful to Maureen Milner for the line illustrations for this chapter and to the photographers credited for their contributions and to Julie Braund-Allen, photo librarian at AEIDC, University of Alaska, Anchorage.

References

Alexander V, Van Cleve K (1983) The Alaska pipeline: A success story. Annu Rev Ecol Systematics 14:443–463.

Anderson GS (1970) Hydrologic reconnaissance of the Tanana Basin, Central Alaska. U.S. Geological Hydrological Investigations Atlases, HA-319. Fairbanks, AK.

Barnes DF (1990) Gravity, gravity-change and other geophysical measurements in Glacier Bay National Park and Preserve. In Milner AM, Wood JD Jr (Eds) Proc Second Glacier Bay Sci Symp. USDI National Park Service, Alaska Regional Office, Anchorage, AK, 12–18.

Barnosky CW, Anderson PM, Bartlein PJ (1987) The northwestern U.S. during deglaciation: Vegetational history and paleoclimatic implications. In Ruddiman WF, Wright HE Jr (Eds) North America and adjacent oceans during the last deglaciation, Vol K3. Geol Soc of Am, Boulder, CO, 289–321.

Beltaos S, Calkins DJ, Gatto LW, et al. (1993) Physical effects of river ice. In Prowse TD (Ed) Environmental aspects of river ice. National Hydrology Research Institute Science Report No. 5. Saskatoon, Saskatchewan, Canada, 3–74.

Benke AC (1990) A perspective on America's vanishing streams. J North Am Bentholog Soc 9:77–88.

Benson C, Harrison W, Gosink J, Bowling S, Mayo L, Trabant D (1986) The role of glacierized basins in Alaskan hydrology. In Kane DL (Ed) Symposium: Cold Regions Hydrology. American Water Resources Association, Bethesda MD, 471–483.

Billings WD, Peterson KM (1980) Vegetational change and ice-wedge polygons through the thaw-lake cycle in arctic Alaska. Arctic Alpine Res 12:413–432.

Black RF (1969) Thaw depressions and thaw lakes: A review. Biuletyn Peryglacjalny 19:131–150.

Brown CS (1989) A description of the United States' contribution to the world glacier inventory. In Oerlemans J (Ed) Glacier fluctuations and climate change. Kluwer Academic Publishers, New York, 103–108.

Brown J, Kreig RA (1983) Guidebook to permafrost and related features along the Elliott and Dalton Highways, Fox to Prudhoe Bay, Alaska. Guidebook No. 4. Fourth International Conference on Permafrost, University of Alaska, Fairbanks.

Brown J, Sellman PV (1973) Permafrost and coastal plain history of arctic Alaska. In Britton ME (Ed) Alaskan Arctic tundra. Arctic Institute of North America, Technical Paper 25, 31–47. Calgary Canada.

Calkin PE (1988) Holocene glaciation of Alaska. Quaternary Sci Rev 7:159–184.

Cannon PJ (1977) Meteorite impact crater discovered in central Alaska with Landsat imagery. Science 196:1322–1324.

Carson CE, Hussey KM (1962) The oriented lakes of Alaska. J Geol 70:417–439.

Chapman DL (1982) Daily flow statistics of Alaskan streams. NOAA Technical Memorandum NWS AR-35. National Weather Service, Anchorage, AK.

Cowardin LM, Carter V, Golet FC, LaRoe ET (1979) Classification of wetlands and deepwater habitats of the United States. U.S. Fish and Wildlife Service, Office of Biological Services FWS/OBS-79/31, Department of the Interior, Washington, DC.

Craig PC (1989) An introduction to anadromous fishes in the Alaskan arctic. Biol Pap Univ Alaska 24:27–54.

Czaya E (1981) Rivers of the world. Van Nostrand Reinhold Co., New York.

Ferrians OJ Jr (1969) Permafrost in Alaska. Plate No. 10. In Johnson PR, Hartman CW (Eds) Environmental atlas of Alaska, University of Alaska, Fairbanks.

Ferrians OJ Jr (1994) Permafrost in Alaska. In Plafker G, Berg HC (Eds) The geology of Alaska. Geological Society of America series. The Geology of North America, Vol G-1. Boulder, CO.

Gordon N, McMahon TA, Finlayson BL (1992) Stream hydrology: An introduction for ecologists. Wiley, Chichester, UK.

Hamilton TD (1994) Late cenozoic glaciation of Alaska. In Plafker G, Berg HC (Eds) The geology of Alaska. Geological Society of America series. The geology of North America, Vol G-1. Boulder, CO.

Hansen TF (1980) The effects of physical variables on the distribution and abundance of insects in Porcupine Creek, southeast Alaska. M.S. thesis, University of Idaho, Moscow, ID.

Hobbie JE (1961) Summer temperatures in Lake Schrader. Limnol Oceanogr 6:326–329.

Hutchinson GE (1975) A treatise on limnology, Vol 1, Part 1. Geography and physics of lakes. Wiley, New York.

Irons JG III (1988) Life histories and trophic energetics of Trichoptera in two Alaska (USA) sub-arctic streams. Can J Zool 66:1258–1265.

Irons JG III, Oswood MW (1992) Seasonal temperature patterns in an arctic and two subarctic Alaskan (USA) headwater streams. Hydrobiologia 237:147–157.

Kane DL, Hinzman LV, Woo M-K, Everett KR (1992) Arctic hydrology and climate change. In Chapin FS III, Jeffries RL, Reynolds JF, Shaver GR, Svoboda J (Eds) Arctic ecosystems in a changing climate. Academic Press, New York, 35–57.

Kline TC Jr (1991) The significance of marine-derived biogenic nitrogen in anadromous Pacific salmon freshwater food webs. Unpublished Ph.D. thesis, University of Alaska, Fairbanks.

Koenings JP, Burkett RD, Kyle GB, Edmundson JA, Edmundson JM (1986) Trophic level responses to glacial meltwater intrusion in Alaskan lakes. In Kane DJ (Ed) Symposium: Cold regions hydrology. American Water Resources Association, Bethesda MD, 179–194.

Koski KV (1984) A stream ecosystem in an old-growth forest in southeast Alaska: Part I Characteristics of Porcupine Greek, Etolin Island. In Meehan WR, Merrell TR Jr, Hanley TA (Eds) Proc Symp on Fish and Wildlife Relationships in Old-Growth Forests. Amer Inst Fish Res Biol Juneau, AK.

Likens GE, Johnson PL (1966) A chemically stratified lake in Alaska. Science 153:875–878.

Livingstone DA (1954) On the orientation of lake basins. Am J Sci 252:547–554.

Mayo LR (1986) Annual runoff rate from glaciers in Alaska: A model using the altitude of glacier mass balance equilibrium. In Kane DL (Ed) Symposium: Cold regions hydrology. American Water Resources Association, Bethesda MD, 509–517.

McPhail JD, Lindsay CC (1970) Freshwater fishes of Northwest Canada and Alaska. Fisheries Research Board of Canada Bulletin 173, Ottawa.

Milner AM, Petts GE (1994) Glacial rivers: Physical habitat and ecology. Freshwater Biol 32:295–307.

Muller SW (1947) Permafrost or perenially frozen ground and related engineering problems. Edwards Bros, Ann Arbor, MI.

Nakao K, Ryvichi R, Oike T, LaPerriere JD (1981) Water budget and lake-level stability of Harding Lake in The Interior Alaska. J Fac Sci, Hokkaido Univ, Ser VII (Geophysico) 7:13–25.

Naiman RJ, Johnston CA, Kelley JC (1988) Alteration of North American streams by beaver. BioScience 38:753–762.

National Climate Center (1982) Climate of Alaska. NOAA Environmental Data Service, Asheville, NC.

Oswood MW, Irons JG III, Schell DM (1996) Dynamics of dissolved and particulate carbon in an arctic stream. In Reynolds JF, Tenhunen JD (Eds) *Landscape function and disturbance in arctic tundra.* Springer–Verlag, New York.

Oswood MW, Miller LK, Irons JG III (1991) Overwintering of freshwater benthic macroinvertebrates. In Lee RE, Denlinger DL (Eds) Insects at low temperature. Wiley, New York, 360–375.

Pielou EC (1991) After the ice age: The return of life to glaciated North America. University of Chicago Press, Chicago.

Pissart A (1988) Pingos: An overview of the present state of knowledge. In Clark MJ (Ed) Advances in periglacial geomorphology. Wiley, New York, 279–297.

Plafker G (1969) Tectonics of the March 27, 1964 Alaska earthquake. U.S. Geological Survey Professional Paper 543-I.

Post A, Mayo LR (1971) Glacier dammed lakes and outburst floods in Alaska. Hydrologic Investigation Atlas HA-455, U.S. Geological Survey, Washington, DC.

Post A, Meier MF (1980) World glacier inventory. Proceedings of the Riederalp Workshop, September 1978. IAHS-AISH Publication No. 126, 45–47, Paris.

Power G, Cunjak R, Flannagan J, Katopodis C (1993) Biological effects of river ice. In Prowse TD (Ed) Environmental aspects of river ice. National Hydrology Research Institute Science Report No. 5. Saskatoon, Saskatchewan, Canada, 97–125.

Sellman PV, Brown J, Lewellen RI, McKim H, Merry C (1975) The classification and geomorphic implications of thaw lakes on the Arctic coastal plain, Alaska. U.S. Army Cold Regions Research and Engineering Laboratory (USA CRREL) Research Report 344, Hanover, NH.

Scrimgeour GJ, Prowse TD, Culp JM, Chambers PA (1994) Ecological effects of river ice break-up: A review and perspective. Freshwater Biol 32:261–276.

Stone KH (1963) Alaskan ice-dammed lakes. Assoc Am Geographers Ann 53:332–349.

University of Alaska (1989) Alaska climate summaries, 2nd ed. Alaska Climate Center Technical Note No. 5. Arctic Environmental Information and Data Center, University of Alaska, Anchorage.

USGS (1986) Water resources data, Alaska, water year 1986. U.S. Geological Survey Water-Data Report AK-86-1, Anchorage, AK.

Vannote RL, Minshall GW, Cummins KW, Sedell JR, Cushing CE (1980) The river continuum concept. Can J Fish Aquat Sci 37:130–137.

Van Cleve K, Dyrness CT, Viereck LA, Fox J, Chapin FS III, OeChel W (1983) Taiga ecosystems in interior Alaska. BioScience 33:39–44.

Viereck LA, Dryness CT, Batten AR, Wenzlick KJ (1992) The Alaska vegetation classification. USDA Forest Service General Technical Report PNW-GTR-286.

Wahrhaftig C (1965) Physiographic divisions of Alaska. U.S. Geological Survey Professional Paper 482. U.S. Government Printing Office, Washington, DC.

Washburn AL (1980) Geocryology: A survey of periglacial processes and environments. Wiley, New York.

Wetzel (1983) Limnology. 2nd ed. WB Saunders, Philadelphia, PA.

Wiles GC, Calkin PE (1994) Late Holocene, high-resolution glacial chronologies and climate, Kenai Mountains, Alaska. Geolog Soc Am Bull 106:281–303.

Zasada (1976) Alaska's interior forests. J Forestry June:334–337.

2. History of Limnology in Alaska: Expeditions and Major Projects

John E. Hobbie

Limnology in Alaska began with the natural history expeditions of the late 1800s by boat and foot and continues a century later with airplanes, helicopters, and synthetic aperture radar (SAR) for satellite investigations of lake ice duration. In the intervening century, the science itself changed drastically in many ways. The other chapters in this book are concerned with the knowledge about Alaska's lakes and streams that has been collected and synthesized by innumerable scientists. In this chapter, we will concentrate on equally interesting information about who some of these scientists were, what their motivation was and who paid for their research, how they traveled to the lakes and streams, and some of their adventures. Needless to say, the bear stories alone would fill a book, but because no limnologist was actually eaten, had to be dropped.

There have been three types of limnological research in Alaska; these are all being pursued today. The first was the expeditionary research where freshwater samples, mainly for the biota, were collected in the few hours spent at a lake or river. These samples range from collections of zooplankton made in northern Alaska as a part of the 1882 First International Polar Year (Wiggens, 1957) to intensive surveys of the biota and chemistry along the entire Noatak River carried out for the National Park Service (e.g., O'Brien et al., 1975). Another type of research is observational in which scientists spend weeks or months at a site making measurements with instruments, collecting samples for chemical analysis, and measuring rates of biological processes with oxygen changes or isotope incorporation. These limnologists may carry out research alone or as part of an integrated team. In a number of sites, the observational type of research has been extended to experiments, a third type of research, in which a single factor was changed during measurements of limnological processes and organisms in bottles, enclosures, and even in whole lakes and rivers.

The remoteness, sheer size of the region, and vast natural areas have never stopped limnologists from studying every conceivable freshwater habitat in Alaska. The logistics of how the research was carried out have changed drastically, however, over the century. Not surprisingly, the first limnologists traveled mainly by ship, although dog teams were used in the Arctic. The airplane and helicopter have been widely used for scientific travel both to place camps and scientists at remote lakes and to transport parties to headwaters for canoe or raft transects. Finally, the expanding Alaska road net has allowed cheap access to rivers and lakes throughout the state including those along the Alyeska Pipeline to the Prudhoe Bay oil fields.

Millions of dollars have been spent on Alaskan limnology. To obtain funding for their research, ingenious entrepreneurs have tapped sources from railroad moguls to the Atomic Energy Commission to university departments. The sales pitch has varied over the years from fundamental research, to species and their distribution, to the potential disturbance of freshwaters by an atomic blast, and to methods to increase the productivity of salmon nursery lakes.

All these interesting facts about how scientists traveled, researched, and obtained research funds in different eras are seldom found in the scientific literature. This chapter brings together some of these facts for the curious. One thing stands out—no matter how difficult the conditions and even when they were cold and hungry, limnologists in Alaska have had a wonderful and exciting time while carrying out excellent research.

Early History of Limnology in Alaska: 1880 to 1920

Harriman Expedition to Alaska: Science on a Cruise Ship

In 1899 the railroad financier E.H. Harriman planned a 2 month family trip to Alaska. In his own words "Our comfort and safety required a large vessel and crew, and preparations for the voyage were consequently on a scale disproportionate to the size of the party. We decided, therefore, if opportunity offered, to include some guests. . . ." (Burroughs et al., 1901). The final party was 14 Harriman family members and servants; 25 scientists; 5 artists, photographers, and stenographers; 4 surgeons, nurses, and chaplains; 11 hunters, packers, and camp hands; and 65 crew to man the ship (Fig. 2.1). Prominent scientists included C. Hart Merriam, W.H. Dall, B.E. Fernow, G.B. Grinnell, John Burroughs, and John Muir. They visited Sitka, Glacier

Figure 2.1. The steamship George W. Elder carried the Harriman Alaska Expedition for the 2 month trip (photo from Peck (1982) with permission).

Bay, Kodiak Island, Unalaska, the Pribilof Islands, and as far north as Port Clarence on the Seward Peninsula as well as Siberia. Their mostly marine and terrestrial collections, reported upon in some 13 volumes, included birds, mammals, land plants, algae including freshwater forms (Saunders, 1910), and insects. An account of the salmon industry by G.B. Grinnell (1901), astounding by modern standards, describes the very common practice on many of the streams of building dams or barricades designed to prevent any spawning fish from ascending. The returning fish piled up at the mouth of the stream and eventually the entire spawning run for that stream was captured.

Canadian Arctic Expedition: Science on a Shoestring

This expedition to northern Alaska and Canada, which took place between 1913 and 1918, was organized by Vilhjalmur Stefansson, an explorer and anthropologist (Fig. 2.2). Funding came from a wide variety of sources including private donations, museums, geographic societies, and governments. Compared with the Harriman expedition, it was a patched-together operation with travel by commercial shipping, small boats, and dog sleds. Much of the collecting was carried out by Fritz Johansen. In one publication (Johansen, 1922) he described the crustaceans of some arctic lagoons, lakes, and ponds collected near Teller, Alaska (on Seward Peninsula) as well as along the coast of northern Alaska. In addition to describing the types of water bodies, he mentions water depths, temperature, ice conditions, sedi-

Figure 2.2. The Canadian Arctic Expedition traveled by ship and dog team. (photo of a dog team in white-out conditions is from Stefansson, 1921).

ments, species of insects, permafrost, and the timing of freeze-up and melt. Specialists were called upon to describe other specimens and so it was that Chancey Juday of the University of Wisconsin wrote up the Cladocera (Juday, 1920). Juday describes the various species and where they were collected (e.g., Teller, Pt. Barrow, Camden Bay). He also took the opportunity to publish information on other arctic collections. One specimen of *Daphnia pulex* he described from the U.S. National Museum collection came from Pt. Barrow in 1882 (presumably on the First International Polar Year Expedition).

The many year lag in publication exasperated Johansen. For example, he sent Professor Sars in Copenhagen an ostracod collection that included specimens from Ottawa. When Sars wrote up the lot and promptly published it in the regular literature, the expedition report republished it in 1926 but with accompanying stern words. "The local Ottawa specimens were sent to Professor Sars without proper explanations by the collector, in his zeal for early publication, independently of the Arctic Publication Committee, and as a comprehensive report was prepared by Professor Sars. . . ." (Sars, 1926: 1).

Southcentral Alaska

Karluk Lake: The First Salmon Limnology Study

One of the first studies of a lake in coastal Alaska, and one of the first detailed studies of the bathymetry, temperatures, and plankton of any Alaskan lake, was of Karluk Lake on Kodiak Island by Willis Rich. Given the coastal location, it is not surprising that a ship was used for all logistics. Some of the scientific information on the lake and lake physics were published by Birge and Rich (1927) and by Juday et al. (1932); but it appears that Birge and Juday of the University of Wisconsin outfitted Rich and helped write up the data but did not travel to Alaska themselves (Frey,

1966; Richard F. Juday, 1994, personal communication). The techniques and sampling instruments were state-of-the-art. Rich was a professor at Stanford and also an employee of the U.S. Bureau of Commercial Fisheries. He was sent to Karluk Lake for the summers of 1926 to 1930 to investigate the fish habitat. One interesting calculation from the study was the amount of organic matter and nitrogen added to the watershed by the millions of dead salmon. This study illustrates how a federal agency sponsored basic research on lakes as a part of the answer to a very practical question, that is, was the lake a good habitat for fish. This is the first of a long series of studies that can be called "salmon limnology."

Bare Lake: Experimental Limnology Begins in Alaska

Twenty years later, another limnological investigation took place on Kodiak Island, this time on Bare Lake (Nelson and Edmondson, 1955). The background story illustrates how funding often chased the scientists in the 1950s. One day in 1950 Tommy Edmondson, a newly appointed assistant professor at the University of Washington, was approached by Phil Nelson of the U.S. Fish and Wildlife Service, who had a theory about the declining commercial harvest of salmon in Alaska. The harvest caused a decrease in the number of spawning fish returning to the headwater streams that led to a decreased input of phosphorus. Primary productivity became lower and eventually this affected the size of the salmon smolts leaving these lakes for the sea. To test a part of the theory, they planned a 1 year series of limnological measurements as a baseline, then a 3 year experimental addition of phosphorus to an entire lake. They chose Bare Lake (49 ha, 4.0 m average depth) for the experiment. For the first fertilization (of seven) they added 2500 lb of superphosphate fertilizer and 6500 lb of nitrate fertilizer. The towed a six-man life raft decked with planks around the lake and "Two men with brooms on the raft swept the mix into the water." Edmondson helped with the planning and interpretation but did not actually travel to Alaska. This was the first whole-lake fertilization study anywhere.

Afognak Island: The First ^{15}N Study in Aquatic Ecology

Another scientific first came from Afognak Island, close to Kodiak. Here, Richard Dugdale, Vera Dugdale (Vera Alexander), John Neess, and John Goering carried out the first nitrogen fixation measurements with ^{15}N in aquatic systems (Dugdale et al., 1959) in Little Kitoi and Upper Jennifer Lakes. In a later paper, Richard Dugdale and Vera Dugdale (Vera Alexander) (Dugdale and Dugdale, 1961) reported on phosphate and nitrogen sources to these lakes. The nitrate measurements were the first made with modern chemical methods. One of their results, the finding that nitrate was highest in streams draining the steepest watersheds, suggested to them that the source of the nitrogen was likely fixation in alder thickets, whose distribution was also correlated with steepness.

Southwest Alaska

Salmon Lakes

A long-term data set on the limnology of the salmon waters of the Bristol
Bay region has been collected by workers from the Fisheries Research
Institute of the University of Alaska. This work began in 1946 under the
leadership of W.F. Thompson. Professor Ole Mathisen summarizes the
work up to 1960 as "fisheries limnology." Then, a large federal appropria-
tion triggered intensive studies of water chemistry, primary and secondary
production, and studies on fish. These continued until 1980 and produced a
dozen M.S. degrees and several Ph.D. degrees with limnological topics; only
a few have been published.

The importance of alder nitrogen fixation for lake nitrogen was also a
conclusion of the work of Charles Goldman, whose Ph.D. thesis work was
carried out on three large lakes (Brooks, Naknek, and Becharof Lakes) of
the Alaska Peninsula (Goldman, 1960). He was supported in the field by the
U.S. Fish and Wildlife Service. The Brooks Lake Research Station was
supplied mainly by amphibious aircraft.

Amchitka Island: Nuke the Aleutians

The Atomic Energy Commission (AEC) initiated detailed studies of the
environmental setting of Amchitka Island, in the Aleutians, in 1966. The
purpose was pragmatic; data were needed for planning and assessing
an underground nuclear test of a warhead larger than could be safely tested
at the Las Vegas site. An environmental study was needed because
Amchitka is a part of the Aleutian Islands National Wildife Refuge and also
because the geology and biota were quite different from those of other
nuclear test sites (Merritt and Fuller, 1977). The island lies at 51°N and
179°E, some 2160 km west–southwest of Anchorage; it is more than halfway
to Asia. Two extensive freshwater studies were carried out, one from
Battelle Columbus Laboratories (Burkett, 1977) on limnology and one
from the Ecology Center, Utah State University on aquatic ecology (Valdez
et al., 1977). Of these, the limnology report deals mostly with extensive
surveys of biota and chemistry; the aquatic ecology report contains
more of the dynamics of animal growth as well as a compartmental stream
model.

The major effect of the several explosions that took place was on the
surface geomorphology. The most powerful nuclear device, named
Cannikin, was 5 megatons in yield; it was detonated at a depth of 1791 m
below the land surface. The explosion cavity formed a vertical rubble
chimney extending to the land surface. It took 10 months for this rubble
chimney to become saturated with water and for a lake to begin to
form. Cannikin Lake is now 12 ha in area and has a maximum depth
of 10 m.

Arctic Alaska

Big Science at Cape Thompson: Nuking the Arctic

The forerunner of the Amchitka project was Project Chariot. It began in the 1950s when the AEC planned up to six atomic explosions to excavate a harbor at Cape Thompson (125 miles northwest of Kotzebue). "We looked at the whole world," said Teller, "and tried to pick a spot where we could demonstrate the peaceful uses of nuclear energy" (O'Neil, 1994). As a part of Project Chariot, a detailed environmental study in 1959 to 1961 with 40 subprojects and over 100 scientists investigated the biological and physical environment. Everything from the geology to the climate to the soils, biology, and limnology was studied. Transportation and field support were all provided for the scientists and the resulting project was one of the first all-out environmental studies anywhere. The objectives were (Wilimovsky and Wolfe, 1966: iii):

1. "to permit: a) estimates of the biological cost of the excavation operation and b) judgments as to whether or not there would be effects that could result in widespread damage or major disruption of ecological systems."
2. to produce findings that "could be used as base lines in studies to be conducted in postexcavation time."

One of the chapters in the gigantic report of the project (Wilimovsky and Wolfe, 1966) covers the limnology of small lakes and a stream (Watson et al., 1966). The authors describe the physical and chemical events in the lakes and streams as well as the algae and animals found. There is no attempt to describe any ecological interactions and no comparison of their findings with any other lakes or streams. There is excellent descriptive information on the geology, climate, soils, stream chemistry, and vegetation. Overall, the results are documented well enough to serve as a baseline; the site should be revisited to look for long-term changes in biota.

From the post-Chernobyl viewpoint of the 1990s, it is incredible that Project Chariot was ever considered. A recent book (O'Neil, 1994) describes the AEC and Alaskan enthusiasm for the project and the caution and disillusion of some of the scientists involved. The author states that a number of the scientists became active environmentalists, came to oppose the project, and lost their jobs as a consequence.

The "Big Laboratory": The Naval Arctic Research Laboratory at Pt. Barrow

A different kind of big science, one could call it the Big Laboratory, existed at Barrow (71°N) from 1947 to the late 1970s. During these three decades, the Naval Arctic Research Laboratory (NARL) supported the research of hundreds of scientists from physicists to biologists. A large number of graduate students, including the author of this chapter and four or five other limnologists, produced theses. The younger scientists working in the Arctic at this time took the NARL for granted. Upon reflection, it stands out as a unique undertaking with high scientific productivity. There has been no comparable laboratory for research in the tropics or temperate regions or in any other habitat or region with the single exception of the McMurdo Laboratory (Antarctic) of the National Science Foundation (NSF).

Why was the military supporting basic research on arctic ecology and was this a good thing? In 1946, the Office of Naval Research (ONR) had just been established and was looked on as "a channel for the flow of federal money to university laboratories for the support of basic research" (England, 1982: 62). In 1946, ONR became interested in setting up some biological research at Barrow in connection with exploration for petroleum in the Naval Petroleum Reserve No. 4, an immense area south of Barrow set aside in the 1920s to ensure that the Navy would have adequate fuel for its ships when war came. The solution to the need for biological research, which seemed reasonable in 1946 and amazing to us today, was to invite a biological scientist at Swarthmore, P.F. Scholander, to set up a new laboratory at Pt. Barrow. The first research followed Scholander's interests: physiological studies of the adaptations of mammals, birds, fishes, and invertebrates to arctic life. Scholander et al. (1953) reports many details about how animals and plants survive freezing and provides the definitive data ending forever the myth that the Alaska blackfish can survive freezing. In fact, even freezing a part of the tail fin was fatal.

Over time, research at NARL came to deal with the whole spectrum of the arctic environment including soils, permafrost, beach formation, microbiology, taxonomy, marine biology, and atmospheric concentrations of carbon dioxide. The ONR support that made it all possible was certainly based on the enlightened attitude stated by Britton (1973: 11) that to the Navy ". . . fell a first responsibility, and what might be considered a moral commitment, to learn both how to use the environment and how to protect it."

By 1960, NARL had evolved to include a large laboratory building, animal quarters, several dormitories, family living quarters, a fleet of jeeps and tracked vehicles (weasels), a small air force of some five planes, and administrators, pilots, custodians, carpenters, and mechanics. All the buildings were quonset huts (Fig. 2.3). There was a research station floating on the Arctic ice and another station at Lake Peters in the Brooks Range. Small field parties were placed at various locations across the North Slope for days or weeks.

Dan Livingstone recalls with enthusiasm his two summers at Barrow in 1951 and 1952. The scientific atmosphere at the laboratory was intense and many prominent scientists, including George Lauff and George Schaller, were then graduate students and research assistants so scientific arguments and discussions cut across all fields and disciplines. Livingstone spent many

Figure 2.3. Quonset huts housed the support facilities for the U.S. DEW Line radar network (Distant Early Warning) as well as the Naval Arctic Research Laboratory (NARL). The NARL is located in the large buildings in the center background (1961 photo by J. Hobbie).

weeks exploring the larger lakes of the Brooks Range utilizing the NARL airplanes, but recalls the apprehension of knowing that at least one pilot was killed each year. There was always the additional risk that a field party would be forgotten or at least picked up very late. One time he and a companion went to the field with 3 weeks of food, but told the pilot to come back in 2 weeks. They ate all the food in 2 weeks, but the pilot knew they had 3 weeks worth so he did not return for 3 weeks. From his work (Livingstone et al., 1958) came the first real concepts of regional limnology of the Arctic. That is, how do the environmental features of the Arctic influence both the present limnology and the development of lakes over time? He concluded that the Arctic affected lakes mostly through the effect of the climate on lake presence and form of development. He called it lake physiography. In relatively flat areas, such as the Coastal Plain, the climate dictates that permafrost is present and thus controls the very existence of lakes. Climate also strongly affects the nature and amount of materials entering the lakes.

W.T. Edmondson also carried out research at Pt. Barrow, this time as advisor to a graduate student, Gabe Comita. Edmondson reports that he was flown all the way from Seattle to Washington, DC, for a meeting about secret research, presumably concerned with environmental effects of atomic explosions. Nothing came of the secret research but he ended up with an offer for research funding at NARL. Comita, while studying the population dynamics of copepods was able to take advantage of the presence of only a few species of zooplankton to follow young stages.

John Hobbie arrived in Alaska by a circuitous route. He had been a cook and research assistant on a Greenland study funded by the U.S. Air Force. In 1958 he was scheduled to go to northern Ellesmere Island as a limnologist but the expedition destination was changed when the U.S. Air Force decided that scientists would no longer be supported out of Thule Air Force Base in Greenland. Instead, the whole expedition was diverted to a Brooks Range lake, Lake Peters. The expedition almost ended at the start when the wheels of a chartered C-46 transport plane punched through the ice in mid June (Fig. 2.4). Three months later, after propeller replacement, the plane successfully took off from the tundra at the edge of the lake. Beginning in August 1960, Hobbie and his wife spent a year at Lake Peters on his Ph.D. research studying the annual cycle of primary productivity as controlled by physical and chemical factors. All support came in by small ski and float planes from NARL at Barrow. One unexpected finding was that the fuel lines leading to the kerosene stoves blocked up every time the temperature dropped below –48°C (–55°F). Because Barrow, on the coast, does not experience the extreme cold of the mountain valleys, the NARL officials

Figure 2.4. The expedition leader, G.W. Holmes, surveys the C-46 transport floating in Lake Peters during the breakup of June 1958. The plane eventually flew again (photo by J. Hobbie).

refused to believe this and blamed the usual suspect, an incompetent graduate student.

The highest productivity in the plankton occurred in the spring and early summer beneath the nearly 2 m of ice when the thin snow cover, clear ice, and clear water combined to produce some 50% of the total primary productivity. During the open water season of July through September, the lake was turbid from glacial rock flour added by streams (Hobbie, 1964).

By the early 1960s, NSF grants were still relatively easy to obtain. For example, David Frey of Indiana University obtained a small NSF grant to support his student Jaap Kalff at Barrow studying links between the lemming highs and lows and nutrient cycles in the small ponds. Kalff points out now that the idea was too half-baked to make it as a 1990s research proposal. His research on limitations on primary production was the first serious study of nutrients in this region. Travel from the laboratory to the research ponds was by tracked vehicles across the tundra; this was taken for granted in the 1960s but would not be allowed at all today. GAIA's revenge for the inevitable destruction of the tundra vegetation was to create small ponds in the roads by thermal erosion. Some ponds could be 3 m deep, enough to drown an unwary driver. Kalff also made good use of samples of opportunity and with the help of the NARL pilots built up a data base of the chemistry of many lakes of northern Alaska.

International Biological Program (IBP) at Barrow: Big International Science

Biologists looked at the boost in knowledge, prestige, and even funding that other types of scientists reaped from the 1958 International Geophysical Year and decided to create their own international science program with the theme of the biological basis of productivity and human welfare. The U.S. participation in the IBP emphasized gaining an understanding of the structure and function of major ecological or human systems. Five major studies of U.S. biomes were funded for 4 to 5 years and the Tundra Biome study was placed at Barrow and the NARL. Here terrestrial and freshwater projects studied the carbon, nitrogen, and phosphorus cycling with the perspective of an entire system. It was focused, question-based research at a scale of support and facilities that enabled scientists to go far beyond descriptive limnology and investigate the processes and controls of carbon and nutrient flux in an entire aquatic system.

The limnology project built on the studies of shallow ponds by Kalff, Bob Barsdate, and Vera Alexander. The three summers in the field featured up to 29 investigators all studying three small (~50 m diameter), shallow (~0.5 m) ponds. Gravel roads were built and a power line reached a small field laboratory near the study ponds. One pond was kept as a control while other ponds were manipulated with treatments of nutrients and oil. An aerial tramway and wooden walkways were necessary to minimize the effect of investigator sampling on the control ponds. The approaches utilized are summarized in the synthesis volume of the project (Hobbie, 1980: vii). "We concentrated first on measuring the fluxes of carbon, nitrogen, and phosphorus through the ecosystem. Next, we used a variety of manipulations, ranging from changed light conditions for plankton in a small bottle to an increase in phosphate in a whole pond, to investigate the controls of various processes. While the field work was going on, we also constructed a mathematical model of the ecosystem." All investigators were required to make measurements in the control pond; this sounds obvious and trivial but the natural inclination of each of the nine principal investigators was to work on the pond best for zooplankton, or for rooted vegetation, or for benthos. In this case, the modeling effort provided the impetus to concentrate on a single pond. The model was also invaluable as the act of construction led to recognition of what we knew and did not know and thus to new directions and overall coordination. The scientific output was only moderately successful because of lack of information on controls of various processes. For example, algal respiration was an important part of the model of the carbon budget, but practically nothing was ever learned about the regulation of this process.

By 1971 NARL occupied a large new building with many modern laboratories and fancy motel-like living quarters. The IBP project scientists had everything at hand including spectrophotometers, microscopes, a liquid scintillation counter, and even computers. Stable isotopes were measured quickly in Fairbanks. I recall the excitement of the summer of 1973 when most of the principle investigators (PIs), postdocs, and graduate students would meet in a laboratory for beer at the end of the day and argue about the latest finding and about the gaps in our understanding of the carbon or nitrogen cycle. Moreover, the array of techniques and skills available meant that we could often test a hypothesis the next day with, among others, Bob Barsdate using ^{32}P; Vera Alexander measuring the algal numbers and movement of nitrogen with ^{15}N; Mike Miller measuring the response of the primary producers; and Tom Fenchel and Hobbie investigating the populations, production, and predation of bacteria and protozoans.

Results were published in many scientific papers and in a 500 page synthesis book. One highlight was the development of a method for the direct counting of planktonic bacteria and the first measurements of bacterial numbers in waters and sediments of a pond and lake. Another was the discovery that the control of phosphorus concentrations, and through this control of the rate of primary production, lay in the chemical reactions of the surface sediments (Prentki et al., 1980). Phosphorus entering a pond was rapidly sorbed into a hydrous iron complex. The concentration of dissolved reactive phosphorus in the water and the release rate of the sorbed phosphorus was controlled by a chemical equilibrium that was, in turn, controlled by the different amounts or iron and inorganic phosphorus in the sediments. Algal photosynthesis in the water column of a series of tundra ponds could be predicted by measuring the phosphate sorption index of the underlying sediments.

Coordinated Research Projects at Toolik Lake

After the IBP ended in 1974, the limnologists on the project decided that the ponds and lakes of the coastal plain were relatively well known but that little was known about those of the foothills. At this time, the oil pipeline construction was well underway and the accompanying haul road opened for vehicles. Accordingly, limnologists received NSF funding and began sampling at Toolik Lake, in the foothills of the Brooks Range. Over the last 21 years an excellent field camp has been built up, now run by the University of Alaska, providing small laboratories, electric power, and meals and lodging.

The Toolik aquatic project has been characterized by group proposals to NSF and research oriented toward understanding processes. Transportation is by truck on gravel roads from Fairbanks and Prudhoe Bay. Stream research has become a vital part of the project. The project scientists believe that great progress can be made by staying at one site and manipulating conditions in lakes and streams to gain insight into effects of changes and into processes.

The Toolik camp attracted additional researchers on a variety of projects. One important project was funded by the Department of Energy and was called the R4D program (Response, Resistance, Resilience to, and Recovery from Disturbance in Arctic Ecosystems). From 1984 to 1989 an integrated project was aimed at creating an understanding of the controls of water and nutrient availability on plant production, plant–nutrient uptake, and water and nutrient movement in a small watershed, Imnavait Creek, in the foothills near Toolik (Reynolds and Tenhunen, 1996). The project experiments included adding water and nutrients to small watersheds. Limnological studies in the R4D project involved the physics (Oswood et al., 1989; Irons and Oswood, 1992), chemistry (Oswood et al., 1996), and ecology of a small beaded stream draining a watershed that contained peat soils and permafrost. Both the transport of dissolved materials (Everett et al., 1996) and the nature of the stream were controlled by hydrological events; that is, snow melt and summer rainstorms created a rapid flow, streamlike system whereas low flow created a lakelike system.

Since 1987 the Toolik Lake research has received additional funding as a part of the Long Term Ecological Research (LTER) program of the NSF. This program emphasizes the development of long-term data sets and comparative studies that cut across the 18 sites that make up LTER. This funding and the advent of digital recording has allowed the accumulation of a complete climate record at the Toolik site as well as long-term data on lake and stream temperature, water flow, and aquatic chemistry. Also, the results from arctic studies are now beginning to appear in articles that compare findings from arctic, boreal, and temperate lakes and streams. For example, the supersaturated CO_2 concentrations found in arctic lakes have now been found to be a feature of many if not most temperate lakes as well and give insight into the interactions between groundwater and lakes (Cole et al., 1994). In another case, a model for stream nitrogen cycling developed in the Toolik project has been applied to LTER streams in North Carolina and Oregon. The Toolik researchers have found that it is necessary to study the lakes and streams for many years to gain the maximum understanding from manipulations such as the increase or decrease in the numbers of the lake trout, which are the top predators, or the fertilization of large streams where major changes still occurred after 10 years of phosphate addition. Results from the Toolik project are presented in two chapters in this book.

Postscript

Freshwater research in Alaska is more than a century old. Over that time the science has changed from expedition collecting and observational research by single scientists to large experiments involving tens of scientists. The facilities needed reflected the complexity of the research, beginning with tents and ending with well-equipped field stations and laboratories. These progressions have occurred elsewhere in the world of freshwater ecology. What is unique to Alaska is the variety and cost of the transport needed for the research. Where else has a 180 foot privately chartered ship been used as a mobile lab? Or dogsleds used as the standard transport? Or airplanes and helicopters used more frequently than roads and boats for limnological surveys? Alaskan limnology has been well funded by an amazing variety of sources. Fish and Wildlife agencies have directed major projects toward improving salmon stocks. The AEC paid for a limnological study carried out "before" six atomic bombs were detonated. The U.S. Navy sponsored numerous aquatic studies on the North Slope. Major NSF research under the IBP and the LTER Program was carried out in Alaska.

The combination of Alaska's physical beauty, wilderness, and opportunities for freshwater research have attracted high caliber and innovative limnologists from throughout the United States. Only a few of their practical problems of funding and logistics have been described here. The quality of their science speaks for itself in the rest of this book.

References

Birge EA, Rich WH (1927) Observations on Karluk Lake, Alaska. Ecology 8:384.
Britton ME (1973) Introduction. In Britton ME (Ed) Alaskan arctic tundra. Arctic Institute of North America Technical Paper No. 25, Washington, DC, 9–15.

Burkett RD (1977) Limnology. In Merritt ML, Fuller RG (Eds) The environment of Amchitka Island, Alaska. Technical Information Center, Energy Research and Development Administration, 269–286.

Burroughs J, Muir J, Grinnell GB (1901) Alaska. Narrative, glaciers, natives, Vol 1. Doubleday, Page and Company, New York.

Cole JJ, Caraco NF, Kling GW, Kratz TK (1994) Carbon dioxide supersaturation in the surface waters of lakes. Science 265:1568–1570.

Dugdale RC, Dugdale VA (1961) Sources of phosphorus and nitrogen for lakes on Afognak Island. Limnol Oceanogr 6:13–23.

Dugdale RC, Dugdale VA, Neess J, Goering J (1959) Nitrogen fixation in lakes. Science 130:859–860.

England JM (1982) A patron for pure science. The National Science Foundation's Formative Years, 1945–57. National Science Foundation, Washington, DC.

Everett K, Kane DL, Hinzman LD (1996) Surface water chemistry and hydrology of a small arctic drainage basin. In Reynolds J, Tenhunen J (Eds) Landscape function and disturbance in arctic tundra. Springer–Verlag, Berlin, 185–201.

Frey DG (1966) Wisconsin: the Birge-Juday era. In Frey DG (Ed) Limnology in North America, University of Wisconsin Press, 3–54.

Goldman C (1960) Primary productivity and limiting factors in three lakes of the Alaska Peninsula. Ecol Monogr 30:207–230.

Grinnell GB (1901) The salmon industry. In Harriman Alaska expedition, Vol II. Doubleday, Page and Company, New York 337–355.

Hobbie JE (1964) Carbon-14 measurements of primary production in two Alaskan lakes. Verh Int Verein Theor Angew Limnol 15:360–364.

Hobbie JE (Ed) (1980) Limnology of tundra ponds, Barrow, Alaska. US/IBP Synthesis Series No. 13. Dowden, Hutchinson and Ross, Inc., Stroudsburg, PA.

Irons JG III, Oswood MW (1992) Seasonal temperature patterns in an Arctic and two subarctic Alaskan headwater streams. Hydrobiologica 237:147–157.

Johansen F (1922) The crustacean life of some arctic lagoons, lakes, and ponds. In Report of the Canadian Arctic expedition 1913–1918, Vol VII, Crustacea. Part N, 1–31. FA Acland, Printer to the King's most excellent majesty, Ottawa.

Juday C (1920) The Cladocera of the Canadian Arctic expedition, 1913–1918. In Report of the Canadian Arctic Expedition 1913–1918. Vol VII, Crustacea, Part H, Cladocera. 1–8. T. Mulvey, Printer to the King's most excellent majesty, Ottawa.

Juday C, Rich WH, Kemmerer GI, Mann A (1932) Limnological studies of Karluk Lake, Alaska. Bull Bur Fisheries 12:407–434.

Livingstone DA, Bryan K Jr, Leahy RG (1958) Effects of an arctic environment on the origin and development of freshwater lakes. Limnol Oceanogr 3: 192–214.

Merritt ML, Fuller RG (Eds) (1977) The environment of Amchitka Island, Alaska. Technical Information Center, Energy Research and Development Administration.

Nelson PR, Edmondson WT (1955) Limnological effects of fertilizing Bare Lake, Alaska. U.S. Fish Wildlife Serv Fish Bull 56:415–436.

O'Brien WJ, Huggins D, deNoyelles Jr F (1975) Primary productivity and nutrient limitations of phytoplankton in the ponds and lakes of the Noatak River basin, Alaska. Arch Hydrobiol 2:263–275.

O'Neill D (1994) The firecracker boys. St. Martin's Press, New York.

Oswood MW, Everett KR, Schell DM (1989) Some physical and chemical characteristics of an arctic beaded stream. Holarct Ecol 12:290–295.

Oswood MW, Irons, III JG Schell D (1996) Dynamics of dissolved and particulate carbon in an arctic stream. In Reynolds J, Tenhunen J (Eds) Landscape function and disturbance in an arctic tundra. Springer–Verlag, Berlin, 275–289.

Peck RM (1982) A celebration of birds. Walker Publishing Co., Inc., New York.
Prentki RT, Miller MC, Barsdate RJ, Alexander V, Kelley J, Coyne P (1980) Chemistry. In Hobbie JE (Ed) Limnology of tundra ponds, Barrow, Alaska. US/IBP Synthesis Series No. 13. Dowden, Hutchinson and Ross, Inc., Stroudsburg, PA, 76–178.
Reynolds J, Tenhunen J (Eds) (1996) Landscape function and disturbance in arctic tundra. Springer–Verlag, Berlin.
Sars GO (1926) Freshwater ostracods from Canada and Alaska. In Report of the Canadian Arctic expedition 1913–1918, Vol VII, Crustacea, Part I, 1–23. FA Acland, Printer to The King's most excellent majesty, Ottawa.
Saunders DA (1910) The algae of the expedition. In Harriman Alaska series, Vol V, Cryptogamic botany. The Smithsonian Institution, Washington, DC, 153–250.
Scholander PF, Flagg W, Walters V, Irving L (1953) Climatic adaptations in arctic and tropical poikilotherms. Physiol. Zool. 26:67–92.
Stefansson V (1921) The friendly arctic. Macmillan Co., New York.
Valdez RA, Helm WT, Neuhold JM (1977) Aquatic ecology. In Merritt ML, Fuller RG (Eds) The environment of Amchitka Island, Alaska. Technical Information Center, Energy Research and Development Administration, National Technical Information Service US Department of Commerce, Springfield Virginia, 287–313.
Watson DG, Hanson WL, Davis JJ, Cushing CE (1966) Limnology of tundra ponds and Ogotoruk Creek. In Wilimovsky NJ, Wolfe JN (Eds) Environment of the Cape Thompson region, Alaska. U.S. Atomic Energy Commission, Division of Technical Information, Oak Ridge, TN.
Wiggens IL (1957) The Arctic—Its discovery and past development. In Hansen HP (Ed) Arctic biology. Oregon State College, Corvallis, OR, 3–11.
Wilimorskyal, Wolfe (Eds) (1996). Environment of the Cape Thompson region, Alaska. U.S. Atomic Energy Commission, Division of Technical Information, Oak Ridge, TN.

3.   The Limnology of Toolik Lake

W. John O'Brien, Michele Bahr, Anne E. Hershey,
John E. Hobbie, George W. Kipphut, George W. Kling,
Hedy Kling, Michael McDonald, Michael C. Miller,
Parke Rublee, and J. Robie Vestal

The scientific study of the Arctic is recent with the exception of various collections and cataloging of plants and animals in the 19th century. The limnological investigation of the Arctic is even more recent with the first review paper of arctic limnology listing only seven papers that dealt with arctic lakes (Rawson, 1953). However, after World War II, research stations developed in arctic Europe, Greenland, and at Point Barrow in Alaska. There was considerable research activity at the Naval Arctic Research Laboratory (NARL) in Barrow (Livingstone et al., 1958; Hobbie, 1964, Chapter 2; Stross and Kangas, 1969), much of which is reviewed by Hobbie (1973).

Further impetus to arctic research developed through the International Biological Program (IBP). The IBP was patterned after the spectacularly successful International Geophysical Year of the late 1950s. The goal of the IBP was worldwide ecological studies of natural communities. In many countries, such studies were of small scale and fragmented. However, in the United States it was decided to fund large-scale studies of five major biomes found in the United States, one of them the tundra biome located near Pt. Barrow. The approach of the U.S. IBP was to study and model the flows of carbon and other major nutrients through these major ecosystems.

The arctic IBP began modestly in 1970 with some preliminary limnological investigations of some of the low center polygonal ponds in the Barrow area. These studies intensified in 1971 and ran through 1973 with the summer of 1974 spent developing reports of the research results. Much of this work is reviewed in Hobbie (1980). The study of Toolik Lake was an outgrowth of the Tundra IBP at Barrow. Although much was learned from the Arctic IBP (Hobbie, 1980) about the limnology and biogeochemistry of small, shallow, fishless ponds, these ponds represented just one of many lake types (see Chapter 1) and just one of several arctic areas. Hence as the IBP came to an end in 1974, the National Science Foundation (NSF) Division of Polar Programs encouraged the continued study of other lake types including deeper lakes that contained fish. Thus in 1975 the study of Toolik Lake was initiated. Toolik Lake was chosen for two reasons. First, the initial research group headed by John Hobbie realized that the Dalton Highway, the road built to facilitate the construction of the trans-Alaska pipeline, traversed areas hitherto unstudied and that logistical costs could be much reduced by studying a lake accessible by road. Second, of the several deep lakes near the road, only Toolik Lake had an Alyeska pipeline construction camp nearby that could provide initial logistic backup. Furthermore, Toolik

Table 3.1. List of Principal Investigators Involved in Research in Toolik Lake Region

V. Alexander Institute of Marine Science University of Alaska Fairbanks, Alaska	G. Kling Department of Biology University of Michigan Ann Arbor, Michigan
R. Barsdate Retired	M. McDonald Department of Chemical Engineering University of Minnesota
P. Colinvaux Smithsonian Tropical Research Institute Panama	Duluth, Minnesota
J. Haney Department of Zoology University of New Hampshire Durham, New Hampshire	M. Miller Department of Biological Sciences University of Cincinnati Cincinnati, Ohio
A. Hershey Department of Biological Sciences University of Minnesota Duluth, Minnesota	S. Mosley Department of Zoology North Carolina State University Raleigh, North Carolina
J. Hobbie The Ecosystem Center Marine Biological Laboratory Woods Hole, Massachusetts	J. O'Brien Department of Systematics and Ecology University of Kansas Lawrence, Kansas
G. Kipphut Department of Geosciences Murray State University Murray, Kentucky	R. Vestal Deceased

Figure 3.1. The Toolik Field Station in 1975. In the background is the Toolik Lake pipeline construction camp, which was active at the time. The pipe in the foreground was the water supply pipe for the pipeline construction camp.

Lake is the deepest lake easily accessible from the Dalton Highway north of the Brooks Mountains.

The initial studies of Toolik Lake began in the summer of 1975 with a series of surveys. The initial principal investigators, coordinated by John Hobbie, consisted of Vera Alexander, Robert Barsdate, James Haney, Michael Miller, Samuel Mosley, and John O'Brien. This team was assembled to cover specific taxonomic groups or types of processes. This type of organization has remained throughout the project even though many of the personnel have changed. A list of all the principal investigators who have been involved in lake investigations in the Toolik Lake region is given in Table 3.1. The initial studies focused almost exclusively on the limnology of Toolik Lake, studying both biomass and various processes within the lake. The initial laboratory facilities were humble, with a single, small camper in 1975 (Fig. 3.1) ultimately growing to a sophisticated field station (Fig. 3.2) that can now accommodate 50 to 70 scientists and that is slated to continue to grow.

Up to the early 1980s, limnological research at Toolik Lake consisted largely of process and biomass measurements with each principal investigator operating rather independently. However, in 1983 we began a series of integrated whole system manipulations where all principal investigators and their students focused on the manipulated systems. From 1983 to 1985 we studied six large (60 to 100 m³) enclosures, often termed limnocorrals,

Figure 3.2. A view of the southern end of Toolik Lake and the Toolik Field Station in the late 1980s. In the background is the Dalton Highway. Structures for terrestrial vegetation manipulation are shown on the hillside.

placed in Toolik Lake to address the general question of the impact of bottom up or top down control of various components of arctic lentic ecosystems. That is, are arctic lakes regulated both in density and structure of biota by the amount of available nutrients or by the type and density of top predators or a combination of both depending upon the level in the ecosystem? In these studies we took Toolik Lake to be the reference and used the manipulated systems to understand the mechanisms operating in the lake (O'Brien et al., 1992).

Limnocorrals proved to be poor surrogate systems for studying the impact on the benthos and fish. Furthermore, maintenance of the limnocorrals for extended periods of time was difficult. Hence in 1985 we began a series of whole lake manipulations, beginning with the partitioning of a small lake and addition of nitrogen and phosphorus nutrients to one side. The overall project also grew to include stream and terrestrial components. In 1987 the Toolik Lake area was designated an NSF Long Term Ecological Research (LTER) site. As such, long-term whole lake manipulations were begun with long-term monitoring of the results. These experiments have grown to include the slow and rapid removal of lake trout from several lakes, nutrient addition to a divided lake that has grayling and sculpin and to a lake that has both lake trout and grayling, and lake trout addition to a small lake with grayling and sculpin. Both LTER funding and NSF Polar Program funding were involved up to 1994.

Site Description of Toolik Lake

Toolik Lake (68°38′00″N, 149°36′15″W) (Fig. 3.3) is a multiple basin kettle lake (Fig. 3.4) and the largest of a number of kettle lakes (Fig. 3.5) set in a terminal moraine (Fig. 3.6) of the Itkillik II glaciation, a local glaciation during the Wisconsin glacial epoch. It is located at 720 m elevation, about 20 km north of the Brooks Mountains (Fig. 3.7) and just off the Dalton Highway. The lake is about 12,400 years old (Colinvaux, personal communication). The morainal material around the lake is composed mostly of Devonian conglomerates of the Chandler formation and Cretaceous sandstones (Keller et al., 1961). Tussock and upland heath tundra dominate the vegetation in the watershed and vegetation cover is nearly 100% (Miller et al., 1986).

The lake has a surface area of 149 ha and a volume of $10 \times 10^6 \, \text{m}^3$ with a maximum depth of 25 m, and a mean depth of 7 m (Fig. 3.8). The lake is composed of five basins that are somewhat isolated by a series of shoals, some of which come within 1 to 2 m of the surface of the lake. The Toolik watershed covers 65 km², yielding a ratio of catchment to surface area of 43 : 1. Seventy percent of the water entering the lake comes from the south inlet stream that drains 10 lakes upstream of Toolik Lake. A secondary inlet provides 9% of the water and several ephemeral streams account for the remainder of the inflow (Whalen and Cornwell, 1985). The major inflow comes in the spring from snowmelt from mid-May to early June (Hobbie et al., 1983). Ice-out in the lake has varied from June 15 to July 2 over the last

Figure 3.3. A view of Toolik Lake looking west from the Toolik Field Station. Snow-covered Jade Mountain is shown in the background.

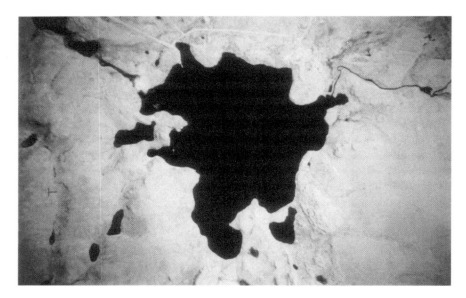

Figure 3.4. An aerial photograph of Toolik Lake and some of the ponds and lakes close to Toolik Lake. The major inlet is shown flowing into the southernmost bay of the lake (at right) and the outflow flowing out of the northernmost bay (at left).

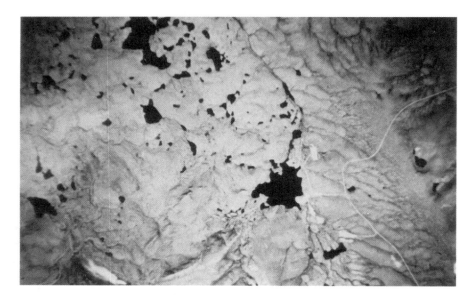

Figure 3.5. An aerial photograph of Toolik Lake showing more of the surrounding area. Many of the lakes and ponds shown have been surveyed. The Dalton Highway and the access road to the Toolik Field Station are shown on the east side of the lake (at right).

Figure 3.6. A view of Toolik Lake from the hills on the southwest side of the lake. Notice the morainal hills and small kettle basins in the foreground.

Figure 3.7. A view of the Brooks Mountains looking due south from Toolik Lake.

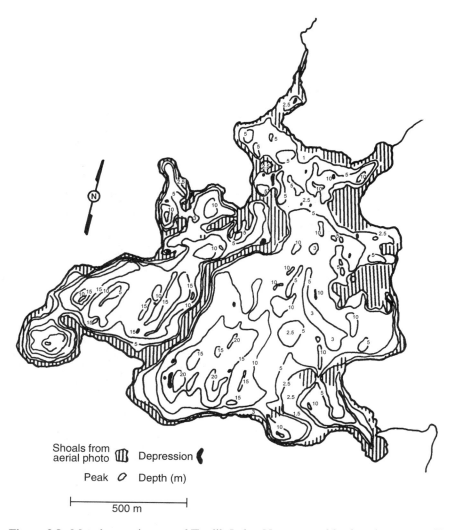

Shoals from
aerial photo 𝕝𝕝𝕝 Depression **❨**

Peak *𝒪* Depth (m)

|———————————|
        500 m

Figure 3.8. Morphometric map of Toolik Lake. Note several basins almost cut off by the shoals (map courtesy of John Miller, North Carolina State University).

19 years. The lake level, which fluctuates by as much as 59 cm, is generally highest in the late spring at the time of major snowmelt inflow, and then declines throughout the summer.

Water Budget

The most detailed information on the annual water budget for Toolik Lake comes from a flow monitoring study conducted from May to September 1980 by Whalen and Cornwell (1985). The main features of the water budget are summarized in Table 3.2. In 1980 (Whalen and Cornwell, 1985) approximately 40% of the annual inflow to Toolik Lake occurred while the lake was ice covered. However, considerable year to year variability in flow exists and high flow may also occur in response to heavy summer rainfall.

Even though the main features of the water budget have been fairly well defined (Table 3.2), there are components of the water cycle that are not well understood but may have an influence upon biological and chemical processes within the lake. The extent to which water within each basin mixes with water in other basins is not known, but there is evidence that mixing between basins may be restricted. No study has been undertaken to determine whether there are consistent chemical differences within the water column of the separate basins. However, routine summer measurements of temperature, conductivity, and chlorophyll show no significant differences between the main basin and limnobay, a small isolated basin at the northern part of Toolik Lake (M. Miller, unpublished data). In May 1985, dissolved oxygen concentrations were measured under ice in the main basin, and in the deep basin in the westernmost part of the lake. These measurements were taken at the end of the long winter season when chemical differences between basins would be expected to be most pronounced. Dissolved oxygen distributions within the two basins were very similar to a depth of approximately 8 m. At depths greater than 8 m, O_2 concentrations were 50 to 90 $\mu mol\, L^{-1}$ less in the deep basin than at similar depths in the main basin (Fig. 3.9). The depth of the shallows that separate the deep basin from the rest of the lake is approximately 8 m. These limited data suggest that interbasin chemical variability is small at shallow depths, probably because water can circulate relatively freely at these depths. However, deep waters within separate basins clearly can develop chemical differences, at least during the winter. In the case of O_2, such differences almost certainly reflect the bathymetric characteristics of each basin.

A 1979 study followed the flow of stream water under the ice by injecting rhodamine dye into the primary inlet stream and determining its distribu-

Table 3.2. Hydrologic Data for Toolik Lake

	Flow ($\times 10^6\, m^3$)	Percent
Inlet 1	13.7	70.6
Inlet 2	1.7	8.8
Undefined flow	4.0	20.6
Total inflow	19.4	100.0
Outlet	19.4	
Theoretical water renewal time (yr)		0.5

Data are based on measurements for the 1980 season and are from Whalen and Cornwell (1985). This study found that direct precipitation and evaporation from the surface of the lake were negligible. Flow was monitored from May 13, to August 31.

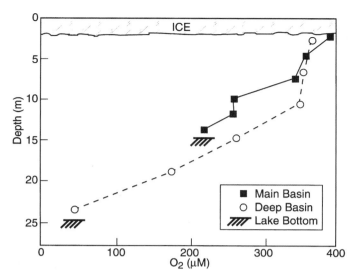

Figure 3.9. Toolik Lake dissolved oxygen profiles beneath the ice, May 1985. Sites included the main basin of the lake and the deep basin in the northwest corner of the lake. To convert micromoles to milligrams per liter multiply by 0.032.

tion under the ice within the lake (Hobbie et al., 1983). During this study much of the stream water flowed as an identifiable layer at depths from between 2 and 5 m located primarily along the eastern shore. The dyed stream water reached the lake outlet in 5 days, suggesting that much of the stream water entering the lake under ice transits the lake rather quickly, and does not mix uniformly. Thus the water renewal time calculated for Toolik Lake (0.5 years, Table 3.1), which assumes complete mixing, does not accurately describe the more complex processes of hydraulic flushing occurring in the lake. This observation may be of significance in explaining the extreme oligotrophic nature of Toolik Lake. The main source of nutrients that support primary production is stream input (Whalen and Cornwell, 1985). The dye tracer results suggest that at least some of the nutrient-rich stream water actually cannot be considered a source of nutrients for the lake, because it does not mix with most of the lake and is flushed from the lake before ice-out.

Physical Limnology

Toolik Lake is ice bound for 9 months, from early October to mid to late June. The lake thermally stratifies within about a week of ice-out, with the thermocline forming around 5 m. The epilimnion deepens during July and August to around 7 m before the lake completely mixes in late August. Ice may form in protected regions by mid-September and forms a complete cover by early October. The ice may reach 155 cm in thickness by the end of winter.

Interestingly, the average temperature of the epilimnion in July has, on average, steadily increased during the time of study. The intercept of the regression of average epilimnetic July temperature against year is 11°C for July 1975 and 14°C for July 1992, the last data used in the regression (Fig. 3.10). As can be seen from the figure there were some very cool summers during this time, but the general trend of warming is unmistakable. The hypolimnion, of course, remains at 4°C throughout the summer and has not changed during the time of study.

Light penetration into the lake is not as deep as might be expected given the extremely oligotrophic nature of the lake. Secchi disk depths of 2 to 4 m at ice-out are common and increase to 6 to 7 m by late July. These shallow secchi depths are mostly due to the high levels of dissolved humic organic matter in the water coming from tundra runoff in the spring (Miller et al., 1986). Hence the compensation depth of 1% light level is often 8 to 10 m and thus the deeper basins of Toolik Lake are not illuminated sufficiently for photosynthesis throughout the entire depth of the basin.

Pelagic Water Chemistry

The chemistry of the Toolik Lake water column is typical of oligotrophic arctic lakes (Kling, O'Brien et al., 1992). Dissolved oxygen is generally at or near saturation during the open water season, and serious oxygen depletion

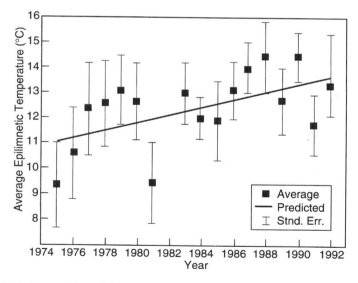

Figure 3.10. Regression of the average July (four measurements) epilimnetic temperature for the years 1975–1992.

is not observed even during the long period of ice cover (Fig. 3.9). Toolik Lake water is moderately dilute, has near neutral pH, and contains relatively large amounts of dissolved organic matter. Average chemical conditions for the water column are summarized in Table 3.3. The concentrations of major cations (Ca^{+2}, Mg^{+2}, Na^+, K^+) are considerably lower than one would expect given the size of the Toolik Lake watershed. Cornwell (1983) has computed the watershed chemical export rates for a number of chemical species. For the major cations, watershed export rates are among the lowest observed for any freshwater ecosystem, probably due to the low rates of chemical and physical weathering in the watershed.

The chemistry of three well-studied lake areas in arctic and subarctic North America and Toolik Lake show a close relation of water chemistry to physical and geological environment, watershed vegetation, and water renewal rate. Char Lake (75°N), the best-studied high arctic lake in North America, is even more oligotrophic than Toolik Lake, but concentrations of total dissolved ions are about four times higher than Toolik Lake (Schindler et al., 1974). Chemical inputs to Char Lake are quite high during the very brief summer season and probably reflect both the presence of easily weathered limestones and the lack of vegetation in the watershed. During the

Table 3.3. Typical Chemical Characteristics of Toolik Lake Water Column During Open Water Season

pH	7.1
Alkalinity (μequiv L^{-1})	500
Conductivity (μmhos)	50
Secchi (m)	5.5
Chlorophyll a (μg L^{-1})	1.8
NO_3^-	0.1
NH_4^+	0.1
Dissolved Si	30
Soluble reactive PO_4	<0.1
Dissolved organic carbon	575
Particulate carbon	18
Dissolved organic nitrogen	17
Particulate nitrogen	2
Total dissolved phosphorus	0.1
Particulate phosphorus	0.1
Ca^{+2}	250
Mg^{+2}	55
Na^+	20
K^+	10
SO_4^{-2}	12
Cl^-	9

All data are reported as μmol L^{-1}, except for pH, alkalinity, conductivity, secchi, and chlorophyll a.
Data from Kling, O'Brien, et al. (1992), Whalen and Cornwell (1985), Whalen and Alexander (1986a), Cornwell (1983), and M.C. Miller (unpublished data).

long period of winter ice cover, ions become further enriched in the water column by cryoconcentration as the ice grows to thicknesses exceeding 2 m. Outflow from the lake in spring is comprised primarily of the more dilute surface waters. As a result, the Char Lake water column is always more saline than its inflow waters.

The Saqvaqjuac lake area on Hudson Bay is subarctic (63°N), but it is underlain by permafrost and the vegetation is typical of barren ground tundra environments. The chemistry of these lakes has been summarized by Welch and Legault (1986). The Saqvaqjuac region is underlain by Precambrian granitic rocks that are not easily weathered, and very dilute waters would be expected. However, the total ionic concentrations in Saqvaqjuac lakes are very similar to those found in Toolik Lake. This situation may result from both input of sea salt from nearby Hudson Bay and from longer water renewal times for Saqvaqjuac lakes (1 to 3 years) in comparison to Toolik Lake (<1 year).

The Experimental Lakes Area, northwestern Ontario, is not in an arctic environment, but it is the best-studied site in North America from a chemical perspective (Armstrong and Schindler, 1971). The chemistry of lakes in this region clearly shows the effect of geological environment. Despite much higher precipitation rates and longer water renewal times than the arctic sites, weathering rates of the underlying Precambrian rocks are very low, and chemical input to lakes is further decreased by the heavily vegetated watersheds. The result is lakes that have average total ionic concentrations less than half those of Toolik Lake.

Summer season measurements of dissolved oxygen, pH, alkalinity, conductivity, dissolved inorganic nutrients, and chlorophyll have been made on Toolik Lake on a weekly basis since 1975. Under-ice data are not nearly as numerous, but sufficient measurements have been made so that annual chemical cycles can be discussed with some confidence. There is no indication of significant changes in water column chemistry during the period 1975 to 1991. Detailed discussions of dissolved nutrient cycles within the water column have resulted from the thesis studies of J.C. Cornwell and S.C. Whalen (Whalen and Cornwell, 1985; Whalen and Alexander, 1986a,b; Whalen et al., 1988) that clearly show dissolved inorganic nitrogen and phosphorus present at concentrations of $0.1 \mu mol L^{-1}$ or less during the open water season. Dissolved reactive phosphate concentrations also remain at extremely low levels during the ice-covered season. However, total inorganic nitrogen concentrations measured under the ice in the spring of 1993 were considerably greater than they had been the previous summer and reached concentrations of approximately $6 \mu mol L^{-1}$ deep in the lake (Fig. 3.11). As discussed elsewhere in this chapter, this increase is due to diffusion of NH_4^+ into the water column from the sediments, followed by rapid nitrification of NH_4^+ to NO_3^-. Nitrate concentrations decrease rapidly due to algal uptake when sufficient light penetrates the ice cover in the spring.

Because the rate of supply of dissolved inorganic nitrogen from the sediments to the water column is much smaller than in typical temperate lakes (Kipphut, 1988), the main source of nitrogen for planktonic growth is allochthonous inputs and recycling within the water column. Whalen et al. (1988) concluded that internal recycling within the water column is the larger of the two sources. However, M.C. Miller (personal communication) has noted that there is some correlation between annual planktonic production in Toolik Lake and discharge rates for the nearby Kuparuk River. If this relationship is valid, then long-term variations in planktonic production in Toolik Lake depend in part on rates of nutrient release from the lake watershed.

An interesting aspect of the inorganic carbon chemistry of Toolik Lake is the observation that Toolik Lake is supersaturated with CO_2 gas, and thus is losing carbon to the atmosphere via gas evasion (Kling et al., 1991). Supersaturation of the water column with respect to atmospheric CO_2 and CH_4 seems to be the usual condition for most lakes and streams in arctic Alaska (Kling et al., 1991, 1992; Kelley and Gosink, 1988). The source of the excess CO_2 is almost certainly related to the shallow groundwater flow of CO_2-rich water draining the arctic tundra. Kling et al. (1991) estimate that the rate of CO_2 lost from Toolik Lake is $35\,mmol\,m^{-2}\,d^{-1}$ during the ice-free season. It is of interest to compare this flux with other inorganic carbon fluxes in Toolik Lake. Calculated fluxes for gas evasion, lake inflow, lake

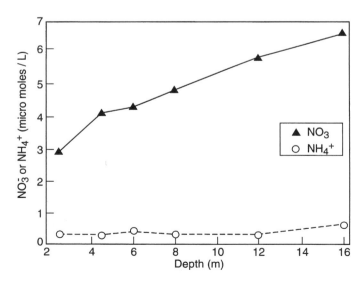

Figure 3.11. Dissolved inorganic nitrogen concentrations ($NO_3^- + NH_4^+$) in the main basin of Toolik Lake, May 16, 1993. Ice thickness was approximately 1.2 m. These represent the annual maximum nitrogen concentrations. To convert micromoles to micrograms per liter multiply by 14.

Table 3.4. Dissolved Inorganic Carbon Fluxes in Toolik Lake Water Column

River inflow[a]	+4600
Shallow subsurface inflow[a,b]	+1400
Lake outflow[a]	−6500
Gas evasion[b]	−3500
Diffusion from sediments[c]	+700
Uptake by phytoplankton[d]	−600

Results are for the entire open water season, which is taken to be 100 days. Results are presented as mmol C m^{-2}. Water column dissolved inorganic concentrations are assumed to be 0.5 mmol L^{-1} (Table 3.1); stream inflow concentrations are assumed to be 0.45 mmol L^{-1} (Kling et al., 1992). Lake surface area is 150 ha. (−) Indicates loss from water column, (+) indicates gain by water column. Fluxes do not balance: it is likely that the subsurface inflow is underestimated.
[a] Using flow data of Whalen and Cornwell (1985).
[b] Kling et al. (1991).
[c] Cornwell and Kipphut (1992).
[d] Whalen and Alexander (1986b).

outflow, diffusion from sediments, and incorporation by phytoplankton are presented in Table 3.4. Clearly, gas evasion is a major component of the dissolved inorganic carbon cycle in Toolik Lake. This has only recently been documented and warrants consideration when investigating carbon cycling in arctic lakes.

Sediment Chemistry

The chemical composition of Toolik Lake sediments is one of the more unusual, perhaps unique, characteristics of the lake (Cornwell and Kipphut, 1992). The sediments are characterized by extremely low rates of organic matter sedimentation and unusually high concentrations of iron and manganese. Toolik Lake contains an unusual type of lacustrine sediment, and in many ways the sediments are more similar to those found in oligotrophic oceanic environments than in lakes. The chemical composition of Toolik sediments has a strong influence on both chemical and biological reactions occurring within the water column.

Approximately 1 m of sediment has accumulated in Toolik Lake since its formation. A ^{14}C date of 12,600 years B.P. (P. Colinvaux, unpublished data) for the bottom sediments is consistent with the time scale of glacial retreat proposed by Hamilton and Porter (1975). The calculated ^{14}C long-term sediment accumulation rate for Toolik Lake is approximately 0.13 mm yr^{-1}. Studies of sedimentation rates covering the last 100 years produced an average rate of 0.16 mm yr^{-1}, which is consistent with the long-term rate (Cornwell, 1985). In terms of mass accumulation, Cornwell (1985) found that the average recent accumulation rate is approximately 27 g m^{-2} yr^{-1}.

Toolik Lake sedimentation rates then are among the lowest reported for lakes anywhere; the only lake with comparable rates is Char Lake (Table 3.5).

Low sedimentation rates in Toolik Lake result from both climatic and biotic factors. Toolik Lake drains a rather large watershed, but export of material from the watershed to the lake via stream flow is small. In fact, for major chemical ions (Na, K, Ca, Mg), export rates from the watershed are among the lowest reported (Cornwell, 1983). Low rates of chemical and physical weathering of watershed material are undoubtedly the result of both low precipitation, the long period during which soils are completely or partially frozen, and the fact that most of the south inlet (Table 3.2) flows through 10 lakes before emptying into Toolik Lake. Biotic production of organic material within Toolik Lake is limited by the short open water season, and by severe nutrient limitation (Whalen and Alexander, 1986a). Cornwell and Kipphut (1992) estimate that 40 to 50% of accumulated sediment is derived from stream input, and approximately 15% is derived from atmospheric input to the surface of the lake. A more complete budget for all sediment sources is not possible at this time, but Cornwell and Kipphut (1992) suggest that direct erosion of the shoreline, especially on steep banks along the northern margin of the lake, may be a significant source of inorganic sediment material.

Toolik Lake sediments are characterized by high concentrations of organic carbon, biogenic silica, and iron and manganese oxides (Cornwell, 1983, 1986, 1987; Cornwell and Kipphut, 1992). Chemical composition of the sediments varies with water depth. The chemical composition of surficial sediments (0 to 1 cm) from a number of water depths is shown in Fig. 3.12. The high concentrations of biogenic silica (5 to 20%) are believed to result from the presence of benthic diatoms, which are most numerous in

Table 3.5. Estimates of Sedimentation Rates for Selected Lakes in Arctic North America

Lake	Sedimentation Rate $g\ m^{-2}\ yr^{-1}$	Reference
Toolik		
Average	27	Cornwell (1985)
Range	13–106	
Char (N.W.T.) range	31–260	de March (1978) and Kipphut (1978)
Imitavik (N.W.T.) range	62–309	Hermanson (1990)
Annak (N.W.T.)	44	Hermanson (1990)

Taken from summary of Cornwell and Kipphut (1992). Estimates are presented only for those lakes to which the relatively reliable ^{210}Pb geochronology method has been applied. Livingstone et al. (1958) reported ^{14}C sedimentation rates for a number of other arctic Alaskan lakes, all of which were greater than for Toolik Lake. Results for Annak are for a single measurement, and should be considered less reliable than for other lakes where multiple measurements were made. N.W.T., Northwest Territories political unit of Canada.

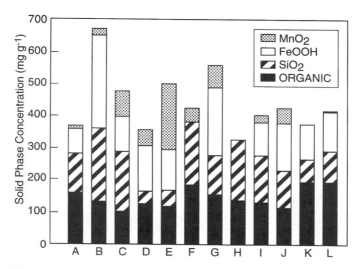

Figure 3.12. Chemical composition of surficial sediments (0–1 cm) at various locations in Toolik Lake, from Cornwell and Kipphut (1992). Letters indicate cores taken at different water depths, ranging from 3.5 to 15 m. The exact location of the cores is given in Fig. 1 in Cornwell and Kipphut (1992). Remainder of sediments consists of uncharacterized material, believed to be aluminosilicates, clay minerals, and quartz. Calcium carbonate is also found in small quantities (1–3%).

the shallower portions of the lake (Cornwell and Banahan, 1992). Unlike biogenic silica, however, the high organic carbon concentrations do not reflect high rates of organic matter production within the lake. Instead, the organic matter is derived from terrestrial detritus. This terrestrial material is quite refractory. If the organic matter within Toolik sediments were more labile, many of the unusual aspects of the sediments would be diminished.

An understanding of the processes controlling the chemical composition of Toolik Lake sediments is not possible without considering the vertical distribution of iron and manganese within the sediments, an example of which is shown in Fig. 3.13. The high concentrations of Mn and Fe oxides, which are present in separate and well-defined zones within the sediments, have not been observed in other freshwater environments. These high metal concentrations within the sediments do not result from either high rates of metal input from the watershed (Cornwell, 1983) or from processes within the water column that would concentrate metals (Johnston and Kipphut, 1988). Rather, the Mn and Fe enrichments are the result of processes responsible for similar Mn and Fe distributions found in certain deep sea sediments (Froelich et al., 1979). These include a very low organic carbon sedimentation rate, low organic carbon reactivity, and an overlying water column that always contains high levels of dissolved oxygen. The essence of the overall process is that soluble Mn and Fe are produced at

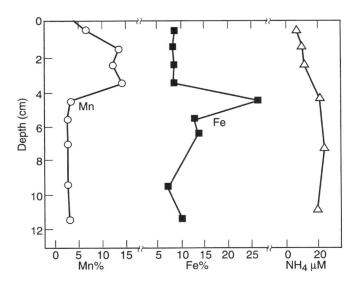

Figure 3.13. Solid phase Mn and Fe, and soluble ammonium distributions in sediments from a deep water core from the main basin of Toolik Lake. Data from Cornwell (1983). The sharply defined metal-rich layers are unique features of Toolik Lake sediments. The ammonium concentrations are very low for an organic-rich sediment, and probably reflect strong adsorption by the metal-rich phases.

depth in the sediments via reaction with organic matter, followed by slow upward diffusion and precipitation of the metals within the oxidized surficial sediments (Cornwell and Kipphut, 1992).

Thus chemical composition of Toolik Lake sediments is the result of a delicate balance among several processes. This balance could be very easily disturbed. Any process that significantly increased the rate of sediment accumulation, or caused oxygen depletion within the water column would result in dissolution of the metal-rich layers of sediment. Such processes could include increased runoff and sedimentation from the terrestrial environment, or an increase in primary production within the lake resulting in increased oxygen consumption rates during the ice-covered season. Such processes could result from human-derived disturbances, or they could result from longer term changes in climatic cycles.

Sediment–Water Chemical Interactions

The chemical characteristics of the Toolik Lake sediments play an important role in maintaining the oligotrophic, oxidizing nature of the Toolik Lake water column. This is accomplished primarily by regulating the fluxes of important chemical constituents between the sediments and the overlying water. Although fluxes of nitrogen and phosphorus from the sediments to the water column provide a significant fraction of the nutrient requirements of the primary producers in many lacustrine environments, this is not the case for Toolik Lake (Kipphut, 1988). Annual mass budgets for N, P, and organic C have been well defined for Toolik Lake by careful monitoring of inflow, outflow, and sedimentation fluxes as well as atmospheric inputs for nitrogen (Whalen and Cornwell, 1985; Whalen et al., 1988). Mass balances indicate the sediments to be minor sources of nitrogen and phosphorus, particularly during the summer season. The small N and P sediment–water fluxes can be satisfactorily explained by the strong adsorption of soluble nitrogen and phosphorus by the metal-rich sediments (see Cornwell, 1987). For P, which is adsorbed more strongly than N, it is likely that the sediments retain virtually all P deposited at the sediment–water interface.

For N, a small but significant flux of NH_4^+ occurs from the sediments. Whalen et al. (1988) estimate that sediments supply only 10 to 19% of the nitrogen required by Toolik Lake phytoplankton on an annual basis. However, this contribution is enhanced because most of the flux occurs during the period of ice cover. Under-ice release of NH_4^+ from the sediments, followed by nitrification, is almost certainly the explanation for relatively high nitrate concentrations found in Toolik Lake at the end of the winter season (Whalen and Alexander, 1986b). Nitrate decreases rapidly when light begins to penetrate the ice in early spring, and Whalen and Alexander (1986b) speculate that the nitrate is utilized by benthic algae.

An important feature of the limnology of Toolik Lake is that significant dissolved oxygen depletion does not occur within the water column, even though the lake may be ice covered for 9 months. Oxygen consumption occurs mainly via reactions with organic matter at the sediment–water interface, and is therefore closely related to the chemical composition of the sediments. Direct measurements of oxygen consumption by Toolik Lake sediments have yielded an average flux of approximately $8\,mmol\,O_2\,m^{-2}\,d^{-1}$ (Cornwell and Kipphut, 1992). This is an extremely low rate of oxygen consumption for a lake sediment. The primary explanation for the low rates is the refractory nature of the organic material found in the Toolik Lake sediments.

Phytoplankton Species Composition

A survey of phytoplankton species from Toolik Lake compiled from live and preserved samples collected during late June of 1991 and early July of 1992 is given in Table 3.6. Sample treatment followed a combination of procedures previously described by Untermohl (1958), Nauwerck (1963), and Fee et al. (1989).

Table 3.6. Phytoplankton From Toolik Lake 1991–1992

1. *Anabaena* sp2 (8.6 μm * 4.2 μm)
1. *Anabaena* sp3 (5.6 μm * 5.6 μm) (*Note*: Species of *Anabaena* can rarely be identified without the presence of akinetes)
2. *Ankistrodesmus falcatus* (Corda) Ralfs
2. *Ankyra judayi* (G.M. Smith) Fott
1. *Aphanocapsa clathrata v. brevis* W. et G.S. West
1. *A. delicatissima* W. et G.S. West
1. *A. elachista* W. et G.S. West
5. *Asterionella formosa* Hassal
4. *Bitrichia chodatii* (Reverdin) Chodat
2. *Botryococcus braunii* Kuetzing
7. *Ceratium furcoides* (Lev.) Langhans
7. *C. hirundinella* (O.F. Mull.) Schrank
5. *Ceratoneis arcus* Kutz
2. *Chlamydomonas* sp.
2. *Chlorella* sp.
1. *Chroococcus limneticus* Lemm.
1. *Chroomonas acuta* Utermohl
4. *Chrysochromulina parva* Lackey
4. *Chrysococcus furcatus* (Dolg.) Nicholls
4. *C. punctiformis* Pascher
4. *C.* sp.
4. *Chrysolykos planktonicus* Mach
4. *C. skujai* (Nauw.) Willen
4. *Chrysosphaerella brevispina* Korshikov emend. Harris and Bradley
4. *C. longispina* Lauterb.
1. *Coelosphaerium minutissimum* Lemm.
2. *Coenococcus fottii* Hindak = *Eutetramorus fottii* (Hind.) Kom.
2. *Collodictyon triciliatum* Carter (colorless flagellate eating greens and centric diatoms)
2. *Cosmarium subtumidum* Nordst.
2. *Crucigeniella rectangularis* Kom.
6. *Cryptomonas cf. obovata* Skuja
6. *C. marssoni* Skuja
6. *C. reflexa* (Marsson) Skuja
1. *Cyanodictyon imperfectum* Cronberg
6. *Cyathomonas truncata* (Fres) Fisch
5. *Cyclotella bodanica* Eulenstein
5. *C. stelligera* Cleve et Grunow
5. *C. tripartita* Hakansson (previously called *C. ocellata* or *C. comensis*)
5. *Diatomae cf elongatum* Agardh
2. *Dictyosphaerium ehrenbergianum* Naegelli
4. *Dinobryon attenuatum* Hill
4. *D. bavaricum* Imhoff + cysts
4. *D. crenulatum* W. and G.S. West (probably the same as *D. acuminatum*)
4. *D. cylindricum* Imhoff
4. *D. njakjaurensis* Skuja (Nordic ultra-oligotrophic species primarily found in nutrient-poor arctic lakes)
4. *D. sertularia* Ehrenberg
4. *D. sociale* Ehrenberg
4. *D. sociale v. americanum* Brunnthaler
4. *D. sociale v. stipitatum* (Stein) Lemm.

Table 3.6. *Continued*

2. *Elakatothrix gelatinosa* Wille
2. *E. genevensis* Hindak
2. *E. spirochroma* (Reverdin) Hindak
4. *Ephiphyxis alaskana* Hilliard and Asmund
4. *E. kenaiensis* Hilliard
1. *Eucapsa alpina* Clem. et Schantz (abundant in small shallow lakes)
2. *Eutetramorus nygaardii* Kom.
7. *Glenodinium edax* Schilling
4. *Gloeobotrys limneticus* (G.M. Smith) Pascher
1. *Gomphosphaeria virieux* Kom. et Hin.
7. *Gymnodinium cf inversum v. elongatum* Nygaard
7. *G. helveticum* Penard
7. *G. lantzschii* Utermohl
7. *G. triceratium* Skuja
7. *G. uberimuum* (Allman) Kofoid and Swezy
2. *Gyromitus cordiformis* Skuja
4. *Ithmochloron trispinatum* (W. and G.S. West) Skuja
6. *Katablepharis ovalis* Skuja
4. *Kephyrion boreale* Skuja
2. *Koliella longiseta* (Visch.) Hindak
2. *K. planktonica* Hindak (often looks like *K. tatrae v. fogararensis*)
1. *Lemmermaniella pallida* (Lemm.) Geitl.
4. *Mallomonas acaroides* Perty
4. *M. akrokomos* Ruttner
4. *M. caudata* Iwanoff
4. *M. crassisquama* (Asmund) Fott
4. *M. deurrschmidtiae* Siver, Hammer and Kling
4. *M. elongata* Reverdin
4. *M. hamata* Asmund
4. *M. pseudocoronata* Prescott
4. *M. tonsurata* Teiling
4. *M. variabilis* Cronberg
1. *Microcystis incerta* Lemm.
2. *Monoraphidium komarkovae* Nygaard = *M. setiforme* Komarkova-legnerova
2. *Mougotia* sp.
2. *Nephrocytium agardhianum* Naeg.
5. *Nitzschia acicularis* G.M. Smith
4. *Ochromonas* spp.
2. *Oocystis borgei* Snow
2. *O. parva* W. et G.S. West
2. *O. solitaria* Wittr.
2. *O. submarina v. variablis* Skuja
4. *Paraphysomonas vistita* (Stokes) de Saedler
2. *Pediastrum boryanum* (Turp.) Menegh.
4. *Pseudopedinella erkensis* Skuja
7. *Peridinium cf apiculata* Penard = *Gonyaulax apiculata* (Penard) Lentz
7. *P. inconspicuum* Lemm.
7. *P. polonicum* Wol.
7. *P. pusillum* (Penard) Lemm.
7. *P. willei* Huetfeldt-kaas
3. *Phacus* sp.
1. *Planktolyngbya subtilis* (W. West) KOM syn. *Lyngbya limnetica* Lemm.

Continued

Table 3.6. *Continued*

2. *Planktosphaeria gelatinosa* (maybe a stage in *Eutetramorus nygaardi*)
2. *Polytomella* sp.
4. *Pseudokephyrion alaskanum* Hilliard
4. *P. entzii* Conrad
4. *P. heimale* Hilliard
4. *P. spirale* (Gerlof) Schmid
4. *P. striatum* Hilliard
4. *Pseudosphaerocystis lacustris* (Teiling) Bourselly (syn *Gemellicystis neglecta* Teiling et Skuja)
2. *Quadrigula closteriodes* (Bohlin) Printz
1. *Radiocystis geminata* Skuja
4. *Rhizochrysis limnetica* G.M. Smith
6. *Rhodomonas minuta* Skuja (3 sizes including *R. minuta v. nanoplanktica* and *R. lacustris*)
2. *Scenedesmus cf opoliensis* P. Richt
2. *Sphaeromantis ochrecea* Pascher
4. *Spiniferomonas coronacircumspina* (Wujek and Kristiansen) Nicholls
4. *S. serrata* Nicholls
4. *S. trioralis* Takahashi
2. *Spondylosium planum* (Wolle) W. and G.S. West
2. *Staurastrum anatinium v. truncatum* Cooke et Wills (taxonomy in flux)
2. *S. avicula* Breb. var *subarcuatum* West
2. *S. turpinii* var *eximium* W. West
4. *Stenokalyx moniliforme* (Gerloff) Schmid
5. *Stephanodiscus alpinus* Hustedt
4. *Stichogleoa doederleinii* (Schmidle) Wille
5. *Synedra acus* Kutz
5. *S. acus var. angustissima* Grun.
5. *S. berolinensis* Lemm.
5. *S. cf nana* Meister
4. *Synura petersenii* Korshikov + various scale formations
5. *Tabellaria flocculosa* (Rothest) Kuetzing
2. *Tetraedron minimum v. tetralobulatum* Reinsch
3. *Trachelomonas* sp.
2. *Troschisca* (resting stage of *Chlorogonium* sp.)
4. *Uroglenopsis cf americanum* Catleins (difficult to distinguish species of Uroglena without seeing statospores) syn. *Uroglena americana* Catleins
1. *Woronichinia robusta* (Skuja) Kom. et Hin. syn. *Gomphosaphaeria robusta* Skuja

Note: The species are arranged alphabetically and the number 1–7 indicate the algal groupings: (1) Cyanophyta, (2) Chlorophyta, (3) Euglenophyta, (4) Chrysophyceae, (5) Bacillariophyceae, (6) Cryptopyceae, (7) Peridineae.

Tooklik Lake phytoplankton is dominated by Chrysophyceae with dinoflagellates and cryptophytes of second and third importance. This phytoplankton assemblage of predominantly chrysophytes puts Toolik Lake in the first group under Holmgren's (1968) lake type classification system for unpolluted arctic lakes. In Toolik Lake, the biomass is made up of 50 to 70% chrysophytes (*Chrysococcus* spp., *Kephyrion* spp., *Chysochromulina parva*, *Spiniferomonas* spp., and *Dinobryon* spp.).

Dinoflagellate (*Gymnodinium* spp., *Peridinium polonicum*, and *Ceratium hirundinella*) biomass ranged from 10 to 25%. Cryptomonads (*Cryptomonas reflexa*, *C. marsonii C. pusilla*, *Rhodomonas minuta*, and *Katablepharis ovalis*) composed between 8 and 28% of the biomass. The other groups on average composed less than 5% of the total biomass. Bacillariophyceae (*Cyclotella bodanica*, *C. tripartita*, *Synedra* spp., *Diatomae elongatum*, and *Tabellaria flocculosa*) biomass ranged between 1 and 3%; green algae composed 2 to 4% and blue-greens were generally less than 1%. Nitrogen fixers and chroococcoid species were present but very rare in Toolik plankton.

Diatoms were found to dominate many of the Colville lakes on the North Slope (Coulon et al., 1972). The same planktonic species were present in Toolik but of low biomass. Dinoflagellates (mainly *Ceratium hirundenella*) were an important component of the plankton of some of the Colville lakes. They are not uncommon in arctic lakes and can occasionally dominate the plankton of some lakes, especially after nutrient additions. Cryptomonads were found to be important components in the Colville lakes (Coulon et al., 1972) as well as many other arctic lakes (Kalff et al., 1975; Welch et al., 1989). Often they dominated under-ice in May to June during periods of low light penetration (mean light 4 to $5 \mu E m^{-2} d^{-1}$) (Welch et al., 1989). Species of this group are motile and migrate in response to light and nutrients, by day concentrating near the surface and at night in the deepest water in the region of anoxia. Cryptophyceae did not dominate in Toolik but composed 25% of the biomass.

Chlorophyte species were relatively numerous but they composed very little of the biomass. This group was more important in the Colville study than in Toolik Lake. Many of the species in the Colville study are not strictly planktonic and would be more prevalent in shallow lakes. Members of the Chlorophyta, especially some of the motile species of the volvocales, have been known to respond positively to nutrient additions in arctic lakes (Schindler et al., 1974; Holmgren, 1968; Welch et al., 1989).

Cyanophyte species were insignificant in abundance and biomass and only the chroococcoid species typical of very nutrient-limited lakes were present in Toolik Lake. This corresponds well with previous studies from arctic lakes in this region (Prescott, 1963; Coulon et al., 1972). However, a recent survey of the plankton of several other lakes immediately surrounding Toolik indicate that the nitrogen-fixing species *Anabaena lemmermannii* was important in the plankton of these lakes (H. Kling, personal communication).

The most important group in Toolik Lake was Chrysophyceae. Species of this group typically dominate nutrient-poor lakes from temperate regions to the arctic. However, Coulon et al. (1972) found chrysophytes to dominate in only 2 of the 10 Colville lakes. In Toolik Lake, the major components of the chrysophyte biomass were several species of the genera *Chromulina*, *Ochromonas*, *Spiniferomonas*, *Pseudopedinella*,

Chrysococcus, Pseudokephyrion, Kephyrion, Mallomonas, and *Dinobryon* plus *Chrysochromulia parva.* The *Chromulina, Ochromonas,* and *Spiniferomonas* are very difficult to identify. *Mallomonas, Synura, Spiniferomonas,* and *Paraphysomonas* need the aid of scanning electron microscopy. *Dinobryon sociale, D. crenulatum* (syn. *D. acuminatum*), and *D. cylindricum* were the most important species of *Dinobryon* in Toolik Lake. Several species of *Mallomonas* reported by Coulon et al. (1972) were found in Toolik Lake. The most common species were *M. crassisquama, M. tonsurata, M. elongata,* and *M. variabilis.* These scaled chrysophytes were similar to those found in the Mackenzie delta and the Tuk peninsula of the Canadian arctic (McKenzie and Kling, 1989).

Primary Productivity

After ice-out, summer primary productivity is quite low and variable in Toolik Lake (Fig. 3.14). Two studies have examined the relationship between seasonal primary production and water column chemistry in Toolik Lake. Miller et al. (1986) summarized the primary production data for the late 1970s and early 1980s. Whalen and Alexander (1986b) and Whalen et al. (1988) reported on very detailed studies of primary production in Toolik Lake during 1980 and 1981. In high latitude aquatic environments, a significant portion of the annual planktonic production typically occurs under the ice (Welch et al., 1989); a phenomenon well documented in Toolik

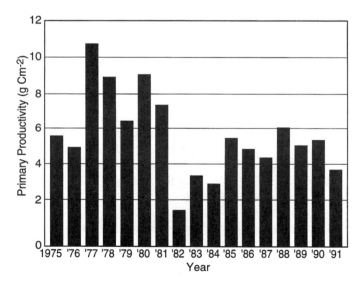

Figure 3.14. Annual total summer primary productivity at main station in Toolik Lake.

Lake. The seasonal trend of planktonic production and chlorophyll during 1980 and 1981 showed that maximum production rates occurred just after ice-out. Production rates then dropped rapidly and remained at low levels for the remainder of the open water season, due to the rapid depletion of winter accumulated dissolved inorganic nitrogen. In other years, the peak period of planktonic production occurred while the lake was still ice covered (Miller et al., 1986). Seasonal variation in chlorophyll concentrations within the water column closely follow the trends in primary production.

The very low dissolved inorganic nitrogen and phosphorus concentrations largely explain the very low rates of phytoplankton productivity in Toolik Lake. Whalen and Alexander (1986a) designed experiments that evaluated the influences of nitrate, ammonia, phosphate, nitrogen + phosphate, vitamins, trace metals, chelates, and common salts on primary production of Toolik Lake plankton. Two important conclusions resulted from this study. First, Whalen and Alexander (1986a) demonstrated that nitrogen and phosphorus are essentially colimiting factors of primary production during the open water season. Second, they concluded that dissolved organic nitrogen (DON) must be utilized to satisfy the nitrogen requirements of the phytoplankton on an annual basis. Further analyses of all available data for Toolik Lake (Whalen et al., 1988) suggest that DON may provide as much as 35% of the total dissolved nitrogen utilized by the plankton, but that recycling of ammonia within the water column was the predominant source of inorganic nitrogen. Riverine and sediment effluxes, which are usually important sources of ammonia in temperate lakes, are only minor sources in the Toolik Lake ecosystem.

Zooplankton

In the mid 1970s the zooplankton community in Toolik Lake consisted of seven species of crustacean zooplankton: *Daphnia middendorffiana* (1.8 to 3.0 mm), *Daphnia longiremis* (0.8 to 1.5 mm), *Holopedium gibberum* (1.4 to 2.2 mm), *Bosmina longirostris* (0.6 to 1.1 mm), *Heterocope septentrionalis* (2.0 to 4.0 mm), *Diaptomus pribilofensis* (1.1 to 1.8 mm), and *Cyclops scutifer* (1.0 to 1.8 mm). All these species are quite typical of arctic lakes and ponds (O'Brien, 1975; Hobbie, 1980). This is the greatest species richness so far reported from the Toolik Lake area (Luecke and O'Brien, 1983; O'Brien, Buchanan, and Haney, 1979). This zooplankton community is unusual in that it is a combination of both large- and small-bodied species. Classically, limnologists have found small-bodied communities in the presence of fish (Brooks and Dodson, 1965) and large-bodied zooplankton species in the absence of fish (Dodson, 1974). This body size pattern is sometimes found in arctic lakes (O'Brien, 1975), although there are exceptions to this both in the arctic (O'Brien, Buchanan, and Haney, 1979; Kling, O'Brien, et al., 1992) and elsewhere (Dodson, 1979).

The zooplankton community in Toolik Lake has changed since 1975 as two of the larger species, *D. middendorffiana* and *H. gibberum*, have all but disappeared (Fig. 3.15). The forces behind the decline and loss of these species are not fully known, but it is thought to be related to the reduced body size and population size of lake trout. Average lake trout body sizes have declined from a median fork length of 388 mm in 1977 to 313 mm in 1986 (McDonald and Hershey, 1989). This decline in size was also thought to be accompanied by a reduction in lake trout density. This reduced body and population size of lake trout is thought to have allowed small zooplanktivorous arctic grayling to become more abundant and also to move into the open portion of the lake. An observation consistent with this is the current common occurrence of fish rises in the middle of the lake during calm evenings, which indicates the presence of small fish in the pelagic region of the lake. In the mid 1970s, fish rises in the open water of the lake were very rare. Furthermore, piscivorous birds such as loons and terns were rarely observed hunting for prey on Toolik Lake in the mid 1970s, whereas in the late 1980s they were commonly observed. Thus it is postulated that as small arctic grayling and lake trout became abundant and common in the open water, their predation on the large zooplankton drove these species to near extinction.

One large zooplankton species, *H. septentrionalis*, is still present in the Toolik Lake zooplankton community, although densities have declined by

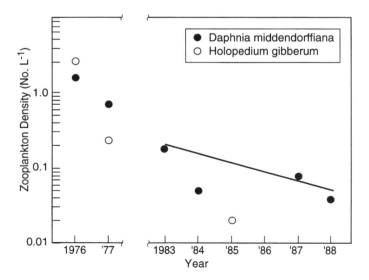

Figure 3.15. Average annual density of *Daphnia middendorffiana* and *Holopedium gibberum* in Toolik Lake from 1976 to 1988.

30 to 50% from 1976 levels. *Heterocope* has been shown to be a voracious predator on small zooplankton (Luecke and O'Brien, 1983) and two species, *D. longiremis* and *B. longirostris*, develop carapace structures shown to minimize predation from *Heterocope* (O'Brien, Kettle, and Riessen, 1979; O'Brien and Schmidt, 1979) when in Toolik Lake.

Although zooplankton were once fairly diverse in Toolik Lake, they have never been abundant by the standards of temperate and more productive lakes. The two currently most abundant cladocerans in Toolik Lake, *Daphnia longiremis* and *Bosmina longirostris*, are rarely more abundant than 1 per liter and often less than 1 per 10 liters. The two copepod species, *Diaptomus pribilofensis* and *Cyclops scutifer*, are commonly more abundant, reaching 10 individuals per liter in late June and early July. However, both species decline throughout the summer commonly to densities less than 1 per liter by late July and early August. This is especially true for *C. scutifer*, which has shown precipitous declines in every year since 1977. This may be due to predation from *Heterocope* since *C. scutifer* is preyed on heavily by *Heterocope* (O'Brien, 1988).

The low abundance and small size of the zooplankton currently present in Toolik Lake means that grazing rates on the phytoplankton are very low. *D. longiremis* has filtering rates in the range of 0.6 to 1.3 mL per individual per hour (Peterson et al., 1978). Typically, copepod filtering rates are considerably lower than cladoceran rates. Thus, assuming *B. longirostris* to have similar filtering rates as *D. longiremis*, the cladoceran daily filtering rate could approach 50 mL d^{-1}. Assuming copepod filtering rates at 20% of *D. longiremis*, then the total copepod filtering rate could be 100 mL d^{-1} for a total filtering rate of 150 mL L^{-1} d^{-1} in later June and early July. Thus, at the greatest, the zooplankton in Toolik Lake might turn over the epilimnetic water once a week at this time but decline to turnover times of once a month by August. Thus it is not likely that zooplankton grazing plays much of a role in nutrient cycling or mortality on the phytoplankton of the lake.

Pelagic Microbial Community

The planktonic microbial community in Toolik Lake includes bacteria (picoplankton, 2 μm in size), small algae and heterotrophic flagellates (nanoplankton, 2 to 20 μm in size), and protozoans and rotifers (microplankton, generally 20 to 200 μm in size). Total bacterial numbers have been determined throughout the year (Hobbie et al., 1983; Johnston and Kipphut, 1988; J. Hobbie and M. Bahr, unpublished data). The seasonal range (Fig. 3.16), from 0.1 to 3.1 × 10^6 cells mL^{-1}, and the average summer numbers, between 1 and 2 × 10^6 mL^{-1}, are comparable to numbers from unpolluted temperate waters. The numbers were usually lowest in the winter and highest in the summer. Depth profiles have shown an irregular

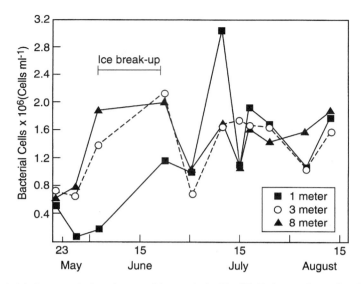

Figure 3.16. Seasonal abundance of bacteria in Toolik Lake at three depths (1, 3, and 8 m).

distribution, with maximum cell numbers in the water column found at various depths. In 1979, the numbers of cells increased in early June in conjunction with the spring discharge of Toolik Inlet stream water and then fluctuated between 1 and $2 \times 10^6 mL^{-1}$ for the remainder of the summer. Hobbie and Bahr counted 1.0×10^6 cells mL^{-1} consistently throughout the summer of 1992. During spring 1993, on May 16 they found (Table 3.7) low numbers of $0.9 \times 10^6 mL^{-1}$ in the winter water beneath the ice. Nine days later, on May 25 the spring runoff into the lake had reached this station and the bacterial numbers jumped to $1.4 \times 10^6 mL^{-1}$. The differences in conductivity of the lake water and stream water at this time of year allowed the plume of inflowing runoff water to be identified. These data confirm a consistent picture of low numbers beneath the ice in winter followed by an increase in numbers when water from the spring runoff entered the lake. Numbers of bacteria remained relatively high and constant during the open water season of Toolik Lake.

Table 3.7. Bacteria Abundance and Productivity, Toolik Lake Main Station, 3 m, 1993

Date	$\times 10^6 mL^{-1}$	$\mu g C L^{-1} d^{-1}$
May 16	0.9	0.3
May 25	1.4	22.4
June 8	1.5	1.6

The planktonic bacteria in Toolik Lake use two food sources, the DOC entering the lake in runoff from the land and the DOC produced during algal growth and decomposition. As noted above, high numbers of bacteria are found in the lake beneath the ice during spring runoff. Additional evidence of the use of runoff DOC comes from an enclosure experiment in which DOC leached from plant litter on the tundra surface caused a tremendous increase in bacterial growth (J. Hobbie and G. Kling, unpublished data). High numbers are also produced in enclosures and high growth rates are produced in lakes that have increased rates of photosynthesis as a result of fertilization with inorganic nutrients. In Lake N-2, the bacterial activity doubled in the fertilized half of the lake while the algal photosynthesis increased 10-fold (Hullar et al., 1986). In a $60\,m^3$ fertilized enclosure, bacterial numbers reached $8 \times 10^6\,mL^{-1}$ 16 days after fertilization began (O'Brien et al., 1992) (Fig. 3.17).

The biomass of microbes at typical summer cell concentrations illustrates the oligotrophic nature of the plankton of Toolik Lake. Using a conversion factor for bacteria of $20 \times 10^{-15}\,g\,C\,cell^{-1}$, bacterial biomass is calculated to be about $20\mu g\,C\,L^{-1}$. The mean values for nanoflagellates, protozoans, and rotifers are about $1\mu g\,C\,L^{-1}$ each, and about $3\mu g\,C\,L^{-1}$ for zooplankton nauplii (Hobbie and Helfrich, 1988; Rublee, 1992).

Bacterial productivity has been estimated using the incorporation of radioactively labeled thymidine, leucine, and acetate. The 1984 enclosure control values of approximately 10pmol thymidine $L^{-1}hr^{-1}$ (Fig. 3.18, limnocorral A) reflect those of the normal lake population; the nutrient

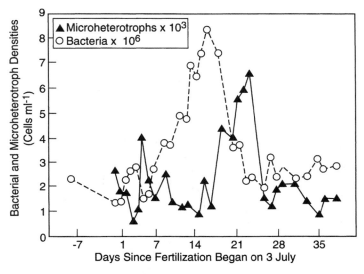

Figure 3.17. Abundance of bacteria and microheterotrophs in a fertilized limnocorral in 1984.

addition (Fig. 3.18, limnocorral B) demonstrates the bacterial response to increased carbon from algal photosynthate. The largest increase in production occurred as bacterial numbers declined, likely due to nanoflagellate grazing (shown in the productivity in corral B in Fig. 3.18 and the nanoflagellate data from Fig. 3.17). While the cause of this stimulatory effect is not known, the close coupling between the bacteria and flagellate activities is clear. Production measured throughout the spring/summer season in 1993 showed an order of magnitude increase following the start of stream flow in May, while cell numbers increased but did not even double (Table 3.7). Based upon the leucine incorporation data from 1993, the bacterial productivity in the plankton is estimated to be 3 to $8\,g\,Cm^{-2}yr^{-1}$. The maximum value for particulate growth (bacterial productivity) is some 66% of the primary productivity, a very high proportion compared with the usual 14% value for temperate lakes (Cole et al., 1988). Although the evidence is not conclusive, it is likely that the bacteria and the microbial food web are using both the DOC from land runoff and the DOC from algal photosynthesis. Whalen and Cornwell (1985) measured an input to Toolik Lake of $98\,g\,DOCm^{-2}yr^{-1}$ and an export of $86\,g\,DOC$. These numbers are not accurate enough to attribute the difference to microbial utilization. However, there certainly is a large amount of DOC moving through the system and only a small proportion would have to be used to be provide this additional input to the bacteria of DOC from runoff.

The nonphotosynthetic nanoflagellates have not been enumerated very frequently in the plankton of Toolik Lake. The existing data show numbers around 1000 cells mL^{-1} or 1000-fold less than the numbers of bacteria.

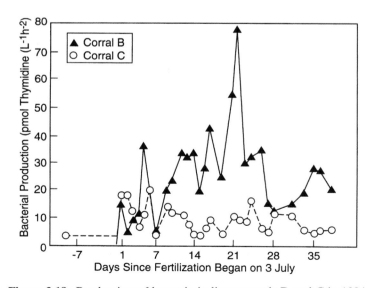

Figure 3.18. Production of bacteria in limnocorrals B and C in 1984.

Table 3.8. Microplankton Abundance and Biomass in Toolik Lake

	Protozoa		Rotifers		Nauplii	
	Number	Biomass	Number	Biomass	Number	Biomass
1989						
June 27	125.4	0.47	22.9	0.44	24.1	5.42
July 3	31.0	0.34	18.5	0.36	32.9	7.41
July 10	40.7	0.13	29.8	0.51	53.0	11.92
July 17	22.9	0.10	19.8	0.36	28.5	6.42
July 24	44.2	0.08	53.3	1.24	18.5	4.16
Aug 1	47.2	0.06	40.7	0.91	33.9	7.64
Aug 8	56.6	0.03	18.0	0.32	23.3	5.25
1990						
June 26	332.3	1.08	16.9	0.31	2.9	0.66
July 3	47.2	0.13	38.6	0.70	12.7	2.85
July 10	20.1	0.09	41.2	0.83	26.5	5.98
July 17	24.7	0.10	25.1	0.43	15.9	3.58
July 24	50.1	0.20	21.0	0.41	38.3	8.61
July 31	124.8	1.05	18.6	0.31	49.4	11.12
Aug 7	55.6	0.43	13.1	0.27	65.9	14.84

Values represent mean number per liter or micrograms carbon per liter, averaged over the upper 16 m of the water column.

While these flagellates are not a major part of the pelagic biomass, it is possible that their grazing can control the numbers of the bacteria. This grazing control is easy to demonstrate in a laboratory flask where the cycles of bacteria and flagellates show the expected predator–prey oscillations. These oscillations are rarely measured in natural lakes and oceans, perhaps because the patches of water in which the bacteria and flagellates are in a lagged synchrony may be only meters across. Our solution in Toolik Lake was to enclose a large quantity of water, 60 to 100 m^3, to insure that the same patch was sampled each day. We found a bacterial cycle of 14 to 16 days, with a nanoflagellate cycle lagging by 6 to 7 days (Hobbie and Helfrich, 1988) (Fig. 3.17). The close correspondence between bacterial troughs and flagellate peaks suggests that the flagellates control bacterial numbers. An estimate of the individual nanoflagellate clearance rate of 0.001 to 0.02 μL h^{-1} was calculated from bacterial production and changes in numbers in this same experiment. At the abundance of nanoflagellates we found, this rate means that 60% of the experimental enclosure was processed by nanoflagellates each day.

Heterotrophic microplankton are represented by ciliated protozoans and rotifers, but these are found in very low abundance in Toolik Lake (Table 3.8). Ciliates are mostly oligotrichs and vorticellids. Eight species of rotifers are present, dominated by *Keratella cochlearis*, *Conochilus unicornis*, and *Polyarthra vulgaris*. Total microplankton biomass averaged over the whole water column is about 1 μg C L^{-1} over the summer, with highest values in the epilimnion. Ciliate abundance is high in the early summer following the peak in algal biomass and may peak again in late summer; rotifer biomass is generally highest during mid- to late summer (Rublee, 1992).

Ciliates, rotifers, and nauplii of crustacean zooplankton graze bacteria, algae, and flagellates. Fluorescent food analogues were used by P. Rublee (unpublished data) to measure grazing rates of an assortment of rotifer species. Clearance rates ranged from 0.01 to 5.5 µL per individual per hour, with differing size and taste preferences exhibited by the various organisms studied. Although the microplankton in Toolik Lake grazed some bacteria, they generally appeared selective for small algae and flagellates. For example, the rotifer *Conochilus unicornis* had highest grazing rates on 4 to 6 µm particles. In a separate study, P. Rublee (unpublished data) measured ciliates, rotifers, and nauplii in 5 m³ enclosures under light and dark regimes augmented with nutrients. Microzooplankton abundances were significantly greater in light enclosures, even though bacteria grew well under both light and dark conditions. Based on the present evidence, we believe that in Toolik Lake most of the grazing on bacteria is carried out by nanoflagellates with much smaller amounts of grazing by microplankton. The microplankton, however, probably play a controlling role in grazing of nanoflagellates.

In conclusion, bacteria are relatively abundant in Toolik Lake. Their abundance and productivity is linked to the supply of available DOC which comes both from inflowing stream water and from algal exudates. Because we know that some of the inflowing DOC is available to microbes and because the extent of microbial activity exceeds the expected 10 to 15% of primary productivity, we conclude that the inflow DOC is at least as important for microbial growth as the algal exudates. Bacterial abundance is controlled by the nanoflagellates and the rate of production is controlled by the supply of substrate. The nanoflagellates, in turn, appear to be controlled by the grazing of the microzooplankton, that is, the ciliates, rotifers, and nauplii. The whole microbial food web is internally controlled and does not appear to be affected by the larger crustacean zooplankton.

Benthic Community

The benthic zone of Toolik Lake can be subdivided into two major habitat types. The rocky littoral zone, comprising approximately 25% of the lake area, is generally less than 2 m in depth and is scoured by ice each year. The remaining 75% of the lake bottom consists of soft silty substrate that may be sparsely covered with diminutive macrophytes at depths up to about 6 m. Both rocky and soft sediment habitats support invertebrate communities dominated by larval chironomids (Diptera: Chironomidae) and mollusks, although the trichopteran *Grensia* sp., and several species of aquatic mites and benthic microcrustaceans are also present (McDonald et al., 1982). The insects and mollusks serve as the major food base for the fish community of

Toolik Lake, coupling benthic productivity to the pelagic zone (Hershey, 1985b, 1990). *Hydra* sp. is also a conspicuous component of the rocky littoral invertebrate community, serving as an additional avenue for benthic–pelagic coupling (Cuker and Mozley, 1981).

At least 25 chironomid species representing 21 genera have been collected from the soft sediments of Toolik Lake (Table 3.9). Bare sediment and macrophyte-covered sediments support different relative abundances of the common taxa (Fig. 3.19). *Stictochironomus rosenschoeldi* (subfamily Chironominae) clearly dominates the macrophyte habitat. This species has a 4 year life cycle in Toolik Lake (Hershey, 1985a). Although *S. rosenschoeldi* is also common on bare sediments, that habitat is dominated by members of the subfamily Orthocladiinae, especially *Heterotrissocladius*, *Parakiefferiella*, and *Zalutschia*.

Table 3.9. Chironomidae Collected From Toolik Lake During Late June Through Early August 1980–1983

Chironomini	
Stictochironomus rosenschoeldi (Zett.)	++
Phaenopsectra sp.	–
Parachironomus sp.	–
Tanytarsini	
Tanytarus spp.	+
Paratanytarsus spp.	+
P. sp. (undescribed)	5 males
Stempellinella minor (Edw.)	1 male +
Constempellina brevicosta (Edw.)	+
Microspectra sp.	–
Tanypodinae	
Procladius spp.	+
P. nr. *vesus* Roback	4 males
P. sublettei var. *grandis* Roback	1 male +
Ablabesmyia sp.	–
Acrtopelopia sp.	–
Diamesinae	
Protanypus saetheri Wied.	4 males –
Prodiamesinae	
Monodiamesa bathyphila Kieff	1 male +
Orthocladiinae	
Heterotrissocladius maeaeri Brund	5 males ++
Parakiefferiella sp.	7 males ++
Zalutschia zalutschicola Lip.	+
Z. tatrica (Pag.)	2 males –
Psectrocladius sp.	–
Cricotopus spp.	–
C. sp. (undescribed)	14 males
Paracladius quadrinodosus Hirv	–
Abiskomyia virgo Edw.	6 males –

Abundance of larvae is indicated as follows: (–) <20 larvae sampled; (+) <200 larvae sampled; (++) >200 larvae sampled.

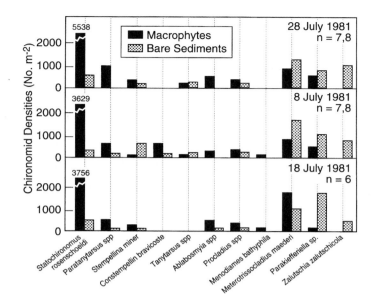

Figure 3.19. Distribution of Chironomidae by habitat. Median density (m²) of chironomid taxa are presented for macrophyte and bare sediment habitats on July 18, 1980, July 8, 1981, and July 28, 1981.

Total chironomid abundance on any given date is about twice as high in macrophytes as in bare sediments. Densities range from about 4000 m⁻² to about 11,000 m⁻² in bare sediments, and from about 8000 m⁻² to about 20,000 m⁻² in the macrophyte beds. In the bare sediment habitat, chironomid abundances are controlled by the benthic-feeding slimy sculpin (*Cottus cognatus*) during the summer, but sculpin are ineffective predators in the macrophyte-covered habitat (Hershey, 1985b; Fig. 3.20). However, because macrophytes are distributed only to a depth of about 6 m, sculpin predation is quite important to the chironomid community below this depth. Their effect is magnified by the fact that at winter temperatures sculpin feed and digest food, but at only 50% the rate occurring at summer temperatures. Because winter temperatures persist for 8 to 10 months, sculpin effect on chironomids during this time is probably as great as during the summer (Hershey and McDonald, 1985). Most other lakes in the region lack macrophytes except around the lake periphery, and sculpin appear to determine chironomid abundance on a regional scale (Goyke and Hershey, 1992).

The mollusk component of the Toolik Lake benthic fauna consists of four snail species (*Lymnaea elodes*, *Valvata lewisi*, *Gyraulus* sp., and *Physa* sp), and two genera of fingernail clams (*Pisidium* and *Sphaerium*). The snails *Gyraulus* and *Physa* are common in the rocky littoral zone, but do not occur in deeper water, whereas *Lymnaea* and *Valvata* are distributed across all habitats. *Lymnaea* and *Valvata* are both preyed upon heavily by lake trout and round whitefish (Merrick et al., 1992). Lake trout predation on *Lymnaea* is intense enough to restrict adults to the lake periphery; adult *Lymnaea* are abundant on the open sediments of lakes lacking lake trout (Merrick et al., 1991). In the presence of adult *Lymnaea*, *Valvata* fecundity is significantly reduced (Hershey, 1990) and its density is correspondingly low (approximately $10 \, m^{-2}$). However, on the soft sediments of Toolik Lake, where adult *Lymnaea* are uncommon, *Valvata* density is approximately $1000 \, m^{-2}$ (Hershey, 1990). Because *Valvata* is a considerably smaller genus than *Lymnaea*, snail biomass is dominated by *Lymnaea* even when *Valvata* density is high (Fig. 3.21).

In the rocky littoral zone, the snail *Lymnaea* is clearly the dominant grazer and controls algal community structure (Cuker, 1983). The diatom-dominated community on the rocks is maintained by the low density of *Lymnaea* grazers. At experimentally high grazer density, the algal community shifts toward small green and blue-green coccoid cells (Cuker, 1983). Nutrients from snail feces may be important to the nutrient-limited epilithic community. Microcosm experiments indicate that photosynthetic activity stimulated by snail feces is approximately 80% greater per unit mass that of ungrazed algae. However, nutrients released by grazing are tightly recycled rather than transferred to the overlying water (Cuker, 1983). *Lymnaea*

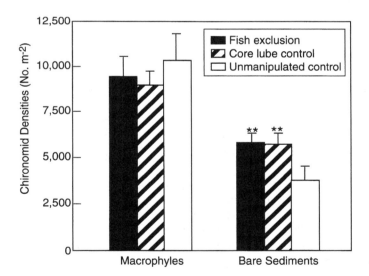

Figure 3.20. Chironomid densities in a fish exclosure experiment in Toolik Lake. Fish exclosures were core tubes covered with 6 mm mesh, core-tube controls were identical tubes with no covering, and unmanipulated controls were core samples from areas accessible to fish, taken at the end of the experiment. Asterisks indicate which treatments differed significantly from the corresponding unmanipulated controls ($p < 0.01$, Tukey's LSD test).

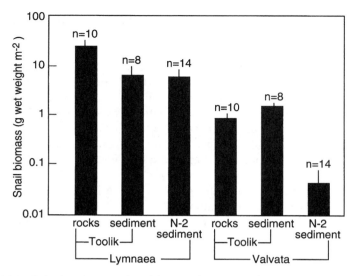

Figure 3.21. Calculated wet-weight biomass (mean ± SE) of *Lymnaea* and *Valvata* on rocky and soft-sediment habitats on July 8, 1982 and August 8, 1982, respectively.

growth rate itself is food limited in Toolik Lake, as evidenced by decreased growth rates at elevated population density (Cuker, 1983).

Fish Community

The Toolik Lake fish community is relatively simple, consisting of lake trout (*Salvelinus namaycush*), burbot (*Lota lota*), arctic grayling (*Thymallus arcticus*), round whitefish (*Prosopium cylindraceum*), and slimy sculpin (*Cottus cognatus*) (Merrick et al., 1992). Other nearby lakes contain these five species or some combination of them (Hanson et al., 1992). Typically, the smaller lakes have fewer species, often only grayling and slimy sculpin. All of these species undergo ontogenetic feeding shifts (Werner and Gilliam, 1984) and, depending on the composition and abundance of species in the fish community, they may variably impact the zooplankton, invertebrate benthos, and small fish communities.

Lake trout and burbot are typically piscivores, but in the lakes of the Toolik region there is no pelagic forage fish. Lake trout in Toolik Lake rely heavily on mollusks, and primarily on the snail *Lymnaea elodes* (Merrick et al., 1992; O'Brien, Buchanan, and Haney, 1979). Young-of-year (YOY) lake trout in Toolik Lake feed on zooplankton (Kettle and O'Brien 1978), likely switching to benthic insects by the end of their first year (McDonald et al., 1982). Juvenile lake trout feed on caddisflies (*Grensia*), chironomids, and zooplankton (Merrick et al., 1992; O'Brien, Buchanan, and Haney,

1979). Adult burbot feed on small lake trout, grayling, and round whitefish; juveniles feed on slimy sculpin (M. McDonald, personal observation). YOY burbot are pelagic and feed on zooplankton, and then move to the bottom and feed on benthic insects (Scott and Crossman, 1973). Burbot are relatively rare in Toolik Lake, perhaps due to their more obligate piscivory or lake trout predation on their pelagic juveniles.

The fishes in Toolik Lake are long lived and slow growing. A lake trout in Toolik Lake has been aged at 34 years and weighed 3400 g (Fig. 3.22). Lake trout growth may be energetically constrained by a diet of just benthic invertebrates (M. McDonald, unpublished data). The median size lake trout in Toolik Lake in 1979 was 578 g and the minimum reproductive size observed was 490 g (McDonald and Hershey, 1989). However, the median weight of lake trout declined significantly to 318 g in 1986 due to angling pressure (McDonald and Hershey, 1989). Burbot have not been sampled in Toolik Lake sufficiently for age and growth determinations, but one 7700 g speciment was 32 years old. Grayling from a nearby lake have been aged at 17+ years old and 815 g. The median size for grayling in Toolik Lake in 1977 was 380 g and the minimum observed reproductive size was 305 g; by 1986 the median size had declined to 185 g (McDonald and Hershey, 1989). Round whiterfish ages are similar to grayling, reaching 20+ years old and 850 g in Toolik Lake. No significant change in the size of round whitefish occurred in Toolik Lake between 1977 and 1986 (median weight 760 g and 680 g, respectively); the minimum observed reproductive size was 300 g (McDonald and Hershey, 1989). Slimy sculpin in Toolik Lake are one of the slowest growing populations of this species described (McDonald et al.,

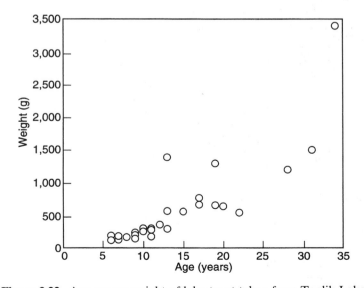

Figure 3.22. Age versus weight of lake trout taken from Toolik Lake.

1982; Hanson et al., 1992), living up to 8 years and attaining a mean weight of 3.2 g (McDonald et al., 1982). However, growth of sculpin in the Chena River approaches this minimum (Oswood et al., 1992). Slimy sculpin growth in Toolik Lake increased after their access to the more food-rich soft sediment habitat improved due to a reduction in lake trout predation (McDonald et al., 1992).

Toolik Food Web

With all the information we have collected over almost 20 years, it is possible to develop a detailed food web for Toolik Lake (Fig. 3.23). We know from the work of Whalen and Cornwell (1985) and through the limnocorral experiment (O'Brien et al., 1992) that the phytoplankton of Toolik Lake are extremely nutrient limited. Several bioassays and the low levels of nitrogen and phosphorus in the lake indicate that both nutrients may be limiting phytoplankton productivity. Hence both nutrients must be taken into account for any thorough understanding of the dynamics of

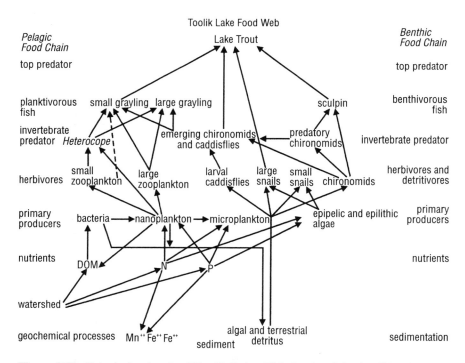

Figure 3.23. Pelagic food web of Toolik Lake. This food web is simplified in many ways but certainly in omitting the two fish species, round whitefish and burbot. Much less is known of their feeding than the species included.

Toolik Lake. Given the low level of inorganic nitrogen it is surprising that more nitrogen fixation does not occur in the lake. One likely reason for low levels of nitrogen fixation is the low levels of phosphorus. Once organisms begin alleviating their nitrogen deficit, they almost immediately run into a phosphorus deficit.

Figure 3.23 oversimplifies the algal community. As the work of H. Kling in this manuscript shows, the number of phytoplankton species that occur in Toolik Lake is quite large. According to niche theory, each must be playing a different role, but we do not know what these various roles might be. Hence we have collapsed this diverse community into one functional group. The importance of the links between zooplankton and phytoplankton is not really known. As mentioned above, the density of grazing zooplankton in Toolik Lake is so low that grazing mortality on phytoplankton is probably of little consequence to their ecology or to nutrient cycling within the pelagic zone. In fact, our current thinking is that most of the energy flow to the two major pelagic fish, arctic grayling and lake trout, does not go through the plankton food chain but rather the benthic food chain. The plankton food chain depicted in Fig. 3.23 is the food chain that existed in the 1970s and early 1980s. Since that time two of the three large zooplankton (*H. gibberum* and *D. middendorffiana*) have gone extinct or are extremely rare. In reported diet composition of Toolik Lake arctic grayling and lake trout (O'Brien, Buchanan, and Haney, 1979), *D. middendorffiana* was an important prey item. Out of eight grayling reported, six contained *D. middendorffiana* while only three had traces of the small-bodied *D. longiremis*. Of the 16 lake trout under 400 mm that were collected (O'Brien et al. 1979a), 9 had some and 7 had many *D. middendorffiana* included in their diet. With *D. middendorffiana* now in very low abundance and *Heterocope* declining, there may be no significant link between the plankton food chain and the two pelagic fish species.

The arrow linking small grayling to lake trout may also be more theoretical than actual. Of the 26 lake trout collected (O'Brien, Buchanan, and Haney, 1979) only 1 (470 mm in length) had a small grayling in the gut, and it was the second largest fish collected. However, it must be borne in mind that we hypothesize that small grayling and lake trout were rare in the late 1970s when these diet data were collected. The overall size structure of lake trout in Toolik Lake has declined from a median fork length of 388 mm in 1977 to a median fork length of 313 mm in 1986 (McDonald and Hershey, 1989). Thus there may no longer be large trout left in Toolik Lake. As shown in Fig. 3.22, fish in the 400 mm size range are around 1000 g and are estimated to be 20+ years old. However, Merrick et al. (1992) found that 30.9% of the large lake trout captured from Toolik Lake in the summer of 1986 had fish in the diet, although most of the fish were slimy sculpin rather than grayling. Interestingly, in the 1979 study, one lake trout had a grayling and one fish had a vole in its gut. One could argue from these data that at that time there was as much linkage to the terrestrial food chain as the planktonic food chain.

Thus the benthic food web of Toolik Lake is very important in understanding the overall trophic dynamics and energy flow within the lake. The two most important components of the benthic food web of Toolik Lake and other lakes in the region are snails and chironomids. We know less about the base of the benthic food chain than the planktonic food chain. However, it is clear that detritus is a very important component of the diets of most benthic invertebrates with the exception of snails. While both common species of snails feed on some detritus, they mainly feed on epipelic and epilithic algae that is likely nutrient limited like the phytoplankton. In most temperate lakes epipelic algae get some or most of their nutrients through sediment nutrient regeneration. However, in Toolik Lake nitrogen and especially phosphorus are so tightly bound that they are unavailable even to algae in close proximity to the sediments. Cuker (1983) showed that epilithic algae in Toolik Lake increased in biomass and primary productivity with added nitrogen and phosphorus. Likewise under intensive snail grazing in experimental containers, algal biomass declined. There was also some indication that nutrients released by grazing snails were rapidly taken up by epilithic algae rather than being released to the overlying water (Cuker, 1983).

Snails are heavily fed upon by lake trout in the lakes in the Toolik region and in Toolik Lake itself (Merrick et al., 1991). Merrick et al. (1992) found that 79.7% of the lake trout sampled had *Lymnaea* in their gut while only 36.6% of round whitefish and 10% of grayling contained snails. Feeding on snails was more pronounced in larger lake trout. It appears that predation from lake trout has considerable impact on snail densities and sizes, especially the large snail *Lymnaea*. It thus seems quite likely that lake trout control the abundance and size distribution of *Lymnaea* (Hershey, 1990; Merrick et al., 1991). In comparing three lakes with lake trout and three without, Merrick et al. (1992) found a 22 times greater density of *Lymnaea* in lakes without lake trout. The snails were also significantly larger in these lakes as well. Thus lake trout derive most of their food from snails, other benthos, and a benthic feeding fish, the slimy sculpin. Especially large lake trout do not feed much on zooplankton or grayling and are thus not part of the pelagic food chain.

Another important part of the benthic food chain is the slimy sculpin that feeds mainly on chironomids. Gut analysis of 51 sculpin captured from Toolik Lake showed that 84% of the diet items were larval chironomids with most of the rest chironomid pupae. Only 2.4% of the diet items were crustaceans (Hershey and McDonald, 1985). It is quite likely that predation by sculpin controls the population density of chironomids in these arctic lakes, especially where lake trout and burbot are absent or in low abundance. In three lakes where sculpin occurred in the absence of lake trout, average chironomid biomass was around $0.4\,g\,m^{-2}$. However, in four lakes, including Toolik Lake, where sculpin occurred in the presence of lake trout and burbot, average chironomid biomass was 3 times higher at $1.2\,g\,m^{-2}$ (Goyke and Hershey, 1992). However, sculpin predation is less effective in macrophyte beds (Hershey, 1985a). In Toolik Lake significantly more chironomids dwell in the macrophyte zone than the bare sediments. Both laboratory and field experiments show that sculpin feed more efficiently on chironomids in bare sediments than in macrophytes. Furthermore, laboratory experiments show that chironomids have no preference for macrophyte areas versus bare sediment regions (Hershey, 1985a). Sculpin also prey on benthic invertebrates in the rocky near-shore areas of Toolik Lake (Cuker et al., 1992). In contrast to the fairly nonselective feeding in the bare sediments, sculpin feeding in the rocky zone are size selective, mainly feeding on the larger, more predaceous chironomids leaving animals smaller than 3.5 mm free from sculpin predation.

Regulation of chironomid populations on bare sediments could occur throughout the year because it is known that sculpin can feed in the dark utilizing their lateral line to locate prey (Broadway and Moyle, 1978; Van Vliet, 1964). Hershey and McDonald (1985) showed that slimy sculpin take more than twice as long to digest prey at 4° as 16°C; but still they could effectively feed during the long arctic winter and thus have considerable impact on chironomid prey populations during the long, cold, dark arctic winter.

Sculpin themselves are of course a source of food for lake trout and burbot. Little is known about burbot feeding but a fair amount is known about lake trout predation on sculpin in the Toolik region. In the late 1970s, when lake trout were quite large in Toolik Lake (McDonald and Hershey, 1989), the distribution of sculpin appears to have been limited by lake trout. At this time sculpin were largely confined to the interface between the rocky littoral zone and the bare sediment deeper in the lake (McDonald et al., 1982). It was hypothesized that the rocky zone offered a refuge from lake trout where the sculpin could remain near the bare sediment where they can feed on chironomids most efficiently. However, as lake trout have been reduced in size, and likely in density as well, the distribution of sculpin has changed. In 1978 the percentage of sculpin caught in the bare sediments was 25% but this increased to 39.5% by 1987 (McDonald and Hershey, 1992). The importance of lake trout predation in restricting the distribution of sculpin is shown in comparing their distribution in lakes with and without lake trout (Hanson et al., 1992). In lakes without lake trout a greater percentage of sculpin are found on the soft sediments than in lakes with lake trout. This habitat shift of sculpin may be why sculpin are not more effectively preyed upon by lake trout. When lake trout are present sculpin are mostly in the rocky littoral zone and are more difficult to capture.

Thus Toolik Lake displays several unusual features when contrasted to other lakes in Alaska. It is extremely oligotrophic, and primary producers are colimited by low levels of both phosphorus and nitrogen. The low levels of nutrients are due both to the low input from the watershed and to the high nutrient sorbing sediments. As previously mentioned much of the spring meltwater, which brings in most of the annual load of nutrients, exits the lake without much mixing, thus delivering little nutrients. The sediments bind nutrients so tightly that there is virtually no cycling of remineralized nutrients. These conditions lead to very sparse levels of pelagic primary productivity and little phytoplankton biomass. Not surprisingly, zooplankton densities are also low and grazing from zooplankton has little impact on phytoplankton mortality or pelagic recycling of nutrients. Furthermore, zooplankton supply little resources to the pelagic fish community, especially with the decline of *D. middendorffiana*. The fish community, lake trout and grayling in particular, derives most of its food from the benthic food web. This is rare in temperate lakes but seems ot be common in the Toolik region where there is no pelagic forage fish.

Dr. J. Robie Vestal

J. Robie Vestal died in the fall of 1992. He had worked with our group for a number of years (1978 to 1986) and is deeply missed. Over 18 publications resulted from his work at Toolik Lake. As impressive and important as these works are, the truly great aspect of Robie's life was the effect he had on others. Robie touched the lives of everyone who came into contact with him through his generous nature and unflagging good cheer. While the world lost a good scientist in his death, those that knew and loved him lost a wonderful friend. This humble chapter is dedicated to his memory and how he made achieving the facts here presented so much more fun than they would have been without him.

References

Armstrong FAJ, Schindler DW (1971) Preliminary chemical characterization of waters in the Experimental Lakes Area, northwestern Ontario. J Fish Res Board Can 28:171–187.

Broadway JE, Moyle PB (1978) Aspects of the ecology of the prickly sculpin, *Cottus asper* Richardson, a persistent native species in Clear Lake, Lake County, California. Environ Biol Fishes 3:337–343.

Brooks JL, Dodson SI (1965) Predation, body size, and the composition of the plankton. Science 150:28–35.

Cole JJ, Findlay S, Pace ML (1988) Bacterial production in fresh and saltwater ecosystems: A cross-system overview. Mar Ecol Prog Ser 43:1–10.

Cornwell JC (1983) The geochemistry of manganese, iron and phosphorus in an arctic lake. Ph.D. thesis, University of Alaska, Fairbanks.

Cornwell JC (1985) Sediment accumulation rates in an Alaskan arctic lake using a modified ^{210}Pb technique. Can J Fish Aquat Sci 42:809–814.

Cornwell JC (1986) Diagenetic trace metal profiles in arctic lake sediments. Environ Sci Technol 20:299–302.

Cornwell JC (1987) Phosphorus cycling in arctic lake sediments: Adsorption and authigenic minerals. Arch Hydrobiol 109:161–179.

Cornwell JC, Banahan S (1992) A silicon budget for an Alaskan arctic lake. Hydrobiologia 240:37–44.

Cornwell JC, Kipphut GW (1992) Biogeochemistry of mangancese- and iron-rich sediments in Toolik Lake, Alaska. Hydrobiologia 240:45–60.

Coulon C, Holmgren S, Alexander V (1972) A survey of phytoplankton and zooplankton of Colville River drainage. In Kinney PJ (Ed) Baseline data study of the Alaskan arctic aquatic environment. University of Alaska, Institute of Marine Science Report R 72–3.

Cuker BE (1983) Competition and coexistence among the grazing snail *Lymnaea*, Chironomidae, and microcrustacea in an arctic epilithic lacustrine community. Ecology 64:10–15.

Cuker BE, McDonald ME, Mozley SC (1992) Influences of slimy sculpin (*Cottus cognatus*) predation on the rocky littoral invertebrate community in an arctic lake. Hydrobiologia 240:83–90.

Cuker BE, Mozley SC (1981) Summer population fluctuations, feeding, and growth of *Hydra* in an arctic lake. Limnol Oceanogr 26:697–708.

de March L (1978) Permanent sedimentation of nitrogen, phosphorus, and organic carbon in a high arctic lake. J Fish Res Board Can 35:1089–1094.

Dodson SI (1974) Zooplankton competition and predation: An experimental test of the size efficiency hypothesis. Ecology 55:605–613.

Dodson SI (1979) Body size patterns in arctic and temperate zooplankton. Limnol Oceangr 24:940–949.

Fee EJ, Hecky RE, Stainton MP, et al. (1989) Lake variability and climate research in northwestern Ontario: Study design and 1985–1986 data from Red Lake district. Canadian Technical Report of Fisheries and Aquatic Sciences No. 1662.

Froelich PN, Klinkhammer GP, Bender ML, et al. (1979) Early oxidation of organic matter in pelagic sediments of the eastern equatorial Atlantic: Suboxic diagenesis. Geochim Cosmochim Acta 43:1075–1090.

Goyke AP, Hershey AE (1992) Effects of fish predation on chironomid (Diptera: Chironomidae) communities in arctic lakes. Hydrobiologia 240:203–212.

Hamilton TD, Porter SC (1975) Itkillik glaciation in the Brooks Range, northern Alaska. Quaternary Res 5:471–497.

Hanson KL, Hershey AE, McDonald ME (1992) A comparison of slimy sculpin (*Cottus cognatus*) populations in arctic lakes with implications for the role of piscivorous predators. Hydrobiologia 240:189–202.

Hermanson MH (1990) ^{210}Pb and ^{137}Cs chronology from small, shallow arctic lakes. Geochim Cosmochim Acta 54:1443–1452.

Hershey AE (1985a) Littoral chironomid communities in an arctic Alaskan lake. Holarct Ecol 8:39–48.

Hershey AE (1985b) Effects of predatory sculpin on chironomid communities of an arctic lake. Ecology 66:1131–1138.

Hershey AE (1990) Snail populations in arctic lakes: Competition mediated by predation. Oecologia 82:26–32.

Hershey AE, McDonald ME (1985) Diet and digestion rates of the slimy sculpin, *Cottus cognatus*, in an arctic lake. Can J Fish Aquat Sci 42:483–487.

Hobbie JE (1964) Carbon-14 measurements of primary production in two Alaskan lakes. Verh Int Verein Theor Angew Limnol 15:360–364.

Hobbie JE (1973) Arctic limnology: A review. In Britton ME (Ed) Alaskan arctic tundra. Arctic Institute of North America Technical Paper 25, Washington, D.C., 127–168.

Hobbie JE (Ed) (1980) Limnology of tundra ponds. Dowden, Hutchinson and Ross, Stroudsburg, PA.

Hobbie JE, Corliss TL, Peterson BJ (1983) Seasonal patterns of bacterial abundance in an arctic lake. Arct Alp Res 15:253–259.

Hobbie JE, Helfrich JVK (1988) The effect of grazing by microprotozoans on production of bacteria. Arch Hydrobiol 31:281–288.

Holmgren S (1968) Phytoplankton production in a lake north of the Arctic Circle. Filosofie Licentiat Thesis, Uppsala University.

Hullar MAJ, Kaufman MJ, Vestal JR (1986) The effect of nutrient enrichment on the distribution of microbial heterotrophic activity in arctic lakes. Proc. IV ISME, 207–212.

Johnston CG, Kipphut GW (1988) Microbially mediated Mn(II) oxidation in an oligotrophic arctic lake. Appl Environ Microbiol 54:1440–1445.

Kalff J, Kling HJ, Holmgren SK, Welch HE (1975) Phytoplankton, phytoplankton growth and biomass cycles in an unpolluted and polluted polar lake. Verh Int Ver Limnol 19:487–495.

Keller SA, Morris RH, Detterman DL (1961) Geology of the Shaviovik and Sagavanirktok Rivers region, Alaska. U.S. Geol. Surv. Prof. Pap. 303D:169–222.

Kelley JJ, Gosink TA (1988) Carbon dioxide and other trace gases in tundra surface waters in Alaska. In Degens ET, Kempe S, Naidu AS (Eds) Transport of carbon and minerals in major world rivers, lakes and estuaries, Part 5, Mitteilungen aus bem Geologisch-Paleontologischen Institut der Universitat Hamburg, Vol 66. SCOPE/UNEP, Hamburg, Germany, 117–127.

Kettle D, O'Brien WJ (1978) Vulnerability of arctic zooplankton species to predation by small lake trout (Salvelinus namaycush). J Fish Res Board Can 35:1495–1500.

Kipphut GW (1978) An investigation of sedimentary processes in lakes. Ph.D. thesis, Columbia University.

Kipphut GW (1988) Sediments and organic carbon in an arctic lake. In Degens ET, Kempe S, Naidu AS (Eds) Transport of carbon and minerals in major world rivers, lakes and estuaries, Part 5. Mitteilungen aus bem Geologisch-Paleontologischen Institut der Universitat Hamburg, Vol 66. SCOPE/UNEP, Hamburg, Germany, 129–135.

Kling GW, Kipphut GW, Miller MC (1991) Arctic lakes and streams as gas conduits to the atmosphere: Implications for tundra carbon budgets. Science 251:298–301.

Kling GW, Kipphut GW, Miller MC (1992) The flux of CO_2 and CH_4 from lakes and rivers in arctic Alaska. Hydrobiologia 240:23–36.

Kling GW, O'Brien WJ, Miller MC, Hershey AE (1992) The biogeochemistry and zoogeography of lakes and rivers in arctic Alaska. Hydrobiologia 240:1–14.

Livingstone DA, Bryan K, Leahy RG (1958) Effects of an arctic environment on the origin and development of freshwater lakes. Limnol Oceanogr 3:192–214.

Luecke C, O'Brien WJ (1983) The effect of Heterocope predation on zooplankton communities in arctic ponds. Limnol Oceanogr 28:367–377.

McDonald ME, Cuker BE, Mozley SC (1982) Distribution, production, and age structure of slimy sculpin in an arctic lake. Environ Biol Fish 7:171–176.

McDonald ME, Hershey AE (1989) Size structure of a lake trout (Salvelinus namaycush) population in an arctic lake: Influence of angling and implications for fish community structure. Can J Fish Aquat Sci 46:2153–2156.

McDonald ME, Hershey AE (1992) Shifts in abundance and growth of slimy sculpin in response to changes in the predator population in an arctic Alaskan lake. Hydrobiologia 240:219–224.

McKenzie C, Kling H (1989) Scale-bearing Chrysophyceae (Mallomonadaceae and Paraphysomonadaceae) from Mackenzie Delta area lakes, Northwest Territories, Canada. Nord J Bot 9:103–112.

Merrick GW, Hershey AE, McDonald ME (1991) Lake trout (Salvelinus namaycush) control of snail density and size distribution in an arctic lake. Can J Fish Aquat Sci 48:498–502.

Merrick GW, Hershey AE, McDonald ME (1992) Salmonid diet and the size, distribution, and density of benthic invertebrates in an arctic lake. Hydrobiologia 240:225–234.

Miller MC, Hater GR, Spatt P, Westlake P, Yeakel P (1986) Primary production and its control in Toolik Lake, Alaska. Arch Hydrobiol 74:97–134.

Nauwerck A (1963) Die beziehungen zwisehen zooplankton und phytoplankton inm see erken. Symb Bot Uppsal 17:1–163.

O'Brien WJ (1975) Some aspects of the limnology of the ponds and lakes of the Noatak drainage basin, Alaska. Verh Int Verein Limnol 19:472–479.

O'Brien WJ (1988) The effect of container size on the feeding rate of *Heterocope septentrionalis*: A freshwater predaceous copepod. J Plankton Res 10:313–317.

O'Brien WJ, Buchanan C, Haney JF (1979) Arctic zooplankton community structure: Exceptions to some general rules. Arctic 32:237–247.

O'Brien WJ, Hershey AE, Hobbie JE, et al. (1992) Control mechanisms of arctic lake ecosystems: A limnocorral experiment. Hydrobiologia 240:143–188.

O'Brien, WJ, Kettle D, Riessen HP (1979) Helmets and invisible armor: Structures reducing predation from tactile and visual planktivores. Ecology 60:287–294.

O'Brien WJ, Schmidt D (1979) Arctic *Bosmina* morphology and copepod predation. Limnol Oceanogr 24:564–568.

Oswood MW, Reynolds JB, LaPerriere JD, Holmes R, Hallberg J, Triplehorn JH (1992) Water quality and ecology of the Chena River, Alaska. In Becker CD, Neitzel DA (Eds) Water quality in North American river systems. Battelle Press, Columbus OH, 7–27.

Peterson BJ, Hobbie JE, Haney JF (1978) *Daphnia* grazing on natural bacteria. Limnol Oceanogr 23:1039–1044.

Prescott GW (1963) Ecology of Alaskan freshwater algae. II–IV. Introduction: General features (Additional notes). IV. Additional notes on Pseudendoclonium, and a transfer. Trans Am Micros Soc 82:137–143.

Rawson DS (1953) Limnology in the North American arctic and subarctic. Arctic 6:198–204.

Rublee PA (1992) Community structure and bottom-up regulation of heterotrophic microplankton in arctic LTER lakes. Hydrobiologia 240:133–142.

Schindler DW, Welch HE, Kalff J, Brunskill GJ, Kritsch N (1974) Physical and chemical limnology of Char Lake, Cornwallis Island (75°N Lat.). J Fish Res Board Can 31:588–607.

Scott WB, Crossman EJ (1973) Freshwater fishes of Canada. Bull Fish Res Board Can 184:1–966.

Stross RG, Kangas DA (1969) The reproductive cycle of *Daphnia* in an arctic pool. Ecology 50:457–460.

Untermohl H (1958) Zur vervollkommnung der quantitativen phytoplanktonmethodik. Mitt Int Verein Theor Angew Limnol 9:1–38.

Van Vliet WH (1964) An ecological study of *Cottus cognatus* Richardson in northern Saskatchewan. M.S. thesis, University of Saskatchewan, Saskatoon, Canada.

Welch HE, Legault JA (1986) Precipitation chemistry and chemical limnology of fertilized and natural lakes at Saqvaqjuac, N.W.T. Can J Fish Aquat Sci 43:1104–1134.

Welch HE, Legault JA, Kling HJ (1989) Phytoplankton, nutrients, and primary production in fertilized and natural lakes at Saqvaqjuac, N.W.T. Can J Fish Aquat Sci 46:90–107.

Werner EE, Gilliam JF (1984) The ontogenetic niche and species interactions in size-structured populations. Annu Rev Ecol Syst 15:393–425.

Whalen SC, Alexander V (1986a) Chemical influences on 14C and 15N primary production in an arctic lake. Polar Biol 5:211–219.

Whalen SC, Alexander V (1986b) Seasonal inorganic carbon and nitrogen transport by phytoplankton in an arctic lake. Can J Fish Aquat Sci 43:1177–1186.

Whalen SC, Cornwell JC (1985) Nitrogen, phosphorus and organic carbon cycling in an arctic lake. Can J Fish Aquat Sci 42:797–808.

Whalen SC, Cornwell JC, Alexander V (1988) Comparison of chemical and biological N budgets in an arctic lake: Implications for phytoplankton production. In Degens ET, Kempe S, Naidu AS (Eds) Transport of carbon and minerals in major world rivers, lakes and estuaries, Part 5. Mitteilungen aus bem Geologisch-Paleontologischen Institut der Universitat Hamburg, Vol 66. SCOPE/UNEP, Hamburg, Germany, 99–115.

4.   The Kuparuk River:
A Long-Term Study of Biological
and Chemical Processes in an Arctic River

Anne E. Hershey, William B. Bowden,
Linda A. Deegan, John E. Hobbie,
Bruce J. Peterson, George W. Kipphut,
George W. Kling, Maurice A. Lock,
Richard W. Merritt, Michael C. Miller,
J. Robie Vestal,[1] and Jeffrey A. Schuldt

Our studies have focused on carbon and nutrient dynamics, primary productivity and decomposition, and abundance and life histories of the macroconsumers in the Kuparuk River in arctic Alaska. The overall objective of these studies is to understand the processes controlling primary and secondary productivity, nutrient dynamics, and trophic structure.

The Kuparuk River originates in the foothills of the Brooks Mountain Range in the Alaskan arctic and flows north–northeast to the Arctic Ocean (total drainage area is $8107\,km^2$). We have intensively studied the foothill section of the river, which intersects the Dalton Highway (Fig. 4.1; formerly the Trans-Alaska Pipeline Haul Road), but have synoptically sampled the river from its headwaters to its mouth. The upper river to the study section drains an area of about $143\,km^2$, with a mean elevation of 842 m above sea level, a main channel length of 25 km, and a mean channel slope of 3.13% (Peterson et al., 1993). Green Cabin Lake (68°32′N, 149°15′W) is the largest of only a few lakes in the watershed (Fig. 4.1), none of which are fed by glaciers. The Kuparuk, therefore, is a clear-water tundra river (Craig and McCart, 1975). Permafrost is present throughout the study area at a mean depth of about 40 cm in the late season (Peterson et al., 1986). The vegeta-

---

[1] Our dear friend and colleague. J. Robie Vestal passed away prior to completion of the final draft of this chapter. We miss him much.

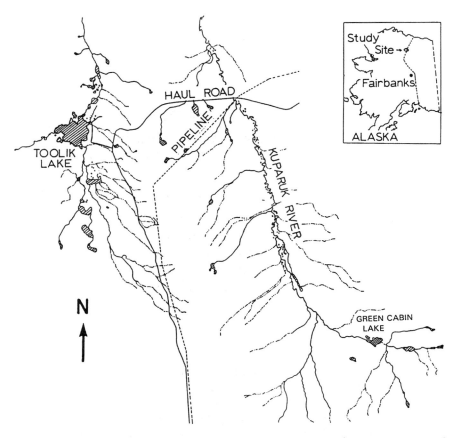

Figure 4.1. Watershed map of the Kuparuk River study area (from Peterson et al., 1993).

tion of the Kuparuk River Valley includes upland heath communities on dry soils, moist tundra communities dominated by the tussock-forming cotton grass *Eriophorum vaginatum*, and wet sedge tundra dominated by *Carex aquatilus*. The riparian zone has large patches of dwarf willows (*Salix* spp.) and birches (*Betula nana*), which rarely exceed 1 m in height. The river meanders through heath peat deposits and the river bed is comprised of large cobbles and boulders. Summer water temperatures average 8 to 10°C, but may range from about 4°C to 14°C on a diel basis, and occasionally reach as high as 20°C at low flow.

Hydrology

The hydrologic regime in the Kuparuk River varies widely during the open water season due to variation in summer rainfall (Fig. 4.2). Average annual total precipitation in this region is about 15 to 25 cm (Selkregg, 1977), of which about half falls as snow from September through May. Snowmelt in late May to early June results in flooding, but during this time large portions of the riverbed are still covered with bottom ice, so there is little bottom scour. Summer discharge records from 1983 to 1990 show that discharge ranged from 0.3 to $30\,m^3s^{-1}$, but mean summer discharge ranges from 0.32 to $3.8\,m^3\,s^{-1}$ (Fig. 4.2). In most years, flow ceases by late September, riffles become dry, and by October pools freeze solid. Flow resumes after snowmelt in May of the following year.

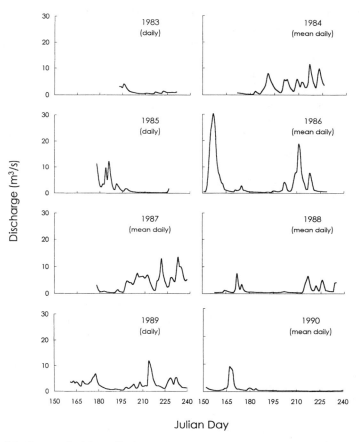

Figure 4.2. Kuparuk River discharge in the study section during the open water season in 1983–1990. During 1983 and 1989 discharge was measured approximately daily. During other years, stage height was measured continuously and discharge was determined from a calibration of discharge against stage height. Mean discharge over each 24-h period is plotted for those years.

Discharge in the study reach is affected by a large spring, approximately 6 km upstream of the road crossing and 2 km downstream of the headwater lakes (Fig. 4.1), which delivers approximately $0.2 \, m^3 \, s^{-1}$. Approximately 50 km downstream of the Dalton Highway is a large aufeis (see Chapter 1). During low flow years, the Kupaurk River goes underground approximately 40 km downstream of the Dalton Highway, then reemerges at the aufeis.

Inorganic Nutrients

Phosphate concentrations in the Kuparuk River are very low, ranging from a summer mean of 0.05 to $0.45 \, \mu M$ PO_4-P in 1983 to 1990 (Fig. 4.3). Lysimeter studies of soil phosphate levels in the riparian zone suggest that soil waters are also low in phosphate ($\leq 0.65 \, \mu M$), but higher than concentrations in the river. Soils associated with the riverside willow zone are higher in extractable phosphate than most upslope soils (Nadelhoffer et al., 1991). However, the mobility of this phosphate into the river is complex and not well understood. On a given day, phosphate (P) concentration can vary at different sites along the river depending on upslope conditions, such that P discharge into the river appears to be a discontinuous process associated with subsurface and riparian flows along the spatially variable river margin. Disturbance processes that may affect phosphate discharge into the river include solifluction, calving of peat soils, and underground erosion into mineral soils within water tracts, which are small, heavily vegetated channels that deliver water and nutrients downslope (Chapin et al., 1978; Kane et al., 1989).

Ammonium concentrations are also typically low, ranging from 0.4 to $1.8 \, \mu M$ NH_4-N (Fig. 4.3). Nitrate concentrations are usually higher than ammonium, ranging from 0.5 to $3.6 \, \mu M$ NO_3-N (Fig. 4.3), and are negatively correlated with discharge.

One source of ammonium to the Kuparuk River appears to be dissolved free amino acids (DFAA) in the riparian zone (Fiebig, 1988) that are mineralized under reducing conditions. Moving downslope toward the river, the riparian zone is often aerobic in the nearest 0.5 m to the river. Lysimeters placed in this narrow zone indicate that soil solutions contain very high concentrations of nitrate (up to $160 \, \mu M$ NO_3-N). In contrast, lysimeters placed in upslope areas show that soil solutions have barely detectable concentrations of nitrate. These results suggest that intense nitrification of the ammonium occurs in the aerated riparian zone. Under aerobic conditions ammonium is readily nitrified by both the riparian microflora and by the epilithic community in the river itself. Rocky steep sided tributaries deliver nitrate to the river through this process. Thus the ammonium concentration in the river is limited by rapid nitrification to nitrate as well as direct algal uptake. Although negatively correlated with discharge,

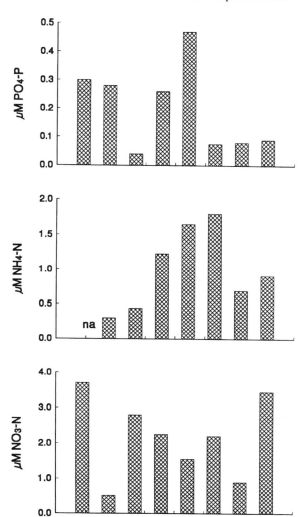

Figure 4.3. Mean molar concentrations of phosphate (PO_4), ammonium (NH_4), and nitrate (NO_3) during the open water season in the Kuparuk River. Nutrient concentrations were measured weekly from 3 to 5 sites.

nitrate levels vary only slightly from station to station suggesting that nitrate uptake and denitrification in the streambed are approximately balanced by nitrate inputs and in-stream nitrification.

Foothills streams generally have low ionic content compared to mountain streams, and the Kuparuk falls at the low end of this range (Table 4.1, see Lock et al., 1989). The conductivity of the Kuparuk River (16 μm hos) was the lowest of foothills and mountain rivers studied (Table 4.1). In contrast to montaine and glacial rivers, the major cation concentrations in the

Table 4.1. Water chemistry of the Kuparuk River Compared with Ranges of Water Chemistry Data for Tundra Foothills and Mountains Streams, Based on Synoptic Sampling of Streams During this Open Water Season

Component	Kuparuk	Foothills ($n = 5$)	Mountain ($n = 5$)
Conductivity (μmhos)	16.0	21–125	30–225
pH	7.1	5.9–8.2	7.4–8.4
Alkalinity (meql^{-1})	0.2	0.05–1.12	0.24–2.61
SRP (μg Pl^{-1})	0.4	0.4–5.0	0.4–4.2
$NO_3 + NO_2$ (μg Nl^{-1})	5.2	4.4–59.0	3.4–110.4
NH_4 (μg Nl^{-1})	18.0	18.0–65.6	39.9–68.8
Calcium (mgl^{-1})	1.7	0.8–16.0	3.5–32.0
Color (250nm 10cm^{-1})	1.5	0.37–2.66	0.08–0.36
DOC (mg Cl^{-1})	2.3	2.3–8.4	0.18–1.84
Amino acids (μM)	6.5	3.7–47.9	12.2–272.5
N:P (molar)	116.0	27–156	72–386

Kuparuk River at low flow are similar to those of subsurface soil water, which has little exposure to unweathered bedrock.

Allochthonous Carbon

The carbon pool of the Kuparuk River is dominated by leaching of dissolved organic carbon (DOC) from the tundra and by inputs of eroding peat, little of which is respired (Fig. 4.4; see Peterson et al., 1986). About 2 to 6% of the net above ground productivity of the watershed appears to be

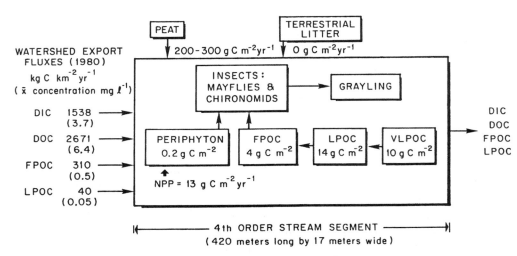

Figure 4.4. Carbon stocks and fluxes for the Kuparuk River watershed upstream of the trans-Alaska pipeline (from Peterson et al., 1986).

exported by the river (Peterson et al., 1986). The total organic carbon (TOC) concentration in the stream water averages 6.8 mg/L (DOC + POC), and the annual export of organic carbon from the watershed is 2 to 3 t km^{-2}yr^{-1}. Both values are similar to those found in temperate systems (Minshall, 1978; Peterson et al., 1986), but relatively high in relation to yearly water flux (see Oswood et al., 1992).

About 88% of the total organic carbon exported from the Kuparuk River is DOC (Fig. 4.4). Although this represents a high DOC/POC ratio (see Richey, 1983), this is a somewhat lower percentage than the 97% of total carbon exported as DOC in nearby Imnavait Creek, a first order beaded tundra stream (Oswood et al., 1989). The Kuparuk River DOC component is predominantly (70%) material with an apparent molecular weight <50,000. This is in contrast to mountain rivers in the region where approximately 90% of the DOC is in this weight range (Lock et al., 1989). Compared to other tundra streams in the region, including Imnavait Creek (Oswood et al., 1989), the Kupaurk also has relatively low DOC concentrations, although typically higher than DOC in mountain streams (Table 4.1). In general, tundra streams exhibit decreasing DOC concentrations with increasing stream order, whereas mountain streams exhibit increasing DOC concentrations with increasing stream order (Lock et al., 1989). Assuming uptake of DOC by the streambed (Lock et al., 1989) and discharge of DOC at stream banks (Kling et al., 1991), the higher bottom area-to-volume ratio and lower bank area-to-volume ratio of larger streams may explain why larger tundra streams have lower concentrations of DOC.

The remaining 12% of TOC exported from the river is in particulate form (POC) (Fig. 4.4). Most of this material is derived from peat, eroded from the stream banks (approximately 200 to 300 g C m^{-2}yr^{-1}). This often occurs as large chunks that are deposited on the stream bottom and subsequently become entrained into the water column as small particles (Peterson et al., 1986). Less than 15% of the POC pool consists of suspended bacteria and algae (Peterson et al., 1986), although this latter fraction of POC is more labile and likely of more importance to invertebrate collectors. The standing stock of POC on the river bottom, including all size fractions, was approximately 20 to 30 g C m^{-2}, whereas suspended POC, which turns over rapidly, was approximately 0.2 g C m^{-2} (Peterson et al., 1986).

The Kuparuk River is supersaturated with CO_2; the CO_2 concentration is approximately 700 ppm, which is about twofold higher than atmospheric equilibrium. The most likely source of this dissolved inorganic carbon (DIC) is tundra soil respiration (Kling et al., 1991). DIC concentration increases as the season progresses ($r = 0.79$), apparently because water enters from deeper in the soil as the frozen peat melts; this water carries more respired carbon (Peterson et al., 1986). The ^{14}C activity of the peat was 70% modern (Peterson et al., 1986), whereas the ^{14}C activity of atmospheric CO_2 is 118% modern (Kling et al., 1991) (1950 atmospheric sample taken as 100% modern; increases are due to nuclear weapons testing). The interme-

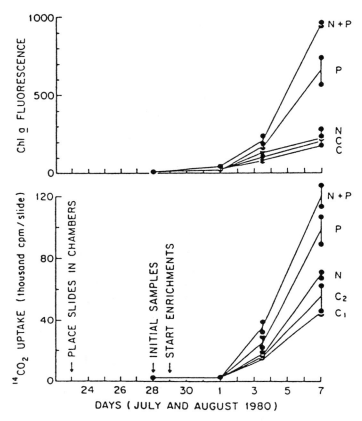

Figure 4.5. Time course of chlorophyll fluorescence and $^{14}CO_2$ uptake on slides in bioassay tubes in the Kuparuk River. C is control; N received added NH_4NO_3; P received added KH_2PO_4; N + P received both (from Peterson et al., 1983).

diate value of 109% for epilithon indicates that the dissolved CO_2 in the river is a mixture of CO_2 from newer terrestrial production and from respiration of older peat in the soils (Kling et al., 1991; Oswood et al., 1992).

Lysimeter studies also suggest that the tundra is the major source of DOC for the river. Concentrations of DOC up to $15.5 \, mg \, L^{-1}$ have been recorded in streamside soils at the Kuparuk River (Lock et al., 1989). Both DOC and DFAA were generally higher at the end of June than later in the season. The highest levels of DOC were 8 to $16 \, mg \, CL^{-1}$ falling to around $4 \, mg$ CL^{-1}; DFAAs were around $10 \mu M$ in late June falling to around $50 \, nM$ by early August. The high concentrations of organic matter may be a consequence of exoenzymatic degradation of organic matter during the winter, which is then flushed from the soils during the spring thaw. Because much of the DOC and DFAA may be taken up and respired in the stream bed (Lock et al., 1989), it is likely that in-stream respiration also contributes to the observed supersaturation of CO_2. Flux of CO_2 to the atmosphere from the Kuparuk River has been estimated at $11.9 \pm 1.5\,\mathrm{mmol\,m^{-2}\,d^{-1}}$, suggesting that the Kuparuk, like other arctic aquatic ecosystems, serves as a conduit that transports terrestrially fixed carbon to the atmosphere (Kling et al., 1991).

Autochthonous Carbon

Net production of epilithic algae is about $13\,\mathrm{g\,C\,m^{-2}\,yr^{-1}}$, two orders of magnitude less than inputs of organic carbon, and quite similar to areal production in nearby Toolik Lake ($14\,\mathrm{g\,C\,m^{-2}\,yr^{-1}}$; Miller et al., 1986). Epilithic production is clearly nutrient limited in the river. Flow-through bioassay tubes, used to evaluate nutrient limitation of algae on glass slides incubated inside the tubes, showed that algal biomass (as chlorophyll *a*) was about threefold higher on P enriched slides than on control slides, and about fourfold higher on P + N enriched slides than control slides (Fig. 4.5; see Peterson et al., 1983). Slides that had experienced N only enrichment were similar to controls (Fig. 4.5). Uptake of $^{14}CO_2$ showed a trend very similar to that of chlorophyll *a* (Fig. 4.5). These data show clearly that P was the primary nutrient limiting algal biomass and production in the Kuparuk River, and that N was of secondary importance (Peterson et al., 1983).

Net photosynthesis of the epilithic community, measured as the difference between rocks in light and dark chambers, always exceeds epilithic respiration in both riffles and pools of the Kuparuk River (Fig. 4.6). Cham-

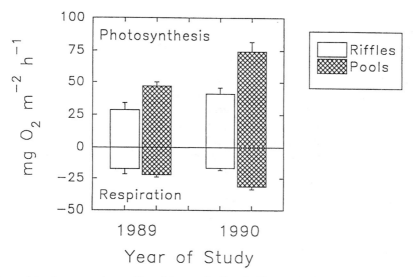

Figure 4.6. A comparison of benthic metabolism in Kuparuk River riffles and pools in 1989 and 1990. Bars show mean ± SE net photosynthesis and respiration of stream rocks in plexiglass chambers.

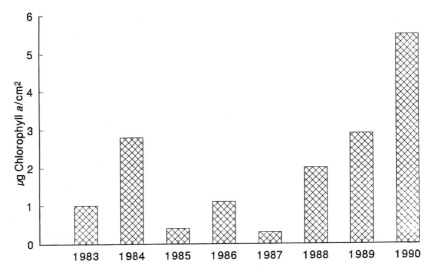

Figure 4.7. Peak epilithic chlorophyll *a* concentrations in Kuparuk River riffles in 1983–1990.

ber measurements made in riffles only in most years between 1983 and 1991, show that chamber net photosynthesis ranges from about 10 to $50\,mg\,O_2\,m^{-2}\,h^{-1}$. Chamber respiration ranges from about 5 to $20\,mg\,O_2\,m^{-2}$ h^{-1}, or about 50% of net photosynthesis. These data suggest that the Kupaurk River epilithic community is autotrophic rather than heterotrophic (but see Peterson et al., 1985); O_2 production in light was 2.5-fold dark consumption. Measurement of metabolic heat output by both autotrophs and heterotrophs was about 2.3-fold higher in experimental biofilm communities grown in the light compared to those grown in the dark (see Lock et al., 1990), also suggesting greater dependence on autotrophy than heterotrophy. However, since chamber and experimental biofilm studies do not account for either decomposition of allochthonous organic matter in transport, or for heterotrophic activity in the hyporheos, these data do not fully represent the trophic status of the river. Studies of invertebrate consumers (see below) show that the largest proportion of secondary production is due to black flies, which use largely allochthonous carbon.

Maximum epilithic algal biomass (as chlorophyll a/cm^2) varied approximately tenfold from 1983 to 1990, ranging from about $0.5\,\mu g\,cm^{-2}$ to about $5\,\mu g\,cm^{-2}$ (Fig. 4.7), and biomass was negatively correlated with mean seasonal discharge ($r = -.78$; Figs. 4.2 and 4.7). The autotrophic community within the epilithic matrix is dominated by diatoms. A total of 187 species of diatoms have been identified over four years, but up to 95% of the individuals enumerated were from only 12 genera, numerically dominated by *Achnanthes*, *Cymbella*, *Hannea*, and *Diatoma*. Species abundance showed greater changes from year to year than between stations in the same year. Filamentous or macroalgae were common, but distributed in isolated patches. These included *Lemonia* (Rhodophyta), *Batrachospermum* (Rhodophyta), and *Calothrix* (Cyanophyta) in fast water. The Chlorophyta *Stigeoclonium*, *Mougotia*, and *Spirogyra* were comonly found associated with moss holdfasts in riffles and pools.

Decomposition

The nutrient content of the various organic substrates determines the rate of carbon processing by the Kuparuk River microbiota. Fresh *Carex aquatilus* leaves, which have a high phosphorus content (Federle et al., 1982) showed 20% (of initial dry weight) biogenic weight loss after 20 days in the river and 30% after 30 days in the river (Hullar and Vestal, 1986). Leaching weight loss was about 20%. These decomposition rates were not affected by experimental incubation with elevated phosphorus (Peterson et al., 1985). Decomposition of autochthonous algal material, which is also high in nutrients, was similarly not nutrient limited (Peterson et al., 1985). In contrast, peat is low in nutrients and does not decay rapidly. However, experimental incubation of peat at high phosphorus concentration resulted in increased rate of mineralization of the relatively refractory peat-derived ^{14}C-lignin and ^{14}C-cellulose (Peterson et al., 1985). Heterotrophic activity, as measured by ^{14}C-bicarbonate incorporation into microbial lipids was significantly lower on glass beads (used as a surrogate for rocks but with known surface area) in darkened flow through bioassay tubes where there was little autochthonous carbon, compared to tubes exposed to ambient light (Peterson et al., 1985). Further evidence that peat is a low quality source of carbon to decomposers was seen during a 4 week experiment in flow-through bioassay tubes, where glucose significantly ($p < 0.05$) stimulated the heterotrophic community on glass beads compared to control beads and beads incubated with tundra peat extract (Table 4.2; see Hullar et al., 1987; Hullar and Vestal, 1989). Heterotrophic activity was very similar on control and peat extract incubated beads (Table 4.2).

Table 4.2. Mean ± SE Heterotrophic Activity as Measured by ^{14}C Acetate Incorporation in Light and Dark Flow-Through Bioassay Tubes. Tubes Were Enriched Either with $300 \mu g l^{-1}$ Glucose, $6.86 mg l^{-1}$ Peat Extract, or Were Unenriched Controls

Treatment	Acetate Incorporation Light (±)		$(dpm \times 10^3 cm^{-2} hr^{-1})$ Dark (±)	
Control	1.0	0.6	1.1	0.5
Glucose	4.1	0.7	2.8	0.7
Peat extract	0.7	0.2	1.1	0.2

The density of suspended bacteria in the Kuparuk was the lowest recorded for any foothills or mountain river in the region (Lock et al., 1989). Peak bacterial populations in the Kuparuk River water occurred during the early spring high discharge period (Hobbie et al., 1983). A similar situation occurs in tundra ponds, where early spring bacterial peaks have been attributed to input of soil microorganisms (Hobbie and Rublee, 1975; Hobbie et al., 1980). However, rising water was not always correlated with bacterial abundance during the remainder of the season, possibly due to flushing of bacteria during previous high water (Hobbie et al., 1983). Activity (^{14}C-acetate incorporation into lipids; McKinley et al., 1982) of the microheterotrophs is generally higher in the foothills streams (56 to 2883 DPM ml^{-1} h^{-1}) than in the mountain rivers (14 to 87 DPM ml^{-1} h^{-1}). However, the Kuparuk falls at the low end of the range for foothills rivers (56 DPM ml^{-1} h^{-1}). The cell specific activity of the Kupaurk was the highest (560 DPH 10^5 cells h^{-1}) of the foothills and mountain rivers studied (Lock et al., 1989).

Photochemical reactions also appear to be important in decompostion of organic matter in the Kuparuk River (Lock et al., unpublished) and in tundra ponds (Strome and Miller, 1978), as well as in marine and other freshwater environments (Kieber et al., 1990). In experiments with dark-grown (heterotrophs) and light-grown (autotroph and heterotroph mixture) biofilm communities from the Kuparuk River, photolysed waters (incubated in UV translucent containers) resulted in higher levels of metabolic heat output (sensu Lock and Ford, 1985; Ford and Lock, 1987) than controls (incubated in UV opaque containers). Dark-grown biofilm communities exposed to photolysed waters averaged 59% higher metabolic heat output than controls (range 10 to 100%, $n = 3$), whereas the single light-grown community showed 80% higher metabolic heat-output than controls. Photolytic changes were confirmed by spectrophotometric examination (sensu Kieber et al., 1990) of the waters showing that photolysis resulted in approximately 10 to 12% reduction in absorption at 250 nm (10 cm path length). Given the 24-hour light regime in the arctic, the residence time of water in the Kuparuk River (≈ 4 days), and the length of the experimental photolysis regime (5 days); these findings suggest that photolysis may substantially increase DOM mineralization and microbial production in the Kuparuk River, as well as in other streams and shallow ponds (Strome and Miller, 1978; Prentki et al., 1980).

Invertebrate Consumers

The macroinvertebrate consumers of the Kuparuk River are dominated by insects. Molluscs are found occasionally in deep pools and in fish stomachs, but are not important in the intensively studied section of the river. The insect community is depauaparate when compared to temperate rivers.

Dipterans domainate both numerically and in terms of secondary production. Ephemeroptera, Trichoptera, and Plecoptera are represented by only a few species (Table 4.3).

The black fly, *Prosimulium perspicuum* Sommerman is the most abundant consumer species in the river, with *Stegopterna mutata* complex also important. These two black flies show very little temporal overlap because *Stegopterna* emerges in early July when the early instars of *Prosimulium* are first sampled. *Prosimulium* emerges in mid-August. Several other black fly species (Table 4.3) are also present at low density. Other conspicuous consumers include the chironomid *Orthocladius rivulorum* (Kieffer), the mayfly *Baetis* sp., and the caddisfly *Brachycentrus americanus* (Banks). Except for *Brachycentrus*, which has a 3 year life cycle in the river, the dominant consumers are all univoltine. A diverse assemblage of chironomids, most very small, have also been sampled from the Kuparuk (Table 4.3). These chironomids all appear to be univoltine, although detailed life history studies have been conducted only for *O. rivulorum*. The silk tubes, several centimeters long, of *O. rivulorum* are conspicuous in the river riffles, and, where they occur, the microflora associated with the tubes constitute 12 to 43% of the epilithic algal biomass (Hershey et al., 1988).

Insect secondary production, calculated using the size-frequency method for 1984 to 1986 data, indicate that black fly production, although variable among years, is the largest component of secondary production (Table 4.4). Total production of the four dominant taxa ranged from 5.72 to 14.38 g dry weight $m^{-2}yr^{-1}$. *O. rivulorum* represented 6 to 20% of the total insect secondary production, but the remainder of the chironomid community contributes <3% to secondary production (estimated for 1984 only), which is very low compared to chironomids in temperate streams (Hauer and Benke, 1991), and reflects their small size and slow turnover.

A 7 year record of abundance of dominant insects collected from three to five sites per year in the river shows that abundance is highly variable among years for most taxa (Fig. 4.8). Significant parameters in a multiple linear regression model predicting black fly abundance were peak spring discharge, algal biomass, and *Baetis* and *Brachycentrus* density. Most of the variability in black fly abundance was related to spring discharge; higher peaks in the spring were associated with higher mean summer density of black flies ($r^2 = 0.77$), probably due to dispersal characteristics of first instar larvae. Algal biomass (as chlorophyll *a*) was negatively correlated with black fly abundance, and when epilithic chlorophyll was included in the regression model, the fit improved substantially ($r^2 = 0.92$). This is consistent with experimental results which show that black fly larvae perfer rock surfaces that are not colonized by algae (Hershey and Hiltner, 1988; Wheeler and Hershey, unpublished data). When *Baetis* density ($r^2 = 0.99$) and *Brachycentrus* density ($r^2 = 1.0$) were included, the model accounted for all the variation in black fly abundance. *Brachycentrus* has also been shown to negatively affect black flies by inducing black fly drift from attachment

Table 4.3. Aquatic Insect Taxa Sampled from the Kuparuk River

Ephemeroptera
  Baetis lapponicus (Bengtsson)
  B. spp.
  Cinygmula atlantica (McDunnough)
  Ephemerella aurivillii Bengtsson
Plecoptera
  Alaskaperla ovibovis (Ricker)
  Nemoura trispinosa Claassen
Trichoptera
  Brachycentrus americanus (Banks)
  Rhyacophila sp.
Diptera
  Tipulidae
    Tipula spp.
  Simuliidae
    Cnephia eremites Shewell
    Metacnephia pallipes (Fries)
    Prosimulium perspicuum Sommerman
    P. neomacropyga Peterson
    Simulium decolletum Alder and Currie
    S. tuberosum cytospecies FGI
    Stegopterna emergens Stone
    St. mutata complex
  Chironomidae
  Tanypodinae
    Natarsia sp.
  Diamesinae
    Diamesa sp.
    Potthastia sp.

Orthocladiinae
  Brillia sp.
  Cardiocladius sp.
  Cricotopus tremulus (L.)
  C. bicinctus (Meigen)
  C. laricomalis Edwards
  C. spp.
  Corynoneura sp.
  Eukiefferiella claripennis group
    (Lundbeck)
  E. spp.
  Eurycnemus sp.
  Krenosmittia sp.
  Nanocladius sp.
  Orthocladius rivulorum (Kieffer)
  Orthocladius spp.
  Parametrionemus sp.
  Paracladius sp.
  Psectrocladius sp.
  Pseudosmittia sp.
  Synorthocladius sp.
  Theinemanniella sp.
  Tvetinia bavarica (Thien.)
Chironominae (Chironomini)
  Chironomus riparius Miegen
Chironominae (Tanytarsini)
  Cladotanytarsus sp.
  Micropsectra sp.
  Rheotanytarsus sp.
  Stempellina sp.
  Tanytarsus sp.

sites (Hershey and Hiltner, 1988). *Brachycentrus* abundance was more stable from year to year than abundance of other dominant taxa, and was negatively correlated with discharge from the preceeding fall ($r^2 = 0.65$). *Orthocladius* and *Baetis* abundances were highly variable from year to year.

Table 4.4. Secondary Production (g dry wt m^{-2}) Estimates (mean ± SE) for the Growing Season for Dominant Insects in the Kuparuk River in 1984–1986

Species	1984	1985	1986
Prosimulium perspicuum	2.62 ± 0.91 (2)	10.26 ± 0.90 (3)	3.2 ± 0.4 (3)
Orthocladius rivulorum	1.30 ± 0.35 (3)	1.42 ± 0.02 (3)	0.37 ± 0.06 (3)
Baetis sp.	0.80 ± 0.39 (2)	2.35 ± 0.71 (3)	1.37 ± 0.23 (3)
Brachycentrus americanus	1.65 ± 0.35 (2)	0.36 ± 0.05 (3)	0.78 ± 0.10 (3)
Total	6.6 ± 0.54 (2)	14.38 ± 1.30 (3)	5.72 ± 0.51 (3)

Number in parentheses indicates the number of stations sampled each year.

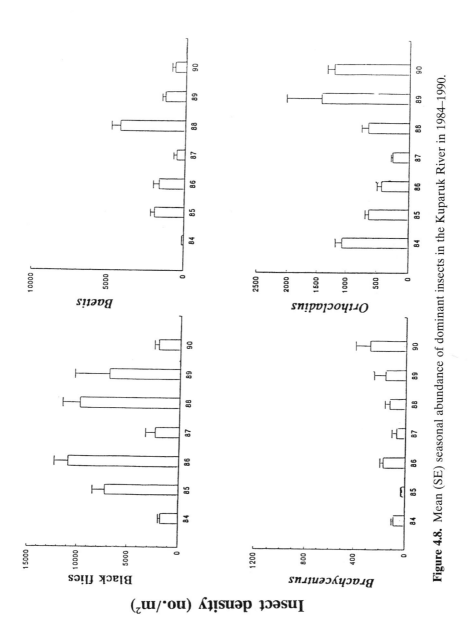

Figure 4.8. Mean (SE) seasonal abundance of dominant insects in the Kuparuk River in 1984–1990.

Neither discharge during any season, abundance of other insect taxa, nor chlorophyll *a* were significant in predicting their abundance in the study section.

Insect drift in the Kuparuk River is very low, but dominated by *Baetis*. Detailed studies of *Baetis* drift behavior (Hinterleitner–Anderson et al., 1992) indicate that *Baetis* drift was not a function of benthic density, but was significantly lower at higher absolute food supply. Drifting *Baetis* showed no diel periodicity during the continuous light of the arctic summer (Fig. 4.9; Hinterleitner–Anderson et al., 1992). However, drift studies in nearby Imnavait Creek, conducted during the arctic fall, showed that diel drift of most taxa increased abruptly when full darkness occured and ice cover was complete (Miller et al., 1986b).

Although algal biomass and productivity is clearly nutrient limited in the Kuparuk River, grazers also act as a control. A nutrient and insecticide releasing substratum experiment showed that algal biomass was approximately twofold higher on discs which excluded grazers (Fig. 4.10; Gibeau, 1992). Discs which diffused nutrients but not insecticide supported approximately fivefold more algal biomass than control discs (Fig. 4.10), suggesting that nutrient limitation is a more important factor than grazing. However, discs that diffused both nutrients and insecticide supported 15-fold higher

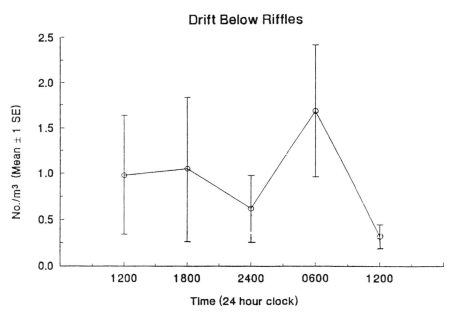

Figure 4.9. *Baetis* diel drift densities below three riffles (two samples per riffle) in the Kuparuk River in mid-July 1988 (from Hinterleitner–Anderson et al., 1992).

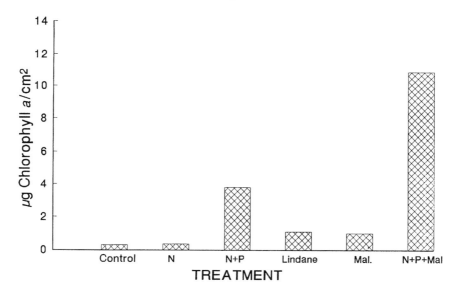

Figure 4.10. Chlorophyll *a* concentrations on diffusing substrates incubated in the Kupaurk River during July 1987. Substrates diffused nitrate (N), nitrate and phosphate (N + P), lindane, malathion (Mal.), and N + P + malathion (see Gibeau, 1990).

biomass than control discs (Fig. 4.10), suggesting that in the unmanipulated stream, nutrients and grazers interact strongly to control algal biomass (Gibeau, 1992).

Fish

The arctic grayling, *Thymallus arcticus*, is the only fish species normally found in the study section of the Kuparuk between the aufeis and the headwater lake. During the summer period, the fish population in the main stem of the Kuparuk is dominated by adults. Based on mark-recapture studies conducted in 1985 to 1990, 7000 and 10,000 adult fish were present in the section between the headwater lake and the aufeis. Adults were between 5 and 12 years old. Total length ranged from 30 to 42 cm, and total weight from 300 to 500 g (Deegan and Peterson, 1992). Young-of-the-year (YOY) grayling emerge from the stream bed by the beginning of July. YOY are most abundant in pools, although they are often seen feeding in riffles and in fast water at the tails of pools by the end of the summer. Juvenile fish have rarely been captured in the study section of the Kupaurk, being principally in tributary streams (Reis, 1988), where they appear to spend several years before entering the river as adults. In recent years (1990 and

1991), however, juveniles appear to have increased in abundance in the study reach, perhaps due to declines in adult density.

Arctic grayling in this section of the river migrate upstream to the headwater lakes and springs to overwinter. The migration begins in mid-August and continues until mid-September. Many of the adult fish tagged in our intensive study reach have been recaptured in Green Cabin Lake at the headwaters in both fall and spring, and several fish tagged at the headwater

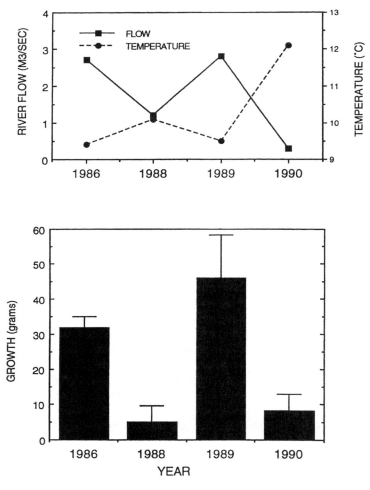

Figure 4.11. Growth of adult arctic grayling and the relationship to river flow and temperature in the Kuparuk River, 1986–1990. Data are the increase in weight for adult grayling based on tagged individuals from the control section during approximately July 1 to August 15 of each year (from Deegan and Peterson, 1992). Data on river discharge and temperature are from recording instruments at the road crossing.

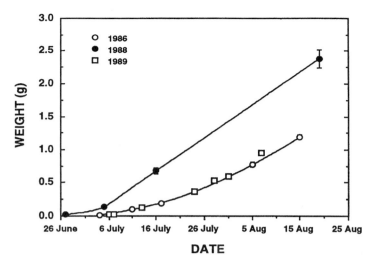

Figure 4.12. Growth of young-of-the-year arctic grayling in the Kuparuk river for high (1986 and 1989) and low (1988) river discharge years.

lake were later recaptured near the aufeis area. Juvenile and young-of-the-year fish may overwinter in the spring several kilometers downstream of the headwater lake. We have observed dense concentrations of juvenile fish in this tributary spring in the fall, while very few appear at the headwater lake. Juvenile fish may use the shallow, spring area to overwinter and thus avoid predation by the large lake trout that are present in Green Cabin Lake. During ice breakup in the spring, fish recolonize the river and tributaries, often returning to the same pools occupied the previous summer.

Growth during the summer period is highly variable for both adults and young-of-the-year grayling. On the average, adults gain about 40 g in weight over the July 1 to August 15 period. However, in poor growth years fish may gain little weight, and some individuals lose weight (Deegan and Peterson, 1992). We have observed poor adult growth in summers of low river discharge and high summer temperatures (for example, 1988 and 1990, Fig. 4.11). In contrast, young-of-the-year grayling grow best in summers of low flow and high temperatures (Fig. 4.12). In 1988, YOY averaged 68 mm in length and nearly 2.5 g in length by mid-August. In contrast, YOY fish were only 43 mm in length and 0.95 g in weight by the end of the summer in 1986 and 1989. These results indicate that growth is controlled by an interaction between temperature and food availability, as determined by river flow. Low flow results in higher water temperatures. For adults, demands of respiration are apparently greater than the food supply at low flow rates. Metabolic demands are lowered because of lower temperature in high flow years, and adult fish may have access to more prey and larger prey items because of higher drift rates. For young-of-the-year, however, food availability exceeds metabolic demands even at low flow, and the higher temperatures promote growth.

Trophic Structure

The macroconsumers of the Kupaurk rely on a combination of allochthonous and autochthonous material. Gut analyses of the major insect taxa (Table 4.5) show that *Prosimulium* and *Brachycentrus* overlap broadly in their diets, whereas *Orthocladius* and *Baetis* have more distinct diets.

Brachycentrus and *Prosimulium* are primarily filter feeders, although *Brachycentrus* also spends some of its time grazing (Gallepp, 1977; Gallepp and Hasler, 1975). When filtering, it attaches the leading edge of its case to a rock surface, and uses the setae on its forelegs as a filtering apparatus. The gut contents of both species are dominated by detrital fine particulate organic matter (FPOM); some of this detrital material is clearly of peat origin, but some is amorphorus, and probably represents sloughed FPOM from the epilithic matrix, flocculated DOM, and feces. Although separation of allochthonous and autochthonous components of filter-feeder diets is not feasible, the high production of black flies compared to that of grazers suggests that allochthonous carbon is important in the river economy. Diatoms and animal prey (instar I chironomids and black flies) were present in >50% of the filter-feeder guts examined (Table 4.5). For *Brachycentrus*, diatoms could be ingested either through filter feeding (of sloughed epilithon) or grazing, but instar I animal prey were probably collected from the drift. The most conspicuous diet difference between *Brachycentrus* and

Table 4.5. Gut Contents of Dominant Aquatic Insects and Arctic Grayling

Insect	N	Peat Detritus	Amorphous Detritus	Diatoms	Filamentous Algae	Animal Prey
Brachycentrus	20	+++	+++	++	+++	++
Prosimulium	20	+++	+++	++	+	++
Orthocladius	17	–	–	+++	–	–
Baetis	19	–	+++	+	–	–

Grayling	N	Mean Dry Wt			Mean (%)	
		Detritus	Terrestrial Insects	Aquatic Insects	*Baetis*	*Brachycentrus*
	12	0.31 g	0.028 g	0.10 g	47	21

Insect gut analyses were qualitative. N = the number of guts examined. Food items scored were peat detritus, amorphous detritus, diatoms, filamentous algae, and animal prey. Possible scores were: (+++) abundant in all guts; (++) present in >50% of guts; (+) present in <50% of guts; (–) absent from all guts. Grayling gut analyses were quantitative. Dry weights of the three components of grayling diets are presented for fish captured in 1985. For the aquatic insect portion of the diet, percent contribution of *Baetis* and *Brachycentrus* is also presented.

Prosimulium, was that filamentous algae were abundant in *Brachycentrus* guts and uncommon in black fly guts. Although these two species overlap broadly in diet, their microdistribution in the river is inversely related. *Brachycentrus* are more aggressive than black flies, and are capable of displacing them or even eating them if they do not drift when encountered (Hershey and Hiltner, 1988). However, *Brachycentrus* has a significant negative effect on black fly populations only when its density is high (Hershey and Hiltner, 1988).

Baetis diets are dominated by detrital FPOM, presumably of epilithic origin. Diatoms are also present in most *Baetis* guts. Although diatoms are probably higher quality food, they are less conspicuous than the detrital fraction (Table 4.5). *Orthocladius* diet is comprised strictly of diatoms (Table 4.5) obtained by grazing the monoculture on their tubes (Hershey et al., 1988).

Aquatic insects comprised only approximately 23% of grayling diets by dry weight (Table 4.5). Of this fraction, approximately 47% consisted of *Baetis* and 21% of *Brachycentrus*. Terrestrial insects comprised an additional 6.3% dry weight, and the remainder of the gut material was detritus (Table 4.5). Presumably, little of the detrital material is assimilated.

Summary

The Kuparuk River is a meandering tundra stream that shows a high degree of temporal variability in terms of discharge, nutrient availability, epilithic primary productivity, and abundance, growth, and production of consumers. Among the major inorganic nutrients, only nitrate delivery to the river is clearly correlated (inversely) with discharge. Epilithic algal biomass and productivity is nutrient limited, primarily by phorphorus, but secondarily by nitrogen. However, grazers are also clearly important in controlling algal biomass. Epilithic heterotrophic and autotrophic processes are both important in fueling the consumers. Insect abundance and production is dominated by black flies, and high spring discharge is the most important factor controlling black fly abundance. Arctic grayling, the only fish species in the river, migrates to upstream lakes and springs for overwintering. During the growing season, adult grayling growth is positively associated with discharge, but young-of-year growth is negatively associated with discharge. Significant factors distinguishing the Kuparuk River from temperate rivers include: (1) ice cover extending into the river bed for approximately eight months of the year; (2) a short growing season with low rates of nutrient delivery; (3) continuous light during the growing season; and (4) low macroinvertebrate diversity and lack of fish diversity. Autotrophy and photolysis may be more important in the clear water Kupaurk River than in arctic rivers fed by glaciers, which are typically turbid from glacial flour.

References

Chapin FS, III, Fetcher N, Kielland K, Everett KR, Linkins AE (1978) Productivity an nutrient cycling of Alaska tundra: Enhancement by flowing soil water. Ecology 69:693–702.

Craig PC, McCart PJ (1975) Classification of stream types in Beaufort Sea drainages between Prudhoe Bay, Alaska and the MacKenzie Delta, NWT. Arctic Alpine Res 7:183–198.

Deegan LA, Peterson BJ (1992) Whole river fertilization stimulates fish production in an Arctic tundra river. Can J Fish Aquat Sci 49:1890–1901.

Federle TW, McKinley VL, Vestal JR (1982) Effects nutrient enrichment on the colonization and decomposition of plant detritus by the microbiota of an arctic lake. Can J Microbiol 28:1199–1205.

Fiebig DM (1988) A study of riparian zone and stream water chemistries and organic matter utilisation at the stream–bed interface. Ph.D. Thesis, University of Wales.

Ford TE, Lock MA (1987) Epilithic metabolism of dissolved organic carbon in boreal forest rivers. FEMS Microbiol Ecol 45:89–97.

Gallepp GW (1977) Responses of caddisfly larvae (*Brachycentrus* spp.) to temperature, food availability, and current velocity. Am Midland Naturalist 98:59–84.

Gallepp GW, Hasler AD (1975) Behavior of larval caddisflies (*Brachycentrus* spp.) as influenced by marking. Amer Midland Naturalist 93:247–254.

Gibeau G (1992) Grazer and nutrient limitation of algal biomass in an arctic tundra river. MS Thesis. University of Cincinnati, Cincinnati, OH.

Hauer FR, Benke AC (1991) Rapid growth of snag-dwelling chironomids in a blackwater river: the influence of temperature and discharge. J North Am Benthol Soc 10:154–164.

Hershey AE, Hiltner AL (1988) Effect of a caddisfly on black fly density: Interspecific interactions limit black flies in an arctic river. J North Am Benthol Soc 7:188–196.

Hershey AE, Hiltner AL, Hullar MAJ et al. (1988) Nutrient influence on a stream grazer: *Orthocladius* microcommunities respond to nutrient input. Ecology 69:1383–1392.

Hinterleitner–Anderson D, Hershey AE, Schuldt JA (1992) The effects of river fertilization on mayfly (*Baetis* sp.) drift patterns and population density in an arctic river. Hydrobiologia 240:247–258.

Hobbie JE, Rublee P (1975) Bacterial production in an arctic pond. Verh Internat Verein Limnol 19:466–471.

Hobbie JE, Corliss TL, Peterson BJ (1983) Seasonal patterns of bacterial abundance in an arctic lake. Arctic Alpine Res 15:253–259.

Hullar MAJ, Johnson CG, Kaufman M (1987) Response of epilithic heterotrophic microbiota to carbon additions in an oligotrophic arctic river. Abstr Ann Mg Am Soc Microbiol 87:189.

Hullar MAJ, Vestal JR (1986) Plant litter decomposition in an arctic stream. Abstr Ann Mg Am Soc Microbiol 86:169.

Hullar MAJ, Vestal JR (1989) The effects of nutrient limitation and stream discharge on the epilithic microbial community of an oligotrophic Arctic stream. Hydrobiologia 172:19–26.

Kane DL, Hinzman LD, Benson CS, Everett KR (1989) Hydrology of Imnavait Creek, an arctic watershed. Holarctic Ecol 12:262–269.

Kieber RJ, Zhou X, Mopper K (1990) Formation of carbonyl compounds from UV induced photodegradation of humic substances in natural waters: fate of riverine carbon in the sea. Limnol Oceanogr 35:1503–1515.

Kling GW, Kipphut GW, Miller MC (1991) Arctic lakes and streams as gas conduits to the atmosphere: implications for tundra carbon budgets. Science 251:298–301.

Lock MA, Ford TE (1985) Microcalorimetric approach to determine relationships between energy supply and metabolism in river epilithon. Appl Environ Microbiol 49:408.

Lock MA, Ford TE, Fiebig DM et al. (1989) A biogeochemical survey of rivers and streams in the mountains and foothills province of arctic Alaska. Arch Hydrobiol 115:499–521.

Lock MA, Ford TE, Hullar MAJ et al. (1990) Phosphorus limitation in an arctic river biofilm—a whole ecosystem experiment. Water Res 24:1545–1549.

McKinley VL, Federle TW, Vestal JR (1982) Effects of petroleum hydrocarbons on plant litter microbiota in an arctic lake. Appl Environ Microbiol 43:129–135.

Miller MC, Hater GR, Spatt P, Westlake P, Yeakel D (1986) Primary production and its control in Toolik Lake, Alaska. Arch Hydrobiol Suppl 74:97–131.

Miller MC, Stout JR, Alexander V (1986) Effects of a controlled under-ice oil spill on invertebrates of an arctic and a subarctic stream. Environ Pollut 42:99–132.

Minshall GW (1978) Autotrophy in stream ecosystems. BioScience 28:767–771.

Nadelhoffer KJ, Giblin AE, Shaver GR, Laundre JA (1991) Effects of temperature and organic matter quality on C, N, and P mineralization in soils from six arctic ecosystems. Ecology 72:242–253.

Oswood MW, Everett KR, Schell DM (1989) Some physical and chemical characteristics of an arctic beaded stream. Holarctic Ecol 12:290–295.

Oswood MW, Milner AM, Irons JG, III. (1992) Climate change and Alaskan rivers and streams. In Firth P, Fisher SG (Eds) Global climate change and freshwater ecosystems. Springer-Verlag, NY.

Peterson BJ, Deegan L, Helfrich J et al. (1993) Biological responses of a tundra river to fertilization. Ecology 74:653–672.

Peterson BJ, Hobbie JE, Corliss TL (1986) Carbon flow in a tundra stream. Can J Fish Aquat Sci 43:1259–1270.

Peterson BJ, Hobbie JE, Corliss TL, Kriet K (1983) A continuous-flow periphyton bioassy: Tests of nutrient limitation in a tundra stream. Limnol Oceanogr 28: 583–591.

Peterson BJ, Hobbie JE, Hershey AE et al. (1985) Transformation of a tundra stream from heterotrophy to autotrophy by addition of phophorus. Science 229:1383–1386.

Prentki RT, Miller MC, Barsdate RJ et al. (1980) Chemistry. In Hobbie JE (Ed) Limnology of tundra ponds. Dowden, Hutchinson, and Ross, Stroudsburg, PA.

Richey JE (1983) Interactions of C, N, P, and S in river systems: A biogeographical model. In Bolin B, Cook RB (Eds) The major biogeochemical cycles and their interactions. Wiley, NY.

Reis R (1988) Foraging behavior of arctic grayling (*Thymallus arcticus*) in a tundra stream. M.S. Thesis, University of Cincinnati, Cincinnati, OH.

Selkregg LL (Ed) (1977) Alaska Regional Profiles: Arctic Region. Arctic Environmental Information and Data Center, Anchorage, AK.

Strome DJ, Miller MC (1978) Photolytic changes in dissolved humic substances. Int Verein Theor Angew Limnol Verh 20:1248–1254.

Stevens PA (1981) Modification and operation of ceramic cup soils solution samples for use in geochemical cycling studies. Institute of Terrestrial Ecology, Occasional Paper No. 8. Bangor, UK.

5. The Limnology of Smith Lake

Vera Alexander and Binhe Gu

Smith Lake (64°52′N and 146°52′W) is located in the Tanana River valley in the interior region of Alaska between two major mountain ranges, the Alaska Range to the south and the Brooks Range to the north (see Fig. 1.1, Chapter 1). Here, rolling hilly terrain is transected by the Yukon–Tanana river system, forming a valley that exceeds 150 km in width. The region experiences large seasonal temperature extremes, with warm dry summers and very cold dry winters, and a mean annual precipitation of only about 30 cm. The Smith Lake watershed is occupied by typical taiga forest, composed of spruce, aspen, birch, and alder, with a ground cover of subarctic muskeg (Fig. 5.1). Black spruce (*Picea mariana*) dominates. The mosses that comprise the muskeg ground cover have low decomposition rates. Research on Smith Lake has focused on the nitrogen cycle, but a considerable body of limnological information has been amassed for this lake in conjunction with that work (Alexander, 1970; Alexander and Barsdate, 1971; Clasby, 1972; Dugdale, 1965; Goering and Dugdale, 1966; Gu, 1992, 1993; Gu and Alexander, 1993a–c; Gu et al., 1994). The initial research was carried out between 1962 and 1967, with a second research program undertaken between 1988 and 1993. The techniques applied to limnological work have changed somewhat in the interim and therefore the results are not exactly comparable; nevertheless this chapter offers an excellent opportunity to make reasonable comparisons to determine whether the lake has changed significantly over the 20 year period.

Figure 5.1. Aerial photograph of Smith Lake.

Smith Lake has a surface area of ca. $7.2 \times 10^4 \mathrm{m}^2$, a volume of ca. $18 \times 10^4 \mathrm{m}^3$ and a mean depth of 1.1 m (Alexander and Barsdate, 1971). The irregular bottom contours of the deeper portions of Smith Lake, together with its location in an area of retransported silts and its proximity to a number of thermokarst pits, suggest that the lake is cryogenic in origin, having been formed by the melting of one or more massive ice bodies (Alexander and Barsdate, 1971). During the Pleistocene, glaciers advanced north from the Alaska Range to within 80 km of the area, and wind-transported silt from the nearby Tanana River flood plain formed thick deposits of loess on the surrounding hills. Much of this loess subsequently was transported into the valleys between the hills. Permafrost is patchy, predominantly occuring in valley bottoms and north-facing shopes. Ice occurs as ice wedges and large subhorizontal masses in the retransported valley-bottom silts.

The water has a strong brown color that severely limits the penetration of light, such that the depth of 1% penetration of light lies at only about 1 to 1.8 m. Although phytoplankton primary production is lower in arctic fresh-waters than at lower latitudes (Hobbie, 1973; Hammar, 1989), Smith Lake exhibits some eutrophic properties. A consistent and dramatic event in the annual phytoplankton cycle is an intense bloom of *Anabaena flos-aquae*, which develops in the epilimnion shortly after the ice leaves the lake and lasts for 2 to 3 weeks. During this time, the weather is often calm with long periods of sunshine. As a result, the stable layer of cold water derived from the ice, initially at 0°C, overlies warmer (3° to 4°C) but denser water below. The initially cold water in the shallow (<1 m) epilimnion warms rapidly and forms a stable thermocline.

Physical Regime

Thermal Cycle

Annual temperature cycles for 3 years are shown in Fig. 5.2. The ice on Smith Lake thawed between May 25 and 29 each year during a 7 year period between 1963 and 1970 (Alexander and Barsdate, 1971), but the thaw occurred 2 to 3 weeks earlier in 1989, 1990, and 1991 (Gu and Alexander, 1993c). This may be a result of climatic factors, because the winters in interior Alaska were very severe during the late 1960s and early 1970s, whereas the more recent winters have been mild.

The warming of the surface waters can be as rapid as 4° or 5°C per day, once the shallow, stable epilimnion develops. Maximum summer temperatures reach 23° to 26°C. Cooling begins early in August, with ice formation by late September or early October.

Heat Budgets

It is difficult to compare heat budgets of lakes that freeze with those that do not, because ice itself significantly affects the heat exchange (Ragotzkie, 1978). Hutchinson (1957) gives annual heat budgets for lakes ranging from about $46,000 \, g \, cal \, cm^{-2} \, yr^{-1}$ for Lake Mead in California to $2000 \, g \, cal \, cm^{-2} \, yr^{-1}$ for Anneks Sø (77°15′N). Most of the thermal information available for lakes, including arctic lakes, is restricted to the summer heat budget. However, Livingstone (1963) pointed out that the annual heat budgets for arctic lakes north of the Brooks Range in Alaska are comparable in size with that of temperate lakes (27,000 to $31,000 \, cal \, cm^{-2}$), but the temperature range through which warming occurs is very different. Smith Lake has a relatively small annual heat budget ($6857 \, g \, cal \, cm^{-2} \, yr^{-1}$), due to its limited depth and small size. Of this, 75%, or $5177 \, g \, cal \, cm^{-2}$, is the winter budget (Alexander and Barsdate, 1971), because latent heat of melting ice accounts for over 60% of the annual heat gain of the lake water. Small lakes often have relatively high sediment heat budgets. At a water depth of 2 m the annual range of surface sediment temperature was 15°C, and thermal gradients on the order of 6° to 8°C were found in the upper meter of sediment following spring ice breakup (Alexander and Barsdate, 1971). The annual heat budget of Ace and Deuce lakes, which are similar to but deeper than Smith Lake, are 11,340 and 7789 $cal \, cm^{-2}$ (Alexander and Barsdate, 1974). These lakes lie only a few kilometers from Smith Lake, and their maximum depths are 8 and 5 m, respectively.

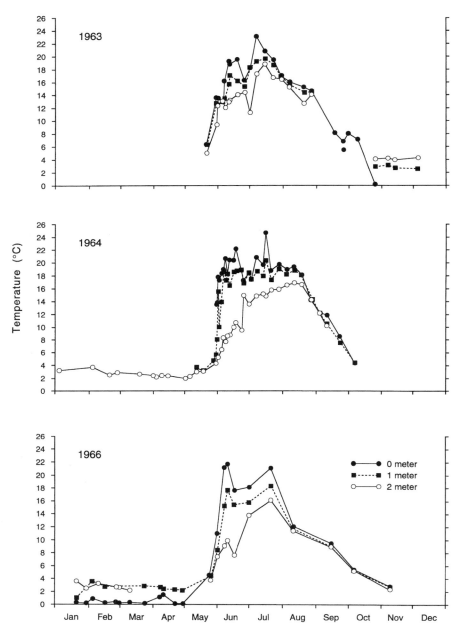

Figure 5.2. Annual temperature cycles in Smith Lake.

Table 5.1. Major Cation Values (mg L^{-1}) from Smith Lake in 1965

Date	Depth (m)	Ca	Mg	K	Na	Fe	Total Cation (meq L^{-1})
July 12	0	5.8	7.2	1.6	8.2	0.08	1.29
	1	5.6	6.0	1.5	6.8	0.18	1.12
	2	5.6	6.1	1.6	6.6	0.24	1.12
Aug. 1	0	5.2	6.6	1.0	6.0	<0.50	1.09

Data for July 12 are taken from Alexander and Barsdate (1971); data for August 1 are taken from Likens and Johnson (1968).

Lake Chemistry

Major Components

The most abundant cations in Smith Lake are Ca^{2+} (25%), Mg^{2+} (32%), and Na$^+$ (36%) (Table 5.1; Alexander and Barsdate, 1971). The calcium concentration is low compared with other Alaskan lakes (Likens and Johnson, 1968). Bicarbonate (87%) dominates the inorganic anions, with SO$_4^{2-}$ (6%) and Cl$^-$ (7%) of minor importance. Total alkalinity in Smith Lake ranges between 24 and 71 mg CaCO$_3$ L^{-1} (Fig. 5.3), dissolved solids between 104 and 192 mg L^{-1}, and conductivity between 87 to 180 μmho cm^{-1} at 25°C (Fig. 5.4). All these variables have seasonal cycles with minima in summer and maxima under winter ice cover. Such seasonal patterns apparently result from exclusion of dissolved substances from the ice during freezing, concentrating the conservative components in the water below.

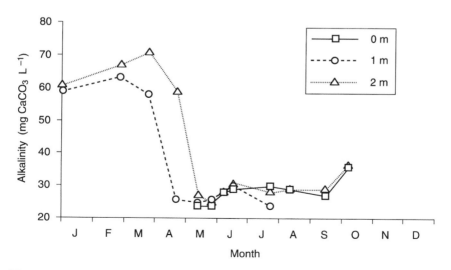

Figure 5.3. Total alkalinity at three depths in Smith Lake during 1964 (data from Alexander and Barsdate, 1971).

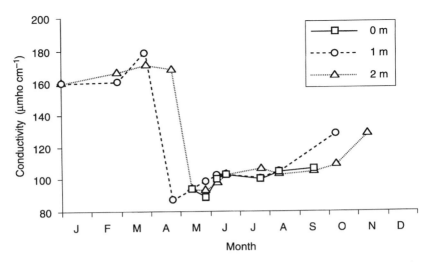

Figure 5.4. Water conductivity at three depths in Smith Lake during 1966 (data from Alexander and Barsdate, 1971).

The dissolved organic carbon (DOC) concentration in Smith Lake water was $43 \, mg \, C \, L^{-1}$ during the summer of 1989 (Satoh et al., 1992); unfortunately the sampling began only in midsummer (4 August 1989) and did not include the spring bloom. This value falls within the range commonly associated with dystrophic lakes. Ballaine Lake, a small lake a few kilometers from Smith Lake, also had high DOC levels (above $46 \, mg \, C \, L^{-1}$).

pH and Dissolved Gases

Smith Lake is not a typical bog lake in that its pH lies close to neutral. Satoh et al. (1992) point out that Alaskan colored waters are unique in their higher, almost neutral pH that is two to three units greater than found in other colored lakes throughout the world. During the winter the pH of Smith Lake water lies between 6.7 and 6.9 (Alexander and Barsdate, 1971). A sudden and temporary increase in the surface water pH, up to 10, was noted on one occasion during the spring bloom (Dugdale, 1965). There appears to be a decrease in the pH on the order of 0.2 to 0.3 units recently compared to the norm for the 1960s (Gu and Alexander, 1993c), so that the current range is approximately 6.4 to 6.6.

Dissolved oxygen (DO) varies seasonally (Fig. 5.5), with high concentrations in summer as a result of phytoplankton photosynthesis and atmospheric input, followed by a progressive decrease in winter. It becomes undetectable in mid and late winter, as ice prevents diffusion from the atmosphere and decomposition continues, especially at the sediment/water interface.

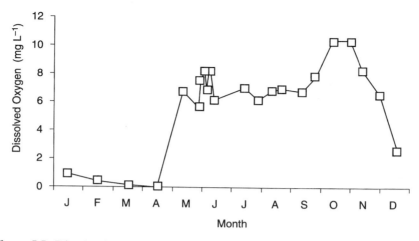

Figure 5.5. Dissolved oxygen in Smith Lake surface water during 1988 (data from Gu, 1993).

Nutrients

The mean concentration of phosphate (PO_4^{3-}) was $2.0 \pm 1.2 \mu mol\,L^{-1}$ ($n = 80$) during the 1960s (1963 to 1966) and $2.0 \pm 1.6 \mu mol\,L^{-1}$ ($n = 166$) more recently (1989 to 1990). PO_4^{3-} accumulates under ice in winter, reaching a maximum in late April. Concentrations decrease as a result of utilization by primary producers by mid-May (Fig. 5.6), and build up again after the spring bloom, reaching a second maximum in mid-July, again followed by a decline.

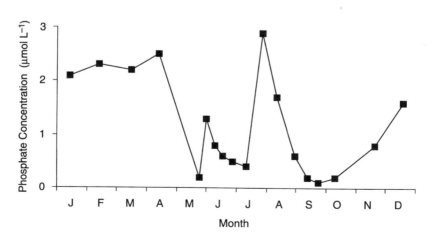

Figure 5.6. Orthophosphate in Smith Lake surface water during 1989.

The concentration of total phosphorus (TP) in Smith Lake during some summer months (August and September) ranges from 1 to $1.6\,\mu mol\,L^{-1}$ (Satoh et al., 1992). This high concentration of TP in September is attributed to the input of nutrients by Canada geese feeding on the macrophytes in the lake. No data are available for the early summer months, when the intense cyanobacterial bloom dominates the lake.

The snowmelt-fed inlet into Smith Lake flows only seasonally. In 1964, the NH^+_4 concentration in this inflow was $8\,\mu mol\,L^{-1}$ on May 5, but by May 27 it had declined to $0.23\,\mu mol\,L^{-1}$. No information exists on the volume of the flow, but it is small. The only other water input, besides rainfall, is through seepage in summer when the ground surrounding the lake has thawed. Under these conditions, the allochthonous supply of inorganic nitrogenous nutrients is probably small, even during the spring thaw. The low decomposition rates in the surrounding watershed probably limit the input of terrigenous nutrients.

Ammonium concentrations in Smith Lake typically increase during the winter to a maximum prior to ice melt, due to the combined effects of decomposition of organic matter, freezing exclusion during ice formation, and a low nitrogen demand by phytoplankton (Fig. 5.7). Concentrations decline rapidly during the spring phytoplankton bloom. In 1964, for example, surface concentrations during the spring melt and early summer declined from $32\,\mu mol$ on April 14 (immediately below the ice) to undetectable levels by May 27; thereafter the concentrations began to rise, although they remained low during the ice-free period. Similar patterns and concentrations were found in 1988 and 1989 (Fig. 5.7).

Nitrate concentrations are extremely low in Smith Lake water at all times during the ice-free season, increase due to nitrification during the fall and early winter, and decrease due to denitrification as the lake water becomes anoxic in late winter (Clasby, 1972; Dugdale, 1965; Goering and Dugdale, 1966; Gu and Alexander, 1993c). Klingensmith (1981) reported a denitrification rate of $35\,ng\,N\,cm^{-3}\,d^{-1}$ based on the C_2H_2 blockage technique (Yoshinari et al., 1977), compared to a previous report of $15\,ng\,N\,cm^{-3}\,d^{-1}$, measured using ^{15}N (Goering and Dugdale, 1966). The latter study did not take into account any denitrification that produced N_2O as the end product.

The concentrations of NO^-_3 increase as oxygen is introduced into the water through photosynthesis with increasing light in spring, allowing the resumption of nitrification. Nitrate concentrations ranged from undetectable at the surface on May 27, 1964 to a maximum of $13\,\mu mol$ immediately under the $1\,m$ thick ice cover on February 3, 1964 (Dugdale, 1965). In 1989, a maximum concentration of about $6\,\mu mol\,L^{-1}$ occurred in early January, and the overall annual pattern was similar (Fig. 5.7).

Total nitrogen (TN) concentrations in Smith Lake, from 1.51 to $1.65\,mg\,L^{-1}$, (107.9 to $117.9\,\mu mol\,L^{-1}$), are fairly constant with depth and time (Satoh et al., 1992). The independence of TN from biological uptake sug-

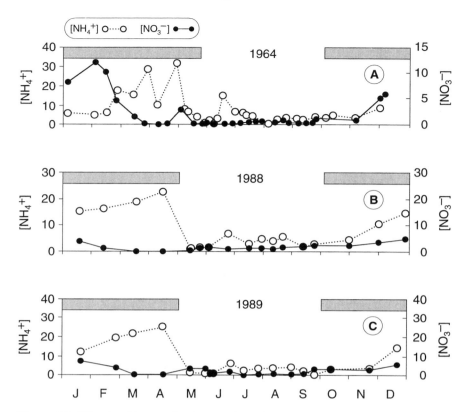

Figure 5.7. Dissolved inorganic nitrogen in Smith Lake during 1964, 1988 and 1989 (data from Alexander and Barsdate, 1971, and Gu and Alexander, 1993c). Concentrations are in micromoles per liter.

gests that phytoplankton utilization of nitrogen only removes a small percentage and does not affect the pool size of TN significantly.

The mean concentration of silicate-silica (SiO_4^{2-}) measured from 1963 to 1966 was 7.7 µmol L^{-1} (SD = 9.7; $n = 38$). Although diatoms are not a major component of the Smith Lake phytoplankton biomass, during the spring of 1964 the SiO^{2-}_4 concentration dropped from 48.7 µmol L^{-1} on May 18 to 1.9 µmol L^{-1} by June 25. This may have been in part due to dilution by water derived from ice. The concentration remained close to the latter figure during the summer.

Trace Metals

Copper, manganese, zinc, and iron were measured in Smith Lake only during 1964 to 1966 (Alexander and Barsdate, 1971), so that no comparison with more recent years is possible. Strong seasonal variations in iron and manganese concentrations were characteristic, with iron (ca. $100\mu g L^{-1}$) more abundant than manganese ($<10\mu g L^{-1}$). Concentrations were higher in winter, and for iron exceeded $600\mu g L^{-1}$ by April. Manganese concentrations reached $90\mu g L^{-1}$ in 1964 and more than $50\mu g L^{-1}$ in 1966. Dissolved iron, much more abundant than particulate iron, amounted to 90% of the total iron at 2m depth. In the case of manganese, dissolved forms were abundant; but even higher concentrations of particulate manganese occurred, particularly during the spring circulation.

Copper concentrations were below $4 mg L^{-1}$, with no seasonal variation; and zinc concentrations were also low except in the hypolimnion during the summer, where, in common with iron and manganese, high concentrations occurred. The mean cobalt concentration in the lake was $1.1\mu g L^{-1}$ in summer, decreased in early fall to $0.7\mu g L^{-1}$, and then increased to $0.9\mu g L^{-1}$ by November. The concentration of molybdenum was low at all times, at about $0.2\mu g L^{-1}$.

Primary Production

Phytoplankton Biomass

The seasonal variations in phytoplankton biomass at four depths, measured as chlorophyll *a* concentration, are shown for 1989 in Fig. 5.8. The bloom starts in the surface water in mid-May, and is followed by a second surface maximum starting in late August. The first peak, dominated by the cyanobacterium *A. flos-aquae*, results in chlorophyll *a* concentrations as high as $40\mu g L^{-1}$ (Gu and Alexander, 1993b). This bloom lasts approxi-

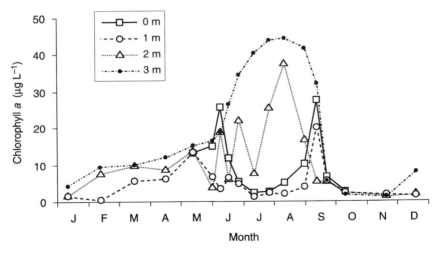

Figure 5.8. Chlorophyll *a* concentrations at four depths in Smith Lake during 1989.

mately 3 weeks and is followed by a summer period of lower chlorophyll *a* concentrations in the surface waters. The second surface chlorophyll *a* peak occurs in September, the fall turnover period. This peak lasts only for a short time due to rapid cooling and formation of ice. A third peak (at the 1 m depth, because the surface layer is composed of ice at this time) sometimes develops in late winter (April) as the photoperiod increases and the ice layer becomes free of snow cover, so that more daylight penetrates the ice, stimulating algal growth (GU and Alexander, 1993c). Depth profiles of chlorophyll *a* show strong variations, ranging from 0.2 to 40μg L⁻¹, over a season. The spring cyanobacterial bloom is mostly confined to the upper 0.5 m, but a chlorophyll *a* maximum develops at the 2 and 3 m depths later during the summer, even though the 1% surface light depth lies at about 1.5 m. These phytoplankton may be adapted to low light intensities, although they may, on the other hand, be inactive and residual from the surface phytoplankton population.

Phytoplankton Carbon Productivity

The annual primary production cycle of Smith Lake is dominated by the spring activity. The sequence of events (Fig. 5.9) typically begins with growth under the ice, primarily composed of green algae, that is followed by the intense bloom of *A. flos-aquae* mentioned above that starts shortly after the ice melts. Maximum photosynthesis rates are very high. In 1963, the

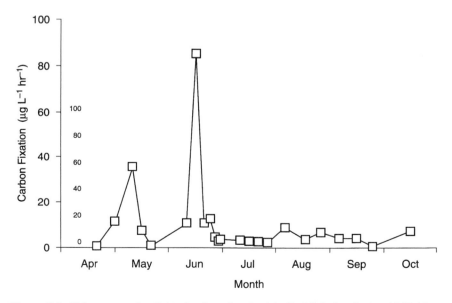

Figure 5.9. Primary productivity (carbon fixation) in Smith Lake during 1966 (data from Alexander and Barsdate, 1971).

maximum in situ surface primary production rate was $175 \mu g\,C\,L^{-1}\,h^{-1}$. In 1964, the maximum surface rate was $90 \mu g\,C\,L^{-1}\,h^{-1}$, and in 1966 it was $468 \mu g\,C\,L^{-1}\,h^{-1}$ (Alexander and Barsdate, 1971). The rates declined rapidly following the peak. Although these rates are spectacular on a volume basis, they do not necessarily imply very high annual production, because the high production takes place over a short period of time and is extremely episodic. The annual planktonic primary production in Smith Lake amounts to about $11\,g\,Cm^{-2}$ (Alexander and Barsdate, 1971). This is not substantially higher than the annual primary production estimate for Toolik Lake, at $7\,g\,Cm^{-2}$ (see Chapter 4). Hammar (1989) gives 4 to $8\,g\,Cm^{-2}$ as the normal range of annual primary production for arctic and antarctic lakes. In the case of Smith Lake, latitude may not be the determining factor, because Ace and Deuce Lakes, only a few kilometers away, have annual primary productions in excess of $60\,g\,Cm^{-2}$.

Phytoplankton Nitrogen Cycling

The seasonal pattern of nitrogen utilization in Smith Lake has been described previously (Billaud, 1968; Gu, 1992). Nitrogen fixation is detectable in the surface waters throughout the entire ice-free season. The combined effects of a low N/P ratio (3.0 prebloom and 6.1 during the bloom), increasing light intensity, water temperature, and water column stratification apparently stimulate the initial *Anabaena* bloom. The short life of the bloom (about 3 weeks) may be attributable to depletion of a trace metal (Billaud, 1968), because other chemophysical conditions remain constant (Gu, 1992). It is also possible that the *Anabaena* act analogously to annual plants, developing akinetes and becoming dormant after a period of activity. The absence of N_2 fixation under the ice is not only due to a high N/P ratio (20.0), but probably also to winter mixing, low light intensity, and low water temperature (Gu, 1992). Nitrogen fixation during the 3 week spring bloom contributed approximately 60 and 57% of the annual N_2 fixation in 1988 and 1989, respectively. The annual fixation amounts to $2.0\,g\,Nm^{-2}$ (Gu, 1992). Similar results were seen in 1963 and 1964 (Billaud, 1968).

Nitrogen fixation is more strongly correlated with light intensity than is NH_4^+ and NO_3^- uptake, which is to be expected because more solar energy is required to reduce molecular nitrogen (Paerl, 1990). Accordingly, 90% of the N_2 fixation rate is confined to the upper 1 m because light penetration is attenuated rapidly due to algal self-shading and the brown water color (Gu and Alexander, 1993b). The seasonal variation in N_2 fixation in Smith Lake is negatively correlated with NH_4^+, but not with the NO_3^- concentration, suggesting that N_2 fixation is inhibited by NH_4^+, which is consistent with the findings in other lakes (Ohmori and Hattori, 1972; Takahashi and Saijo, 1988).

The utilization of NH_4^+ and NO_3^- by phytoplankton showed three peaks in spring, fall, and late winter (Fig. 5.10) that correlated strongly with chloro-

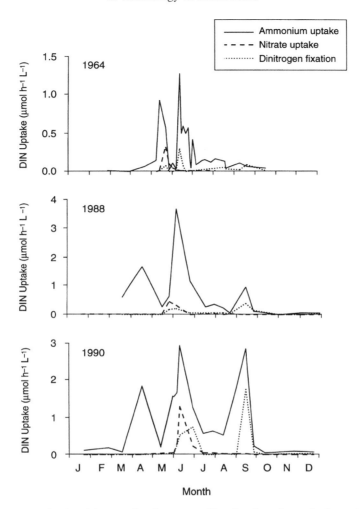

Figure 5.10. Dissolved inorganic nitrogen utilization by phytoplankton in Smith Lake during 1964, 1988, and 1989 (data from Billaud (1968) and Gu and Alexander, 1993b).

phyll *a*, water temperature, and day length (Gu, 1992). Ammonium uptake dominated at all times (Billaud, 1968; Gu, 1992). In 1988 and 1989, uptake of ammonium accounted for 79% of the total nitrogen utilized by phytoplankton during the year; nitrate uptake and N_2 fixation accounted for only 16 and 5%. Dissolved organic nitrogen (DON) in the form of urea was also an important source of nitrogen for the spring *Anabaena* bloom, accounting for 25% of total nitrogen utilization (Gu, 1992). The turnover time for NH_4^+ was less than 3 days (2 year mean), and as short as several hours in summer. This, along with the preference for NH_4^+, supports recycled nitrogen as the primary form of dissolved inorganic nitrogen used by these phytoplankton. The rapid NH_4^+ turnover and low ambient NH_4^+ levels suggest that the supply rate of NH_4^+ is sufficiently rapid to meet the demand by summer phytoplankton. Gu and Alexander (1993b) suggested that this is primarily generated by zooplankton excretion and bacterial remineralization.

Gu and Alexander (1993b) discussed the mechanisms behind the utilization of several forms of dissolved nitrogen by the spring bloom (*Anabaena*). Uptake of NH_4^+, NO_3^-, urea, and N_2 was measured in spring 1990. *A. flos-aquae* is able to utilize all four nitrogen forms simultaneously, even though NO^-_3 and N_2 are expensive energetically compared with reduced forms of nitrogen. NH_4^+ concentrations in the surface water are inadequate to meet the demand for the cyanobacterial growth. Furthermore, an uptake kinetic experiment indicated that cyanobacteria in Smith Lake have a low affinity for NH_4^+, but a high affinity for NO_3^-, allowing for simultaneous utilization of other forms of nitrogen for optimal growth (Gu and Alexander, 1993a).

We have compared the seasonal patterns of nitrogen uptake, along with other limnological variables, for the recent (1988 to 1989) and earlier work (1964), and found no significant differences (Gu and Alexander, 1993b). Nevertheless, it is risky to judge conclusively whether or not the lake nitrogen cycle dynamics over approximately the last 25 years have changed, although all indications are that the nitrogen cycle has remained consistent.

Macrophyte Production

Aquatic macrophytes occupy the majority of the Smith Lake basin from the shore to about 1 to 1.5 m depth. Calculated net carbon fixation for *Ceratophyllum demersum*, *Dreponocladus* sp., *Potamogeton perfoliatus*, *Potamogeton* sp., and *Utricularia vulgaris* at water depths from 0.0 to 0.5, 0.5 to 1.0, and 1.0 to 1.5 m is 21, 57, and 9 mg C m^{-2} h^{-1}, respectively. Macrophyte productivity per unit lake surface area exceeds the phytoplankton productivity (Alexander and Barsdate, 1971).

Benthic Algal Production

Alexander and Barsdate (1971) showed that the epiphytic algae are not important in Smith Lake until late summer and fall, when they become abundant on the stems and lower leaves of submergent macrophytes. In situ slide incubation in 1965 revealed that the average daily rate of carbon accumulation over 30 days on glass substrates was more than $2 \mu g C cm^{-2} d^{-1}$. Vertically suspended slides showed a decrease in carbon: chlorophyll *a* ratios from the surface to a depth of 1 m, probably as a result of shade adaptation with reduced light intensity; below 1 m the carbon and chlorophyll *a* ratio is higher due to inclusion of animals in the *aufwuchs*. Smith Lake has an abundant benthic fauna dominated by aquatic insect larvae, snails, and crustaceans in the littoral zone (see further discussion below).

Application of Natural Abundance of Carbon and Nitrogen Stable Isotopes

Stable carbon and nitrogen isotope analyses are powerful tools for tracking carbon and nitrogen sources and sinks in laboratory and field studies. Isotope effects result from differing metabolic transfer rates for the heavy and light isotopes, so that small and predictable variations in isotope composition are introduced. The isotope ratio is expressed in the conventional delta (δ) notation, defined as the per mill (‰) deviation from the isotope standard. In the case of nitrogen, ($^{15}N/^{14}N$), this standard is atmospheric nitrogen, and Peede Belemnite formation (PDB limestone) is used for $^{13}C/^{12}C$. $\delta X = [R_{sample}/R_{standard}] - 1] \times 10^3$, where R is the appropriate ratio ($^{13}C/^{12}C$ or $^{15}N/^{14}N$) (Peterson and Fry, 1987). Isotope ratio analyses use a mass spectrometer, a highly sensitive instrument designed to separate elements quantitatively by mass. The following sections discuss results using this approach in studying nitrogen dynamics and food webs in Smith Lake.

Nitrogen Fixation

Natural abundance of ^{15}N can be used to estimate N_2 fixation. There is only a small isotope fractionation during N_2 fixation (Delwiche and Steyn, 1970; Hoering and Ford, 1960), so that organisms that fix N_2 have $\delta^{15}N$ ratios close to that (0.0‰) of atmospheric N_2. Non-N_2 fixing organisms should have $\delta^{15}N$ ratios reflecting their nitrogen sources. If the alternative source of nitrogen (i.e., dissolved nitrogen in lake water) has an ^{15}N composition significantly different from that of atmospheric N_2, then the use of $\delta^{15}N$ ratios as an indicator of N_2 fixation is valid. We have measured the $\delta^{15}N$ ratios for a variety of blue-green and green algae and found that $\delta^{15}N$ ratios for N_2 fixing blue-green algae were 0.6 ± 0.8‰ ($\bar{X} \pm$ SD), but six green algae showed substantially higher $\delta^{15}N$ ratios (6.6 ± 4.5‰) (Fig. 5.11). The $\delta^{15}N$ ratio (3.6 ± 0.5‰) for one blue-green alga, *Lyngbya* sp., also was considerably higher, suggesting that it was not fixing nitrogen (Gu and Alexander, 1993a). Analyses of the nitrogen content of N_2 fixing blue-green algae and non-N_2 fixing algae showed that the former consistently had a higher nitrogen content.

Planktonic Food Web

Carbon isotope ratios vary considerably among food sources and can be used to identify the food sources of consumers (Fry, 1991). Isotope fractionation of ^{13}C is small during carbon transfer between trophic levels (DeNiro and Epstein, 1978). In contrast, nitrogen isotopes are useful for defining

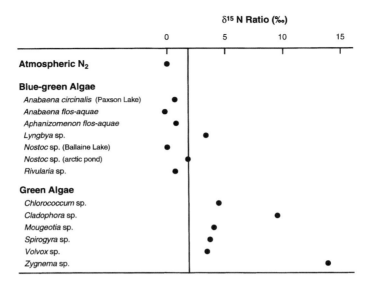

Figure 5.11. Natural abundance of nitrogen stable isotopes (δ^{15}N) in freshwater algae as indicators of nitrogen fixation (data from Gu and Alexander, 1993c).

trophic levels because of a consistent increment of 3 to 4‰ in the ^{15}N content of consumers with each trophic transfer (Minagawa and Wada, 1984; Owens, 1987; Peterson and Fry, 1987).

The summer phytoplankton community in Smith Lake is dominated by *Chlamydomonas*, *A. flos-aquae*, and *Aphanizomenon flos-aquae*, whereas the winter population is composed of green flagellates with low carbon and nitrogen productivity (Alexander and Barsdate, 1971; Gu, 1992). The zooplankton consists principally of several species of crustaceans, dominated by *Diaptomus pribilofensis*, with *Daphnia middendorffiana* and *Heterocope septentrionalis* less abundant. Other zooplankters such as the copepod, *Cyclops scutifer*, and the rotifer, *Conochilus hippocrepis*, appear only briefly. Samples of particulate organic matter (POM, mostly algae) and

Table 5.2. Summer δ^{13}C of Zooplankton and Various Carbon Sources in Smith Lake

Source	δ^{13}C (‰)	SD	n
Zooplankton	−34.4	1.1	3
POM	−33.0	3.6	10
Terrestrial plants	−28.5	2.2	5
Lake plants	−27.0	1.6	7
Lake sediment	−30.5	0.1	12

Data are taken from Gu (1993).

zooplankton were collected during 1989 to 1990 and analyzed for $\delta^{13}C$ and $\delta^{15}N$ ratios to trace nutrient and energy flow pathways in the planktonic food web. The $\delta^{13}C$ ratios of summer zooplankton resemble those of POM, implying an autochthonous source of carbon fueling the food web (Table 5.2). The $\delta^{15}N$ ratios of presumed herbivorous *Daphnia* and *Diaptomus* were 2.2 ± 0.7‰ and 4.7 ± 0.6‰, respectively, with a mean of 3.5 ± 1.8‰, higher than that of POM. The $\delta^{15}N$ (7.0 ± 0.7‰) of *Heterocope* suggests that this presumed carnivore did not prey exclusively on *Diaptomus* ($\delta^{15}N$ = 6.8‰), a dominant species in the lake, but may also have used POM. Similar results have been reported for some arctic lakes (Kling et al., 1992).

Benthic Food Web

The benthic primary consumers in Smith Lake are clearly separated into two carbon-isotopically distinct groups (Fig. 5.12). Group I primary consumers include the conchostracan *Lynceus*, the bivalve *Sphaerium*, and the insects *Cymatia* and *Gyrinus*. These animals had $\delta^{13}C$ ratios from −31.2 to −32.1‰, with a mean of −31.5‰. Based on this, there is little doubt that their energy was derived from phytoplankton, which show a $\delta^{13}C$ of −33.0‰. *Sphaerium* is associated with *Ceratophyllum* or *Typha* communities and filters phytoplankton from the water column. However, other species in this group, including *Cymatia*, *Gyrinus*, and tadpoles, are not filter feeders.

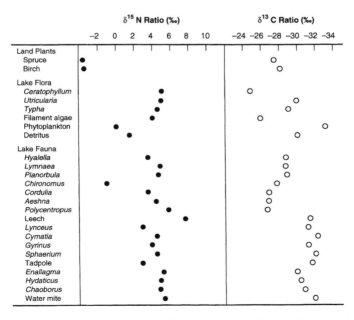

Figure 5.12. Natural abundance of stable carbon and nitrogen isotopes in source organic matter and benthic fauna in Smith Lake (data from Gu, 1993).

These species might be feeding on phytoplankton detritus deposited on surface sediment or other substrates, because phytoplankton detritus possesses lower fiber content and a higher nitrogen content compared to vascular plant detritus (Mann, 1988). This is consistent with the known diets for these organisms except *Gyrinus*, which is thought to be a predator or scavenger on organisms in the surface film (M. Oswood, personal communication).

The group II benthic fauna includes the amphipod *Hyalella*, gastropods *Lymnaea* and *Planorbula*, and the dipteran *Chironomus*. With the exception of *Chironomus*, their $\delta^{13}C$ ratios fell between −28.7 and −27.7‰ with a mean of −28.3‰, indicating a dependence on detritus (−30.0‰) or more probably a mixture of detritus of vascular plants, phytoplankton, and periphyton. *Chironomus* shows a $\delta^{15}N$ ratio of −1.2‰, which is lighter than that (1.8‰) of the detritus mixture (Fig. 5.12). We speculate that *Chironomus* relies on a specific component of detritus in the sediments, possibly leaf litter. The $\delta^{15}N$ of birch leaves supports this.

Benthic predators can also be classified into two groups isotopically. Group I, including *Cordulia*, *Aeshna*, and *Polycentropus*, has a $\delta^{13}C$ of −27.0‰. Group II, with a $\delta^{13}C$ of −31.0‰, includes the damselfly *Archilestes*, the water beetle *Hydaticus*, dipteran *Chaoborus*, and water mites. The distinct $\delta^{13}C$ ratios in these two predator groups indicate that these animals rely on specific and different prey as their major carbon source, although benthic predators have been reported to feed on a large spectrum of prey, typically microcrustaceans, Chironomidae, Trichoptera, Zygoptera, and Ephemeroptera (Blois, 1985; Pritchard, 1964). Our $\delta^{15}N$ values show that Chironomidae are the major nutrient source for *Cordulia* and *Aeshna*. Although *Hyalella* was abundant on the sediment, the $\delta^{15}N$ values suggested that amphipods are not important as food for anisopteran larvae in Smith Lake. *Polycentropus* had the highest $\delta^{15}N$ (6.0‰) among animals in this group and might feed primarily on *Hyalella*. Group II predators all have similar $\delta^{15}N$ (5.2 ± 0.4‰), suggesting that they rely on similar diet sources, such as microcrustaceans. There is no evidence that *Chaoborus* relies on zooplankton such as the dominant *Diaptomus* ($\delta^{15}N$ = 6.9‰), although it could be feeding on *Daphnia*, and has been reported to feed on crustacean zooplankton in other lakes (Rau, 1980). Although water mites have been reported to actively feed on zooplankton (Butler and Burns, 1991; Matveev and Martinez, 1990), the $\delta^{15}N$ ratio in Smith Lake indicates that they rely exclusively on microcrustaceans.

According to the $\delta^{15}N$ value for leeches, they are tertiary consumers. Leeches feed on a variety of animal prey (Davies et al., 1978, 1981). Blinn and Davies (1990) reported that 90% of the diet for the Leech *Erpobdella montezuma* consisted of the amphipod *Hyalella montezuma*. *Hyalella* is abundant in Smith Lake, and its $\delta^{15}N$ ratio of 3.5‰ was 5.0‰ lower than that of leeches. This indicates that, although *Hyalella* may be a significant food source, these leeches probably also feed on gastropods.

Nitrogenous Nutrient Sources of Aquatic Macrophytes

The $\delta^{15}N$ of 14 species of aquatic macrophytes collected from Smith Lake during 1988 to 1990 range from −1.2 to 5.6‰ with a mean of 4.3‰ (Table 5.3). Nonrooted macrophytes have $\delta^{15}N$ values (4.9 ± 0.4‰) significantly higher than rooted macrophytes (2.9 ± 1.8‰). The $\delta^{15}N$ of nonrooted macrophytes (approximately 5.0 to 6.0‰) is close to that of summer DIN in the water column, indicating that these macrophytes derive their nitrogen from lake water. Furthermore, the close $\delta^{15}N$ ratios of macrophytes and DIN suggest little isotopic fractionation during nitrogen assimilation. This would be expected if DIN was the limiting nutrient, in which case essentially all available nitrogen would be assimilated without isotopic fractionation (Peterson and Fry, 1987). Toetz (1974) showed that the stem and leaves of *C. demersum* can take up NH_4^+ from sediment and translocate it to the shoots. However, the high $\delta^{15}N$ in our study suggests that this species obtained most of its nitrogen, if not all, from water. Best (1980) speculated that the nutrients absorbed by modified leaves of *C. demersum* might be translocated only to a slight extent due to the highly reduced vascular system. It is reported that *E. crassipes*, and *Lemna* spp. obtain nitrogen from N_2 fixing symbiotic bacteria (Iswaran et al., 1973; Zuberer, 1982). However, the $\delta^{15}N$ of these macrophytes do not show significant influence of light isotopic nitrogen derived from N_2 fixation. The $\delta^{15}N$ of the moss *Drepanocladus* sp. growing on lake sediments is 4.5‰, similar to those of nonrooted macrophytes, demonstrating that although moss occurs on lake sediment, its nitrogen source derives from the overlying water (Hutchinson, 1975).

Table 5.3. $\delta^{15}N$ (‰) of 14 Species of Aquatic Macrophytes in Smith Lake

Taxon	$\bar{x}$	SD
Nonrooted plants		
Eichhornia crassipes	4.4	1.0
Lemna trisulca	4.3	1.0
L. minor	5.6	0.1
Pistia stratiotes	4.9	0.1
Ceratophyllum demersum	5.0	0.8
Utricularis vulgaris	4.9	1.0
Drepanocladus sp.	5.4	1.5
Rooted plants		
Polygnum sp.	−1.2	0.1
Eleocharis palustris	3.2	0.2
Potamogeton gramineus	4.5	0.3
Nuphar polysepalum	2.7	0.6
Sparganium angustifolium	3.5	1.5
Typha latifolia	4.6	0.2
Nymphaea tetragona	2.9	0.1

Data are taken from Gu (1993).

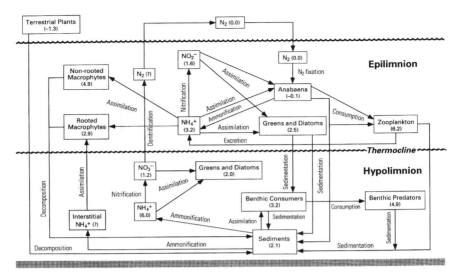

Figure 5.13. The Smith Lake food web based on the analysis of nitrogen isotopic composition (after Gu, 1993).

The $\delta^{15}N$ (2.9 ± 1.8‰) of rooted macrophytes was close to the $\delta^{15}N$ (2.0‰) of sediment nitrogen, suggesting that they utilized nitrogen from sediment rather than from the water column. This agrees with previous studies that showed rooted submergent macrophytes mainly obtained their nitrogenous nutrients from sediments (see review by Barko et al., 1991).

Our current understanding of the Smith Lake food web based on the isotopic information discussed above is reflected in Fig. 5.13.

Lakes of Tanana Valley

The question of how typical Smith Lake is among lakes in the Tanana Valley is difficult to address critically, because less information exists for all other lakes. There are a number of differing lake types in the area, with the most abundant being thermokarst lakes, aggradation lakes, and numerous oxbow lakes that dominate the wetlands adjacent to the river itself. Smith Lake appears to be characteristic of one type of lake that is abundant in the Tanana/Yukon River Valley. LaPerriere (1990) has described a series of lakes in the Tanana Valley. Harding, Little Harding, Birch, Chisholm, and Quartz lakes all lie in tributary valleys that run perpendicular to the river and to the north. These lakes differ substantially from the thermokarst lakes. She describes them as aggradation lakes, except that Harding has an older tectonic basin under the aggradation layer. They differ in their clear water and their biological regimes must be very different, if only because these deeper, well-oxygenated lakes contain fish as an apex predator.

There is an excellent opportunity for comparative limnological work in the interior valleys of Alaska. Lakes of several types, often readily accessible but largely unstudied, occur within convenient distances from each other. The work on Smith Lake has been a good beginning toward an understanding of these lakes, but much more remains to be done in the region.

Acknowledgments. The authors wish to acknowledge the contributions to research on Smith Lake, over the years, by many faculty, staff, and students. In particular, we wish to recognize Dr. R.J. Barsdate's long-term participation and intellectual investment in the limnology of interior Alaskan lakes. The figures have benefited greatly from careful "fine tuning" by M. Billington.

References

Alexander V (1970) Relationships between turnover rates in the biological nitrogen cycle and algal productivity. Proc 25th Ind Waste Conf, Purdue Univ Eng Ext Serv 137:1–7.

Alexander V, Barsdate RJ (1971) Physical limnology, chemistry and plant productivity of a taiga lake. Int Rev Gesamten Hydrobiol 56:825–872.

Alexander V, Barsdate RJ (1974) Limnological studies of a subarctic lake system. Int Rev Gesamten Hydrobiol 59:737–753.

Barko JW, Gunnison D, Carpenter SR (1991) Sediment interactions with submersed macrophyte growth and community dynamics. Aquat Bot 41:41–65.

Best EPH (1980) Effects of nitrogen on the growth and nitrogenous compounds of *Ceratophyllum demersum.* Aquat Bot 8:197–206.

Billaud VA (1968) Nitrogen fixation and the utilization of other inorganic nitrogen sources in a subarctic lake. J Fish Res Board Can 25:2101–2110.

Blinn DW, Davies RW (1990) Concomitant diel vertical migration of a predatory leech and its amphipod prey. Freshwater Biol 24:401–407.

Blois C (1985) Diets and resource partitioning between larvae of three anisopteran species. Hydrobiologia 126:221–227.

Butler MI, Burns CW (1991) Predation by a water mite (*Piona exigua*) on enclosed populations of zooplankton. Hydrobiologia 220:37–48.

Clasby RC (1972) Denitrification in a Subarctic Lake. M.S. thesis, University of Alaska, Fairbanks.

Davies RW, Wrona FJ, Everett RP (1978) A serological study of prey selection by *Nephelopsis obscura* Verrill (Hirudinoida). Can J Zool 56:587–591.

Davies RW, Wrona FJ, Linton L, Wilkialis J (1981) Inter- and intra-specific analyses of the food niches of two sympatric species of Erpobdellidae (Hirudinoidea) in Alberta, Canada. Oikos 37:105–111.

DeNiro MJ, Epstein S (1978) Influence of diet on the distribution of carbon isotopes in animals. Geochim Cosmochim Acta 42:495–506.

Delwiche CC, Steyn PL (1970) Nitrogen isotope fractionation in soils and microbial reactions. Environ Sci Technol 4:929–935.

Dugdale VA (1965) Inorganic nitrogen metabolism and phytoplankton primary productivity in a subarctic lake. Ph.D. thesis, University of Alaska, Fairbanks.

Fry B (1991) Stable isotope diagrams of freshwater food webs. Ecology 72:2293–2297.

Goering JJ, Dugdale VA (1966) Estimates of the rates of denitrification in a subarctic lake. Limnol Oceanogr 11:113–117.

Gu B (1992) Dissolved nitrogen utilization by phytoplankton in a subarctic lake. M.S. thesis, University of Alaska, Fairbanks.

Gu B (1993) Nitrogen isotope cycling in a subarctic lacustrine system. Ph.D. thesis, University of Alaska, Fairbanks.

Gu B, Alexander V (1993a) Dissolved nitrogen uptake by a cyanobacterial bloom (*Anabaena flos-aquae*) in a subarctic Alaska lake. Appl Environ Microbiol 59:422–430.

Gu B, Alexander V (1993b) Estimation of N_2 fixation based on differences in natural abundance of ^{15}N among freshwater N_2-fixing and non-N_2-fixing algae. Oecologia 96:43–48.

Gu B, Alexander V (1993c) Seasonal variations in dissolved inorganic nitrogen utilization by phytoplankton in a subarctic Alaska lake. Arch Hydrobiol 126:273–288.

Gu B, Schell DM, Alexander V (1994) Stable carbon and nitrogen isotopic analysis of the plankton food web in a subarctic lake. Can J Fish Aquat Sci 51:1338–1344.

Hammar J (1989) Freshwater ecosystems of polar regions: Vulnerable resources. Ambio 18:7–22.

Hobbie JE (1973) Arctic limnology: A review. In Britton ME (Ed) Alaskan Arctic tundra. Arctic Institute of North America Technical Paper 25, 127–168. Washington DC.

Hoering TC, Ford HT (1960) The isotope effect in the fixation of nitrogen by *Azotobacter*. J Am Chem Soc 82:376–378.

Hutchinson GE (1957) A treatise on limnology, Volume I—Geography, physics, and chemistry. Wiley, New York.

Iswaran V, Sen A, Apte A (1973) *Azotobacter chroococcum* in the phyllosphere of water hyacinth (*Eichhornia crassipes* Mort. Solms). Plant Soil 39:461–463.

Kling GW, Fry B, O'Brien WJ (1992) Stable isotopes and planktonic trophic structure in arctic lakes. Ecology 73:561–566.

Klingensmith KM (1981) Sediment nitrification, denitrification, and nitrous oxide production in an arctic lake. M.S. thesis, University of Alaska, Fairbanks.

LaPerriere JD (1990) Variations in aggradation lakes of the Tanana Valley: Influence of morphometry and chemistry on phytoplankton photosynthesis. Verh Int Verein Limnol 24:309–313.

Likens GE, Johnson PL (1968) A limnological reconnaissance in interior Alaska. Research Report No. 239, CRREL, Hanover, NH.

Livingston DA (1963) Alaska, Yukon, Northwest Territories, and Greenland. In Frey DG (Ed) Limnology in north America. University of Wisconsin Press, Madison, WI, 559–574.

Mann KH (1988) Production and use of detritus in various freshwater, estuarine and coastal marine ecosystems. 1988. Limnol Oceanogr 33:910–930.

Matveev VF, Martinez CC (1990) Can water mites control populations of planktonic Cladocera? Hydrobiologia 198:227–231.

Minagawa M, Wada E (1984) Step-wise enrichment of ^{15}N along food chains: Further evidence and the relation between ^{15}N and animal age. Geochim Cosmochim Acta 48:1135–1140.

Ohmori M, Hattori A (1972) Effects of nitrate on nitrogen-fixation by the blue-green alga *Anabaena cylindrica*. Plant Cell Physiol 13:589–599.

Owens NJP (1987) Natural variations in ^{15}N in the marine environment. Adv Mar Biol 24:389–451.

Paerl HW (1990) Physiological ecology and regulation of N_2 fixation in natural waters. In Marshall KC (Ed) Advances in microbial ecology, Vol 11. Plenum Press, New York, 305–344.

Peterson BJ, Fry B (1987) Stable isotopes in ecosystem studies. Ann Rev Ecol Syst 18:293–320.

Pritchard G (1964) The prey of dragonfly larvae (Odonata; Anisoptera) in ponds in northern Alberta. Can J Zool 42:785–800.

Rau GH (1980) Carbon-13/carbon-12 variation in subalpine lake aquatic insects: Food source implications. Can J Fish Aquat Sci 37:742–745.

Ragotzkie RA (1978) Heat budgets of lakes. In Lerman A (Ed) Lakes: Chemistry, geology, physics. Springer–Verlag, New York, 1–19.

Satoh Y, Alexander V, Takahashi E (1992) Dissolved organic carbon (DOC) and some other chemical profiles of various Alaskan lakes in summer. Jpn J Limnol 53:207–216.

Takahashi M, Saijo Y (1988) Nitrogen metabolism in Lake Kizaki, Japan V. The role of nitrogen fixation in nitrogen requirement of phytoplankton. Arch Hydrobiol 112:43–54.

Toetz DW (1974) Uptake and translocation of ammonia by freshwater hydrophytes. Ecology 55:199–201.

Yoshinari T, Hynes R, Knowles R (1977) Acetylene inhibition of nitrous oxide reduction and measurement of denitrification and nitrogen fixation in soil. Soil Biol Biochem 9:177–183.

Zuberer DA (1982) Nitrogen fixation (acetylene reduction) associated with duckweed (Lemnaceae) mats. Appl Environ Microbiol 43:823–828.

6. Waterfowl and Wetland Ecology in Alaska

James S. Sedinger

Alaska's waterfowl are receiving more attention as they become an increasingly important component of North American continental populations. Continental populations of ducks are at record low levels (Bortner et al., 1991), largely because of loss of habitat in midcontinent breeding areas (Tiner, 1984) and drought conditions through much of the 1980s (Bortner et al., 1991). Because breeding populations in Alaska have remained stable, or even increased, a larger proportion of continental duck populations are currently recorded in Alaska during breeding pair surveys than ever before (Conant and Dau, 1991). For example, >50% of breeding Northern Pintails (*Anas acuta*) in North America have been recorded in Alaska each year since 1987 (Conant and Dau, 1991). Alaska also contains >90% of the world's Trumpeter Swans (*Cygnus buccinator*), which are classified as endangered in the contiguous 48 states. Many other breeding populations of waterfowl achieve their highest concentrations in Alaska.

Except for waterfowl species that are associated with coastal tundra (e.g., Black Brant, *Branta bernicla nigricans*), Alaska was historically thought to be relatively unproductive for waterfowl (Lynch, 1984), although the presence of local, relatively productive areas was recognized (Smith et al., 1964). This belief reflects the consensus among waterfowl biologists until recently that boreal forest wetlands were unproductive and, therefore, could not support high densities of waterfowl (e.g., Wellein and Lumsden, 1964). The belief in the low productivity of Alaskan wetlands is based, in part, on experience in the forests overlying the granitic Canadian shield, which are relatively unproductive (Wellein and Lumsden, 1964). In Alaska, however, certain habitats are very productive for waterfowl (Conant and Hodges, 1985; King and Lensink, 1971; Lensink, 1965).

In this review, I examine factors affecting the distribution of waterfowl in Alaska and their use of inland freshwater wetlands and coastal wetlands that are infrequently (less than monthly) tidally influenced. Wetlands are generally defined as areas that are at least periodically flooded but do not regularly contain standing water > 2 m deep (Cowardin et al., 1979). I will first briefly review basic requirements of waterfowl during the breeding, molting, and migration periods (few waterfowl winter in Alaskan freshwaters) and how meeting these requirements is influenced by wetland processes. I then consider wetland processes themselves and conclude with a review of waterfowl distribution in Alaska and a discussion of linkages among the various temporal and spatial scales of consideration necessary to understand waterfowl ecology in Alaska.

Waterfowl Requirements

Breeding Season

Requirements for successful reproduction by waterfowl can be separated into three parts: sufficient nutrients for maintenance and productive processes in breeding adults, suitable nesting habitat, and nutrients for growth of young and molt of adults. Numerous studies have shown that egg production in waterfowl is limited by the combination of exogenous nutrients (those available in the environment) and endogenous nutrients (nutrients females store before laying) (Afton and Ankney, 1991; Alisauskas and Ankney, 1992; Ankney and Alisauskas, 1991; Ankney and MacInnes, 1978; Mann and Sedinger, 1993; Raveling, 1979a). Ducks, which have relatively small body size, depend proportionally more on exogenous nutrients during breeding (Alisauskas and Ankney, 1992), although recent evidence for several species suggests that some ducks also rely heavily on previously stored reserves for breeding (Afton and Ankney, 1991; Alisauskas and Ankney, 1992; Ankeny and Afton, 1988; Ankney and Alisauskas, 1991; Mann and Sedinger, 1993). In Alaska, a substantial portion of nutrient reserves required for breeding in Northern Pintails are stored after arrival on the breeding area (Mann and Sedinger, 1993). Control of clutch size in ducks by nutrient availability at the time of (or just before) egg formation is demonstrated by the fact that ducks lay more eggs when fed ad libitum, than in nature (Batt and Prince, 1979; Duncan, 1987; Rohwer, 1984). Female attentiveness to their nests, which is directly correlated with hatching success (Gloutney and Clark, 1991; Harvey, 1971), is a function of the energy reserves of the female (Afton and Paulus, 1992), and therefore, at least indirectly related to food availability. Therefore, the ability of wetlands to provide sufficient nutrients will directly influence the productivity of ducks using wetlands.

In geese, which are large bodied, females rely heavily on endogenous nutrients for egg formation and for meeting their energy requirements during the incubation period (Aldrich and Raveling, 1983; Ankney and MacInnes, 1978; Raveling, 1979a; Thompson and Raveling, 1987). Nevertheless, food availability on breeding areas is still important to some species (Ankeny, 1984; Budeau et al., 1991; Gauthier and Tardif, 1991; Raveling, 1979b). Egg formation and nest attentiveness each depend on the total nutrients available from both exogenous and endogenous sources. Because nest attentiveness and associated nest success (Harvey, 1971) depend on the nutritional status of females, productivity of many goose populations will also reflect local nutrient availability for geese.

Growth of young waterfowl has been directly linked to food availability and quality for numerous species (Cooch et al., 1991; Hunter et al., 1986; Larsson and Forslund, 1991; Pehrsson and Nyström, 1988; Sedinger, 1992; Sedinger and Flint, 1991). Recently, growth of young waterfowl has been shown to influence first-year survival and future fecundity in arctic nesting geese (Cooke et al., 1984; Sedinger, unpublished data). It is likely that similar relationships exist for ducks, but these will be more difficult to establish for most species because of the difficulty of following known individuals. Nevertheless, the advantage of productive wetland habitats to both breeding adults and their offspring should favor active preference for such habitats, resulting in higher densities of breeding waterfowl in productive areas.

While primary and secondary productivity determine the abundance of food for waterfowl, physical attributes of the habitat determine food availability for waterfowl. Most dabbling ducks (tribe Anatini) feed from the surface. They are therefore restricted to feeding in water less than 1 m deep where they reach the benthos or aquatic vegetation by tipping up (Sugden, 1973). Ducklings require even shallower water, <0.3 m deep (Sugden, 1973). Northern Shovelers (*Anas clypeata*) are an exception to this general rule because they feed predominantly at the surface, where they strain plankton. Northern Shovelers, therefore, may feed in deeper water than other dabbling ducks. Diving (Aythini) and sea ducks (Mergini) feed by diving in deeper water, but still predominantly in water < 2 m deep. Wetlands lacking foraging areas of appropriate depth will therefore not support breeding ducks.

Waterfowl exhibit species-specific preference for particular nesting habitats. Most geese prefer open tundra (Owen, 1980), with the exception of some Canada (*Branta canadensis*) and White-Fronted Geese (*Anser albifrons frontalis*) that nest in shrubby or even forested areas (Bellrose, 1980). Ducks also vary among species in their preferences for nesting habitats. The origin of these preferences is unknown but they are not, for the most part, phylogenetically based. Some species (e.g., Northern Pintails [*Anas acuta*], Northern Shovelers, and Greater and Lesser Scaup [*Aythya marila* and *A. affinis*]) prefer open habitats that include either tundra or open meadows within the boreal forest. Others, Eiders (*Somateria* spp.) for example, are tundra specialists (Bellrose, 1980). Another group prefers to nest in shrubby (Mallards [*Anas platyrhynchos*]) or forested habitats (American Wigeon [*A. americana*], Green-Winged Teal [*A. crecca*], and Goldeneyes [*Bucephala islandica* or *B. clangula*]) within the boreal forest (Bellrose, 1980). These habitat preferences influence the nature and distribution of wetlands used by waterfowl in Alaska because of the spatial distribution of these various habitat types.

Molt

The simultaneous wing molt (remigial molt) of waterfowl is the time period when adult waterfowl are most vulnerable to predation. During this period, adults not accompanying young typically move to habitats offering some protection from predators. Some dabbling ducks (e.g. Pintails) move to wetlands with emergent vegetation, thereby reducing their visibility (Bellrose, 1980). Most geese and several species of ducks move to large lakes or rivers to molt (e.g., Derksen et al., 1981).

Whether nutritional requirements for molt represent a substantial nutritional cost is currently controversial (Ankney, 1984; Heitmeyer, 1988). Flight muscles of all species atrophy (Ankney, 1984; Hanson, 1962), likely representing a response to disuse of these muscles immediately before and during the remigial molt. There is no evidence that postbreeding waterfowl draw on stored nutrient reserves for feather growth. Nevertheless, the nutrients required to produce new feathers, or particular foods eaten to meet these requirements, may limit the ability of birds to increase muscle mass or store lipids that will be needed for migration. Such tradeoffs will be exacerbated by the increased wariness of most waterfowl and resulting reduced nutrient intake during molt.

Migration Period

Our understanding of waterfowl ecology in Alaska during the migration period is poor as few studies have been conducted during this period (e.g., Sedinger and Bollinger, 1987; Ward and Stehn, 1989). It is possible to establish plausible hypotheses, however, based on nutrient dynamics in waterfowl, phenological events during waterfowl migration, and knowledge of migration elsewhere.

Before fall migration, waterfowl increase primarily stored lipids (Wypkema and Ankney, 1979). Natural selection probably favors individuals with large lipid stores before fall migration because such individuals survive migration at a higher rate (Owen and Black, 1989). Also, in late summer into fall, proteinacious foods necessary for deposition of lean body mass become less available in wetlands (Butler et al., 1980; Sedinger and Raveling, 1986; Stross et al., 1980; Taylor, 1986), forcing shifts to energy-rich foods like berries (Sedinger and Raveling, 1984) or seeds and tubers (Coleman and Boag, 1987; Prevett et al., 1979).

Energy-rich foods are also important in spring because individuals must replenish lipid reserves depleted during migration (Budeau et al., 1991; Mann and Sedinger, 1993; Thomas, 1983). Relatively high energy foods like seeds tend to be available in spring because they survive the winter dormant period (Burris, 1991). Both ducks and geese feed on tubers of wetland plants in spring (Budeau et al., 1991; Burris, 1991; Prevett et al., 1979), which are excellent sources of energy but may also provide some protein (Thomas and Prevett, 1980). One study of Northern Pintails in interior Alaska indicates that females also increase their lean body mass (protein) after arrival in Alaska (Mann and Sedinger, 1993), suggesting that foods containing sufficient protein for positive nitrogen balance are available at that time.

Marsh Ecology

Freshwater wetlands were relatively neglected by scientists until the late 1970s. Therefore, much of our understanding of these wetlands rests on single studies of particular processes; and what follows should thus be viewed as hypotheses requiring further testing, rather than accepted paradigms, especially with respect to Alaska.

Hydrology fundamentally drives processes in freshwater wetlands. Wetlands with hydrological regimes of intermediate energy are the most productive (Mitsch and Gosselink, 1986). Hydrology regulates nutrient flow into wetlands (Heglund, 1988; LaBaugh, 1989; Murphy et al., 1984; Seppi, 1993); areas that are periodically flooded receive more nutrients from outside the system, thereby increasing primary productivity (Mitsch and Gosselink, 1986). In the prairie potholes of the midcontinent, wetlands lower in hydrological basins or those that receive more groundwater have greater nutrient influx (Swanson et al., 1988 in La Baugh, 1989; Sloan, 1972). Evidence that nutrient input into wetlands influences their primary productivity is provided by the observation that agricultural runoff carrying nitrogen and phosphorus fertilizers significantly increases wetland primary productivity (Neely and Davis, 1985). Periodic drying of prairie wetlands is also believed to increase nutrient availability by increasing oxygen available to decomposers and consequently the rate of decomposition (Mitsch and Gosselink, 1986).

Hydrology also influences the wet meadows and salt marshes used by grazing geese. Grasses and sedges (graminoids) in the arctic and subarctic are nutrient limited (Bazely and Jefferies, 1985; Chapin et al., 1980). Therefore, areas that regularly exchange nutrient-rich water are likely to be more productive and contain plants with greater nutrient concentration than other areas, possibly explaining the high concentrations of geese associated with river deltas and tidally influenced areas.

Primary productivity influences waterfowl use of wetlands directly because waterfowl consume the leaves of plants (Bartonek and Hickey, 1969; Bergman, 1973; Sedinger and Raveling, 1984; Sugden, 1973; Swanson et al., 1979), their seeds (DuBowy, 1985; DeBruyckere, 1988; Burris, 1991; Kirchhoff, 1978; Krapu and Swanson, 1975) or underground storage organs (Bartonek and Hickey, 1969; Bergman, 1973; Budeau et al., 1991; Prevett et al., 1979). Submergent and emergent vegetation, however, also create habitat for aquatic invertebrates (Jacobs, 1992; Krull, 1970; Voigts, 1976) that are important foods for breeding adult waterfowl (Swanson et al., 1979) and growing ducklings (Sedinger, 1992). Detritus from vegetation provides food for numerous aquatic invertebrates (Murkin, 1989). Surfaces of stems and leaves provide substrate for epiphytic algae that is eaten by grazing invertebrates (Hunter, 1980). Aquatic plants also influence aquatic invertebrate abundance less directly by providing escape habitat from predatory fish and by influencing dissolved oxygen concentrations (Nelson and Kadlec, 1984). Dissolved oxygen concentration declines along a gradient from open water into stands of vegetation, likely because of reduced mixing and increased decomposition in stands of vegetation (Kairesdo, 1980). Therefore, aquatic invertebrates experience the optimum combination of conditions at the interface between emergent vegetation and open water (Nelson and Kadlec, 1984). Temperate marshes with equal dispersion of open water and emergent vegetation are, therefore, most attractive to ducks (Weller and Fredrickson, 1973). Two experiments in which emergent vegetation was experimentally removed, verified that a mixture of open water and emergent vegetation maximizes the abundance and diversity of aquatic invertebrates and is heavily used by waterfowl (Kaminski and Prince, 1981; Murkin et al., 1982).

Wetlands in Alaska

Distribution and Hydrology

Wetlands can be separated into three broad categories based on geography and predominant habitat: boreal forest, subarctic coastal tundra, and arctic coastal tundra (Fig. 6.1). Boreal forest includes the forested areas of interior and southcentral Alaska. Upland areas are dominated by spruce (*Picea* spp.) and paper birch (*Betula papyrifera*). This habitat grades into temperate rain forest in southcentral Alaska. Our knowledge is insufficient to differentiate between wetland processes in these two types of forest, but the lower precipitation–evapotranspiration balance in the boreal forest increases the importance of surface flow recharge of wetlands and associated

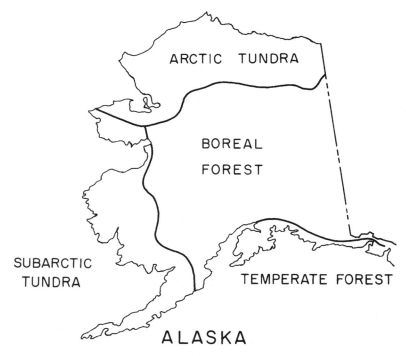

Figure 6.1 General distribution of arctic coastal tundra, subarctic coastal tundra, and boreal forest zones in Alaska. Arctic and subarctic tundra were separated by the southern boundary of the arctic climate zone (Anon, 1973). Subarctic tundra was defined by treeline south of arctic tundra. The boundary between temperate and boreal forest was defined as the northern limit of Sitka spruce (*Picea sitchensis*) and western red cedar (*Thuja plicata*) (Hultén, 1968).

nutrient input in this habitat relative to temperate forests (Ford and Bedford, 1987). Subarctic tundra extends from the Aleutian Islands along the west coast of Alaska to the Seward Peninsula. Coastal portions of this tundra are dominated by graminoid communities. Tidal influence extends far inland in some areas, resulting from low topographic relief and large tidal fluctuations (Dupre, 1980; Tande and Jennings, 1986). More than 30 km inland from the coast this habitat is dominated by *Sphagnum* sp. and several species of prostrate or dwarf shrubs (Maciolek, 1989; Tande and Jennings, 1986). Arctic tundra is also characterized by graminoid communities but tidal influence is less important because of the much smaller tidal fluctuations in arctic waters (Norton and Weller, 1984), compared to those in the subarctic (Dupre, 1980).

Arctic freshwater wetlands have been the most intensively studied in Alaska because of the International Biological Program and permit requirements associated with oil development. These wetlands, predominantly the result of the thaw-lake cycle (Hopkins, 1949; also see Chapter 1), were classified by Bergman et al. (1977) based on their history and plant communities. Bergman et al. (1977) defined eight wetland classes. The first five classes represent freshwater basins ranging in depth (and permanence) from flooded tundra (class I) to deep open lakes (class V). Classes II, III, and IV are relatively permanent ponds with varying amounts of emergent *Carex aquatilis* (class II) or *Arctophila fulva* (classes III and IV). Pond depth ranged from 10 to 30 cm (class II) to >40 cm (class IV). Class III ponds are nearly completely vegetated with *C. aquatilis* in the shallower shoreward zone and *A. fulva* in the pond center. Class VI wetlands are large, partially drained basins containing any of the other seven types. Class VII wetlands are small fluvial systems containing a series of pools, and coastal wetlands are in class VIII.

Arctic tundra is characterized by a high density of wetlands. Between Prudhoe Bay and Barrow, from 42 to 86% of several areas are covered by water (Derksen et al., 1981). Class II to IV wetlands, which are of greatest value to breeding waterfowl, are most abundant near the Beaufort Sea coast (Derksen et al., 1981) and pond density declines east of Prudhoe Bay.

Subarctic coastal tundra of the Yukon–Kuskokwim (YK) delta and the Alaska Peninsula is influenced substantially by tidal hydrology (Tande and Jennings, 1986; U.S. Department of Commerce, 1987). Wetlands within 16 km of the coast of the YK delta are flooded several times each decade by storm surges (Dupre, 1980). Tidal flooding regularly deposits nutrients into wetlands. Introduction of brackish water also influences the species composition of plants in these wetlands (Kincheloe and Stehn, 1991). Pond densities are high on the YK delta; more than 50% of the area is covered by water (Mickelson, 1975). In the coastal zone of the central YK delta ponds are small (<2 ha) and shallow (<1 m deep). Farther inland, ponds increase in size and depth (personal observation) and there is evidence of thaw-lake origins (Maciolek, 1989).

Most lakes in the boreal forest were formed by hydrological processes associated with rivers (e.g., Heglund, 1988). Therefore, oxbow lakes or crescentic levee lakes formed between old river levees and lateral lakes (formed by damming of tributary streams) are important classes of wetlands (Heglund, 1988). Thaw lakes are present but are less important in boreal forest than in tundra areas. Beavers also create wetlands that are used by waterfowl (Kafka, 1988). Finally, in the Minto Flats area (60 km west of Fairbanks) geological subsidence along a fault on the east side of the flats has created a low area that functions as a hydrological sink, holding large quantities of water under many hydrological conditions. Processes responsible for forming wetlands in the boreal forest have produced shallow lakes (<3 m deep) (Heglund, 1988; Jacobs, 1992; Kafka, 1988) with relatively flat bathometry. As a result, wetlands in the boreal forest frequently contain extensive areas suitable for emergent or submergent vegetation.

Dogma among scientists studying wetlands and waterfowl has traditionally held that wetlands in the boreal forest are hydrologically stable;
flooding occurs during the spring melt but otherwise water levels do not
fluctuate substantially (Smith et al., 1964). Interpretation of these early
observations was biased by the belief that the hydrological variation of
principal importance was drought. In the boreal forest of Alaska, however,
spring flooding is of major importance. Flooding occurs along the major
rivers in interior Alaska associated with two types of events: a heavy snow
pack in spring or a late and rapid breakup. Either of these events produces
extensive flooding of wetland basins associated with rivers (e.g.,
Fredrickson and Reid, 1990; Petrula et al., 1992), thereby depositing nutrients into wetlands. Frequency of flooding depends on the elevation of the
wetland relative to the river, presence and height of natural levees, and the
presence of channels directly connecting the wetland to the river. Major
flooding events occur several times per decade in some wetland basins and
more frequently in wetlands directly connected to rivers (Petrula et al.,
1992).

Primary Productivity

Primary productivity at high latitudes is limited by the availability of nutrients, that is, in turn, limited by the rate of decomposition (Alexander et al.,
1980; McKendrick et al., 1978; Tieszen, 1978). Decomposition in wetlands is
limited by low temperatures and oxygen availability (Prentki et al., 1980).
We expect, therefore, that factors that influence the rate of decomposition
or that provide limiting nutrients from outside the system will increase
primary productivity in Alaskan freshwaters.

Annual primary productivity is substantially lower in high latitude
($<200 g m^{-2} yr^{-1}$) (Alexander et al., 1980; Cargill and Jefferies, 1984;
Chapin, 1978; Sedinger and Raveling, 1986), than in temperate wetlands
($>1000 g m^{-2} yr^{-1}$) (McNaughton, 1966; Van der Valk and Davis, 1978; White
et al., 1978; Zedler, 1980). Lower primary productivity in the Arctic has
been attributed to the short growing season, low temperatures, and resulting low decomposition rates (Chapin et al., 1980). The relatively stable
hydrology (among years) of arctic wetlands also is important because the
absence of "drawdown" conditions eliminates the possibility for acceler-
ated release of nutrients from organic substrates associated with unflooded
conditions (Mitsch and Gosselink, 1986: 240). The ultimate importance of
inorganic nutrients is demonstrated by the effect of adding nutrients to
wetlands, which dramatically increases phytoplankton production
(Alexander et al., 1980). The short summer further reduces primary productivity because of the small number of growing days for primary producers (Tieszen, 1978).

Macrophyte production is influenced by pond depth and hydrology. The
two principal emergent plants, *Carex aquatilis* and pendent grass

(*Arctophila fulva*) of tundra wetlands, occur at water depths less than 30 and 50 cm, respectively (Bergman et al., 1977); so shallow ponds support a greater area of emergents than do deeper ponds. The position of ponds in the thaw-lake cycle should also affect primary production. Ponds in drained lake basins are shallower, thereby increasing the proportional area that will support macrophytes (Bergman et al., 1977). Also, the topographical depression created by the original lake may increase water flow and nutrient flux into the small ponds within thaw-lake beds, relative to ponds in more elevated sites (e.g., low-center polygons) (Bergman et al., 1977).

Salt marshes, although poorly studied, may also support higher primary productivity than isolated ponds. These areas receive nutrients from sediment loads carried by associated rivers and tidal actions. Further, the relatively open hydrological regime associated with salt marshes reduces litter accumulation, and therefore increases thaw depth and soil temperatures (Miller et al., 1980).

The YK delta has substantially warmer summer temperatures and a potentially longer growing season than the arctic coast of Alaska (Sedinger, 1986; Webber, 1978). This results in greater thaw depths in soils of the coastal zone on the YK delta (Dennis et al., 1978; Kincheloe and Stehn, 1991) and probably greater rates of decomposition and primary productivity (Flanagan and Bunnell, 1980). These factors, combined with tidal flooding, should increase nutrient availability in wetlands on the YK delta, relative to the arctic.

Wetlands inland from the coastal zone of the YK delta probably experience lower primary productivity than those on the coast because they are larger and deeper, reducing the relative size of the emergent zone. Inland wetlands on the YK delta also are relatively hydrologically isolated, which reduces the availability of nutrients from outside the wetland. Soil thaw depth and corresponding decomposition are potentially retarded around these wetlands because of the preponderance of *Sphagnum*-influenced tundra (Miller et al., 1980).

Primary productivity has been measured in only a few wetlands in the boreal forest (Alexander and Barsdate, 1971, 1974; Heglund, 1988) and these measurements are restricted to phytoplankton. My assessment of primary productivity in the boreal forest will therefore be preliminary. Chlorophyll *a* (an index of phytoplankton standing crop) levels are highly variable among boreal forest wetlands, ranging from 0.4 to $262\,\mu g\,L^{-1}$ (Alexander and Barsdate, 1971, 1974; Heglund, 1988). Sixty-eight percent of the lakes studied on Yukon Flats had chlorophyll *a* levels $>7\,\mu g\,L^{-1}$ (Heglund, 1988), more than three times peak levels in tundra ponds at Barrow (Alexander et al., 1980).

Variation in hydrology, because of its effect on nutrient availability, probably has a strong influence on primary productivity. Minto lakes are flooded frequently during spring and summer by high water in the Chatanika River and Goldstream Creek (Shepherd and Mathews, 1985).

These lakes are characterized by extremely high primary productivity of emergent and submergent plants (although it has yet to be measured) and nutrient availability (Jacobs, 1992). It is likely that further studies will show that other hydrologically connected wetlands in the boreal forest are also highly productive.

Trophic Interactions

The study of trophic interactions in Alaskan wetlands is in its infancy. Flows of carbon between trophic levels are less than in temperate wetlands (Hobbie, 1980; MacLean, 1980), reflecting colder temperatures and shorter growing seasons in the Arctic. The same invertebrate guilds exist in the Arctic as at lower latitudes, but they have lower species diversity and lower biomass in the Arctic (Butler et al., 1980; MacLean, 1980). Detritivores, especially Chironomidae, dominate benthic macroinvertebrate communities in coastal ponds (Hobbie, 1984). Aquatic invertebrates have extended life histories in arctic wetlands (Butler et al., 1980; MacLean, 1980). Reduced secondary productivity results from lower algal production and lower rates of decomposition in the Arctic, relative to lower latitudes.

Densities of aquatic invertebrates in a productive boreal forest wetland were three times greater than in the Arctic but lower than in some temperate wetlands (Jacobs, 1992), suggesting that secondary productivity is higher in the boreal forest than in the Arctic. Taxonomic diversity was also higher in boreal forest wetlands (Jacobs, 1992). Detritivores (primarily Chironomidae) were the most abundant group of macroinvertebrates, but grazers were also an important component of the food web in one boreal forest wetland (Jacobs, 1992). Jacobs (1992) observed that invertebrate abundance was highest in stands of submergent vegetation, likely associated with higher levels of periphyton on submergent plants. More studies of productive boreal forest wetlands are required to assess the generality of this observation.

Waterfowl Foraging Ecology and Nutrient Dynamics in Alaska

Female ducks are in negative energy and sometimes protein balance during egg laying and incubation (Alisauskas and Ankney, 1992). The availability of high energy foods before nesting that allow deposition of lipids is, therefore, important for most species (Krapu and Reinecke, 1992). For long-distance migrants these foods must be available on breeding areas early in spring to allow restoration of lipid reserves depleted during migration. Species that depend on endogenous protein stores to produce eggs must have proteinacious foods available before nesting. Even for species that rely partially on endogenous protein reserves for egg production, however, the production of eggs depends on the total available protein from both exogenous and endogenous sources. Therefore, protein availability in the environment will influence productivity for most duck species.

In response to nutritional requirements for breeding, females of many species of ducks consume high energy foods before nesting and increase the proportion of invertebrates in the diet during egg production (Krapu and Reinecke, 1992). Data from high latitudes, while limited, are not completely consistent with this paradigm. Northern Pintails from three sites in Alaska—Minto Flats in the boreal forest (Burris, 1991), Selawik National Wildlife Refuge (DeBruyckere, 1988), and the Yukon Delta (Kirchoff, 1978)—all ate fewer invertebrates (28 to 58% of the diet) during the breeding season than expected based on studies in the temperate zone (Krapu and Reinecke, 1992). Substantial proportions of both Northern Pintails and American Wigeon have historically nested north of the midcontinent prairies (Bellrose, 1980). Thus, we do not know whether the differences between the diets of these species in the north and other species in the south result from differences in the timing and biomass of secondary productivity of invertebrates in temperate and high latitudes or differences in ecology among species of ducks.

One study has been completed for diving ducks (Lesser Scaup) that nest later in summer in the boreal forest near Yellowknife, NWT (Bartonek and Murdy, 1970). Diet was dominated by aquatic invertebrates and was similar to that for the same species in prairie wetlands (Bartonek and Hickey, 1969), suggesting that invertebrate foods were sufficiently abundant in late summer to allow Lesser Scaup to meet their nutritional requirements while concentrating on these foods.

Sea ducks, an important group in Alaska (see below), are poorly studied. One study of foraging ecology exists: Taylor's (1986) study of Oldsquaw Ducks (*Clangula hyemalis*) on the north slope of Alaska. Individuals ate exclusively aquatic invertebrates, although the diet varied in relation to foraging habitat and time of season. Oldsquaws fed opportunistically on blooms of particular foods (e.g., fairy shrimp). There was apparent competition with fish in deeper (>2 m deep) lakes, where Oldsquaws ate smaller size classes of invertebrates than in shallower lakes, probably because fish removed the larger size classes in deep lakes (Taylor, 1986).

Waterfowl Distribution and Habitat Use in Alaska

Waterfowl managers and ecologists interested in waterfowl have similar goals with respect to understanding ecological factors underlying abundance of waterfowl across Alaska. For managers, these factors allow identification of wetland habitats with high waterfowl values; for scientists, questions of the role of trophic interactions in regulating species distributions have long been of central importance.

It is currently possible to examine waterfowl distribution in Alaska at three spatial scales: the scale of the entire state (relative to the remainder of the continent), regions within the state, and individual wetlands within regions. Breeding pair surveys and other surveys conducted in waterfowl production areas by the U.S. Fish and Wildlife Service provide the best data on distribution at continental and statewide scales (Conant and Dau, 1991). More intensive, ground-based studies are necessary to elucidate patterns at the level of individual wetlands.

Alaska as a Waterfowl Breeding Area

Mean densities of ducks counted on breeding pair surveys are lower in Alaska than in the midcontinent prairie-parkland breeding areas (Pospahala et al., 1974), which was thought to reflect the lower productivity of Alaskan wetlands. Numerous areas within Alaska, however, support densities of breeding ducks comparable to those in the highly productive midcontinent prairie parklands (King, 1985). Minto Flats, near Fairbanks, contains higher densities of breeding ducks than lower latitude areas (Conant and Hodges, 1985). Yukon Flats, north of Fairbanks, supports densities of ducks comparable to midcontinent areas (King, 1985). Some areas on the YK delta in southwestern Alaska also contain higher densities of some species (e.g., Northern Pintails) than some midcontinent areas (personal observation).

Even for wetland areas of lower productivity, the sheer size of many of these areas makes them important for waterfowl production. Many investigators evaluate Alaskan waterfowl production areas from a perspective based on experience in temperate production areas. When we consider the size of Alaska we realize that many wetland areas in Alaska are huge by ordinary standards. For example, Yukon Flats and the YK delta represent areas 15 and 38% as large as the size of the entire state of North Dakota! Minto Flats is a tiny area by Alaskan standards yet it represents an area equivalent to 40% of all the remaining wetlands in California. Minto Flats exceeds the area of all but two of the National Wildlife Refuges in the lower 48 states, and exceeds the total area of National Wildlife Refuges in all but nine states. As a result of the magnitude of Alaskan wetlands and high waterfowl densities on some areas, Alaska supported substantial proportions of the North American breeding populations for several species in 1991. These include Norther Shovelers (21% of the breeding pair index was in Alaska), Canvasbacks (*Aythya valisineria*; 26% in Alaska), American Wigeon (28%), Green-Winged Teal (35%), and Northern Pintails (57%) (Conant and Dau, 1991).

Several other species of waterfowl nesting in Alaska are high latitude specialists that do one nest in temperate areas. These include Tundra Swans (*Cygnus columbianus*), Emperor Geese (*Anser canagicus*), White-Fronted Geese, several subspecies of Canada Geese, Black Brant, Greater Scaup, four species of eiders (*Somateria* spp. and *Polysticta stelleri*), and Oldsquaws. Harlequin Ducks (*Histrionicus histrionicus*) and the three species of scoters (*Melanitta* spp.) nest predominantly, but not exclusively, at higher latitudes (Bellrose, 1980). Alaska is an important, if not exclusive, breeding area for these species.

Variation Within Alaska

Breeding pair surveys divide Alaska into 11 sampling strata (Fig. 6.2) (Conant and Dau, 1991), which can be allocated to the three broad habitat categories: boreal forest, subarctic coastal tundra, and arctic coastal tundra. Six strata—Kenai–Susitna, Nelchina, Tanana–Kuskokwin, Yukon Flats, Innoko, and Koyukuk—are in the boreal forest (Table 6.1). Two strata are in subarctic tundra, YK delta and Bristol Bay; two strata, Seward Peninsula and Kotzebue Sound represent arctic tundra. The Copper River delta, a survey stratum not considered here, is a large alluvial, mostly unforested delta in the temperate rain forest zone. Few historical breeding pair data exist for the north slope of Alaska (King, 1979), although surveys have recently been expanded to that area (Brackney and King, 1993).

Densities of pairs of ducks (estimated as pairs, plus single males who are assumed to be attending incubating females) were comparable in the boreal

Figure 6.2 Location of U.S. Fish and Wildlife Service strata used for the waterfowl breeding pair survey (from Conant and Dau, 1991). Strata 7 (Copper River Delta) and 12 (Old Crow Flats, Yukon Territory) are not shown.

Table 6.1. Mean Densities of Breeding Ducks on USFWS Breeding Pair Surveys of Three Major Habitat Types in Alaska, 1989–1991

Species	Boreal Forest[a]	Subarctic Tundra[a]	Arctic Tundra[a]	North Slope[b]
	Number of Pairs km^{-2}			
Mallard	0.94	0.46	0.52	0.02
Gadwall	0	0.06	0	<0.01
Wigeon	1.70	0.57	1.68	0.02
G-W teal	1.03	1.02	0.75	0.02
B-W teal	0	0	0	0
Shoveler	0.88	0.42	0.85	0.01
Pintail	1.77	1.73	4.00	0.74
Redhead	0.02	0	0.02	0
Canvasback	0.21	0.01	0.13	0
Scaup	1.91	1.74	2.65	0.15
Ring necked	0.12	0.01	0.01	0
Goldeneye	0.24	0.08	0.02	0
Bufflehead	0.25	0.02	0.02	0
Oldsquaw	0.08	0.70	1.45	0.48
Eider	0	0.02	0.01	0.04
Scoter	0.25	1.02	0.59	0.05
Ruddy	<0.01	0	0	0
Merganser	0.02	0.04	0.05	<0.01
Total	9.42	7.90	12.75	1.53

[a] Data from Conant and Dau (1989, 1990, 1991). Arctic tundra in these surveys is on the Seward Peninsula and around Kotzebue Sound.
[b] Data from Brackney and King (1993).

forest (9.4 pairs km^{-2}) and subarctic tundra (7.9 pairs km^{-2}). The two strata in arctic tundra habitat from the long-term breeding pair surveys represent isolated patches of relatively high quality habitat, which support substantially higher densities of breeding ducks than surrounding areas. Recent surveys, covering a large proportion of Alaska's North Slope, show that most species nest there in densities one to two orders of magnitude lower than in boreal forest and subarctic tundra. Northern Pintails and Oldsquaws, which were 40 and 70% as dense on the North Slope as in subarctic tundra were exceptions to this general pattern. The YK delta is the largest stratum surveyed and includes large areas of relatively low nesting density. Coastal tundra segments of this stratum contain much higher densities of nesting ducks than indicated for the entire YK delta (personal observation). Several of the boreal forest strata are also large and heterogeneous with respect to their value for waterfowl. For example, Minto Flats, which represents about 25% of the Tanana–Kuskokwim stratum, contains densities of nesting waterfowl comparable to those in the midcontinent prairies (Conant and Hodges, 1985). Yukon Flats, the largest continuous expanse of waterfowl habitat in the boreal forest, supports the highest total densities of breeding ducks of any of the strata (Table 6.1).

Generally, areas supporting high densities of nesting ducks are hydrologically open systems of intermediate energy. Yukon Flats and Minto Flats are both river basins that are regularly flooded during spring runoff or heavy summer rains; the coastal portion of the YK delta is flooded by storm surges.

Despite variation among strata in total numbers of nesting ducks, it is possible to discern variation among the three major habitat types in nesting densities of different species. Similar patterns to those reported here were described by Lensink and Derksen (1990). Among dabbling ducks, Mallards were most abundant in the boreal forest (Table 6.1), whereas Wigeon and Northern Shovelers were least abundant in subarctic tundra. Over large areas, Northern Pintails were most abundant in subarctic tundra and the boreal forest, although the north slope owing to its size, is an important breeding area for Northern Pintails. Scaup were approximately equally abundant in all three habitats but Lesser Scaup predominated in the boreal forest (>90% of the scaup), while Greater Scaup were the dominant Scaup in the tundra. Not surprisingly, species of the genus *Bucephala* (Goldeneyes and Buffleheads), which are cavity nesters, occurred at the highest densities in the boreal forest. Scoters, Eiders, and Oldsquaws occurred primarily in tundra areas, although White-Winged Scoters (*M. fusca*) were restricted to forested areas.

Consistent with patterns for other waterfowl, Tundra Swans (*C. columbianus*), which are tundra specialists, are four to six times as dense on the YK delta as in other tundra habitats in Alaska (Conant and Cain, 1987). Trumpeter Swans nest at highest density in Minto Flats, in the vicinity of Gulkana and on the Copper River delta (Groves et al., 1991).

Local Variation in Density

Consistent with expectations from temperate zone studies, ducks preferred arctic wetlands with a substantial emergent zone (class II to IV wetlands; Bergman et al., 1977; Derksen et al., 1981). Ducks likely used these wetlands because they contained higher food densities (see above) and appropriate water depths for foraging.

Studies on Yukon Flats and in the Innoko River area found that duck broods used wetlands with higher concentrations of inorganic nutrients and higher productivity of phytoplankton, as evidenced by higher chlorophyll *a* concentrations (Heglund, 1988; Seppi, 1993). Broods near the Alaska–Yukon border also use wetlands hydrologically connected to streams, which explains their higher nutrient concentrations (Murphy et al., 1984). On Yukon flats, soil type was also associated with the availability of some nutrients (Heglund, 1988). Also, on Yukon Flats broods use wetlands more frequently if they support substantial areas of emergent vegetation (Heglund, 1988). These patterns are generally being confirmed in two other studies, neun the Innoko River (Seppi, 1993) and on Minto Flats, where apparent extremely high primary productivity is associated with high densities of nesting ducks there.

Breeding geese achieve their highest densities in, or near, coastal salt marshes because such areas provide the highest concentrations of nutritious plant foods necessary for growth of goslings (Sedinger and Flint, 1991; Sedinger and Raveling, 1984). On the YK delta, which supports high densities of four species of geese, densities are highest in the coastal zone of the central delta, associated with extensive areas of salt marsh in this region relative to other parts of the delta.

Large numbers of waterfowl molt in Alaska. During drought in midcontinent breeding areas, large numbers, especially Northern Pintails, move to high latitude wetlands to molt (Derksen and Eldridge, 1980; Henny, 1973). Molting Northern Pintails on the North Slope prefer deep *Arctophila* wetlands, probably because such wetlands provide both food and cover (Derksen et al., 1981). Large flocks of molting diving and dabbling ducks form on large boreal forest lakes (e.g., Yokum, 1970) because these large lakes provide protection from predators and provide adequate food. Molting geese are also associated with larger and deeper water bodies (lakes or rivers) than is true of other phases of their annual cycle (e.g., Derksen et al., 1981).

Waterfowl Distribution and Marsh Ecology in Alaska

Waterfowl densities are associated directly with nutrient availability at the local level and with hydrological variables likely to affect nutrient availability at the regional scale. Nutrient availability probably regulates primary productivity, particularly of phytoplankton, but this requires confirmation at a number of sites. Productivity of macrophytes is poorly known (but see Alexander et al., 1980), except in the semiterrestrial sites used by geese. The magnitude of secondary productivity and factors influencing this variable are also generally unknown. Finally, only a handful of studies of the foods of waterfowl have been conducted in Alaska. Because such studies tell us how waterfowl use lower trophic levels, they are critical to our understanding of waterfowl and wetland ecology.

If the goal of scientists and managers is to understand factors regulating the distribution and productivity of waterfowl, and I believe it is, then a series of studies in representative wetlands must be conducted. The relationships among hydrology, inorganic nutrients, and use by waterfowl is becoming well established in the boreal forest wetlands. Similar studies are needed in subarctic and arctic tundra wetlands to complete our understanding of these relationships throughout the state. Observational and experimental studies of the relationship between primary production of both macrophytes and phytoplankton and availability of inorganic nutrients are necessary. Finally, we do not know the extent to which the "marsh model"

(Nelson and Kadlec, 1984) developed for temperate wetlands is applicable to Alaska. One study of the relationship between macrophytic vegetation and aquatic invertebrates has been completed (Jacobs, 1992). More studies of the role of macrphytes as food (to shredders), as a substrate for food (for grazers), and as a regulator of dissolved O_2 are needed.

Remote sensing, based primarily on satellite images, of wetlands has been completed for some federally owned lands (e.g., Kempka et al., 1993). Although remote sensing is an important tool, it is unclear whether it can identify features of importance to waterfowl, thereby predicting which wetlands waterfowl will prefer. Such imaging studies must be closely linked to trophic studies and studies of geomorphology and vegetation surrounding wetlands to develop predictive models of wetland importance to waterfowl based on remote images. In summary, the few studies of waterfowl ecology in Alaska have produced tantalizing correlations. To achieve the goal of identifying wetlands of potential importance to waterfowl, a more coordinated approach over a range of habitats is needed.

Acknowledgments. I wish to thank a dedicated group of survey biologists from the U.S. Fish and Wildlife Service who have provided much of what we know about waterfowl distribution in Alaska. B. Conant and R. King kindly granted permission to include unpublished survey data and I thank R. Post, J. LaPerriere, and F. Reid for reviewing an earlier draft.

References

Afton AD, Ankney CD (1991) Nutrient reserve dynamics of breeding Lesser Scaup: A test of competing hypotheses. Condor 93:89–97.

Afton AD, Paulus SL (1992) Incubation and brood care in waterfowl. In Batt BDJ, Afton AD, Anderson MG, et al. (Eds) Ecology and management of breeding waterfowl. University of Minnesota Press, St. Paul, MN, 62–108.

Aldrich TW, Raveling DG (1983) Effects of experience and body weight on incubation behavior of Canada Geese. Auk 100:670–679.

Alexander V, Barsdate RJ (1971) Physical limnology, chemistry and plant productivity of a taiga lake. Int Rev Ges Hydrobiol 56:825–872.

Alexander V, Barsdate RJ (1974) Limnological studies of a subarctic lake system. Int Rev Ges Hydrobiol 59:737–753.

Alexander V, Stanley DW, Daley RJ, McRoy CP (1980) Primary producers. In Hobbie JE (Ed) Limnology of tundra ponds. Dowden, Hutchinson and Ross, Stroudsburg, PA, 179–250.

Alisauskas RT, Ankney CD (1992) The cost of egg-laying and its relation to nutrient reserves in waterfowl. In Batt BDJ, Afton AD, Anderson, MG, et al. (Eds) Ecology and management of breeding waterfowl. University of Minnesota Press, St. Paul, MN, 30–61.

Ankney CD (1984) Nutrient reserve dynamics of breeding and molting Brant. Auk 101:361–370.

Ankney CD, Afton AD (1988) Bioenergetics of breeding Northern Shovelers: Diet, nutrient reserves, clutch size, and incubation. Condor 90:459–472.

Ankney CD, Alisauskas RT (1991) Nutrient-reserve dynamics and diet of breeding female Gadwalls. Condor 93:799–810.

Ankney CD, MacInnes CD (1978) Nutrient reserves and reproductive performance of female Lesser Snow Geese. Auk 95:459–471.

Anonymous (1973) Major ecosystems of Alaska. Joint Federal-State Land Use Planning Commission for Alaska. Anchorage, AK.

Bartonek JC, Hickey JJ (1969) Food habits of Canvasbacks, Redheads, and Lesser Scaup in Manitoba. Condor 71:280–290.

Bartonek JC, Murdy HW (1970) Summer foods Lesser Scaup in subarctic taiga. Arctic 28:35–44.

Batt BDJ, Prince HH (1979) Laying dates, clutch size and egg weight of captive Mallards. Condor 81:35–41.

Bazely DR, Jefferies RL (1985) Goose feces: A source of nitrogen for plant growth in a grazed salt marsh. J Appl Ecol 22:693–703.

Bellrose FC (1980) Ducks, geese and swans of North America, 3rd ed. Stackpole, Harrisburg, PA.

Bergman RD (1973) Use of southern boreal lakes by postbreeding Canvasbacks and Redheads. J Wildl Manage 37:160–170.

Bergman RD, Howard RL, Abraham KF, Weller MW (1977) Water birds and their wetland resources in relation to oil development at Storkersen Point, Alaska. U.S. Fish and Wildlife Service Resource Publ. 129. Washington, DC.

Bortner JB, Johnson FA, Smith GW, Trost RE (1991) Status of waterfowl and fall flight forecast. Office of Migratory Bird Management, U.S. Fish and Wildlife Service, Washington, DC.

Brackney AW, King RJ (1993) Aerial breeding pair surveys of the arctic coastal plain of Alaska: Revised estimates of waterbird abundance 1986–1992. Unpublished Report. U.S. Fish and Wildlife Service, Fairbanks, AK.

Budeau DA, Ratti JT, Ely CR (1991) Energy dynamics, foraging ecology, and behavior of prenesting Greater White-Fronted Geese. J Wildl Manage 55: 556–563.

Burris FA (1991) Foods and foraging ecology of Northern Pintails breeding in interior Alaska. Unpublished M.S. thesis, University of Alaska, Fairbanks.

Butler M, Miller MC, Mozler S (1980) Macrobenthos. In Hobbie JE (Ed) Limnology of tundra ponds. Dowden, Hutchinson and Ross, Stroudsburg, PA, 297–339.

Cargill SM, Jefferies RL (1984) Nutrient limitations of primary production in a subarctic salt marsh. J Appl Ecol 21:657–668.

Chapin FS III (1978) Phosphate uptake and nutrient utilization by Barrow tundra vegetation. In Tieszen LL (Ed) Vegetation and production ecology of an Alaskan arctic tundra. Springer–Verlag, New York, 483–507.

Chapin FS III, Tieszen LL, Lewis MC, Miller PC, McLown BH (1980) Control of tundra plant allocation patterns and growth. In Brown J, Miller PC, Tieszen LL, Bunnell FL (Eds) An arctic ecosystem the coastal tundra at Barrow, Alaska. Dowden, Hutchinson and Ross, Stroudsburg, PA, 14–185.

Coleman TS, Boag DA (1987) Foraging characteristics of Canada Geese on the Nisutlin River Delta, Yukon. Can J Zool 65:2358–2361.

Conant B, Cain SL (1987) Alaska Tundra Swan status report. Unpublished Report. U.S. Fish and Wildlife Service, Juneau, AK.

Conant B, Dau CP (1991) Alaska–Yukon waterfowl breeding population survey. Unpublished Report. U.S. Fish and Wildlife Service, Juneau, AK.

Conant B, Hodges JI (1985) Minto Flats waterfowl resources. Unpublished Report U.S. Fish and Wildlife Service, Juneau, AK.

Cooch EG, Lank DB, Dzubin A, Rockwell RF, Cooke F (1991) Body size variation in Lesser Snow Geese: Environmental plasticity in gosling growth rates. Ecology 72:503–512.

Cooke F, Findlay CS, Rockwell RF (1984) Recruitment and the timing of reproduction in Lesser Snow Geese (*Chen caerulescens caerulescens*). Auk 101:451–458.

Cowardin LM, Carter V, Golet FC, LaRoe ET (1979) Classification of wetlands and deep water habitats of the United States. USFWS/OBS Publ. 79/31 Washington, DC.

DeBruyckere LA (1988) Feeding ecology of Northern Pintails (*Anas acuta*), American Wigeon (*Anas americana*), and Long-Billed Dowitchers (*Limnodromus scolopaceus*) on Selawik National Wildlife Refuge, Alaska. Unpublished M.S. thesis, University of Maine, Orono, ME.

Dennis JG, Tieszen LL, Vetter MA (1978) Seasonal dynamics of above and below ground production of vascular plants at Barrow, Alaska. In Tieszen LL (Ed) Vegetation and production ecology of an Alaskan arctic tundra. Springer–Verlag, New York, 113–140.

Derksen DV, Eldridge WD (1980) Drought displacement of Pintails to the Arctic coastal plain, Alaska, J Wildl Manage 44:224–229.

Derksen DV, Rothe TC, Eldridge WD (1981) Use of wetland habitats by birds in the National Petroleum Reserve–Alaska. U.S. Fish and Wildlife Service Resources Published 141. Washington, DC.

DuBowy PJ (1985) Feeding ecology and behavior of postbreeding male Blue-Winged Teal and Northern Shovelers. Can J Zool 63:1292–1297.

Duncan DC (1987) Nesting of Northern Pintails in Alberta: Laying date, clutch size, and renesting. Can J Zool 65:234–246.

Dupre WR (1980) Yukon delta coastal processes study. National Oceanic and Atmospheric Administration, Outer Continental Shelf Environmental Assessment Program 1988, Final Report 58. Washington, DC, 393–447.

Flanagan PW, Bunnell FL (1980) Microflora activities and deconmposition. In Brown J, Miller PC, Tieszen LL, Bunnell FL (Eds) An arctic ecosystem the coastal tundra at Barrow Alaska, Dowden, Hutchinson and Ross, Stroudsburg, PA, 291–334.

Ford J, Bedford BL (1987) The hydrology of Alaskan Wetlands, U.S.A.: A review. Arct Alpine Res 19:209–229.

Fredrickson LH, Reid FA (1990) Impacts of hydrologic alteration on management of freshwater wetlands. In Sweeney JM (Ed) Management of dynamic ecosystems. North Central Section Wildlife Society, West Lafayette, IN, 71–90.

Gauthier G, Tardif J (1991) Female feeding and male vigilance during nesting in Greater Snow Geese. Condor 93:701–711.

Gloutney ML, Clark RG (1991) The significance of body mass to female dabbling ducks during late incubation. Condor 93:811–816.

Groves DJ, Conant B, Hodges JI (1991) A summary of Alaska Trumpeter Swan surveys 1991. Unpublished Report. U.S. Fish and Wildlife Service, Juneau, AK.

Hanson HC (1962) The dynamics of condition factors in Canada Geese and their relation to seasonal stresses. Arctic Inst. N. Amer. Tech. Paper No. 12.

Harvey JM (1971) Factors affecting Blue Goose nesting success. Can J Zool 49: 223–224.

Heglund PJ (1988) Relations between water bird use and the limnological characteristics of wetlands on Yukon Flats National Wildlife Refuge, Alaska. Unpublished M.S. thesis, University of Missouri, Columbia.

Heitmeyer ME (1988) Protein cost of the prebasic molt of female Mallards. Condor 90:263–266.

Henny CJ (1973) Drought displaced movement of North American Pintails into Siberia. J Wildl Manage 37:23–29.

Hobbie JE (1980) Introduction and site description. In Hobbie JE (Ed) Limnology of tundra ponds. Dowden, Hutchinson and Ross, Stroudsburg; PA, 19–50.

Hobbie JE (1984) The ecology of tundra ponds of the arctic coastal plain: A community profile. U.S. Fish and Wildl. Service FWS/OBS-83125. Washington, DC.

Hopkins DM (1949) Thaw lakes and thaw sinks in the Imurik Lake area Seward Peninsula, Alaska. J Geol 57:119–131.

Hulten E (1968) Flora of Alaska and neighboring territories. Stanford University Press, Stanford, CA.

Hunter ML Jr, Jones JJ, Gibbs KE, Moring JR (1986) Duckling responses to lake acidification: Do Black Ducks and fish compete? Oikos 47:26–32.

Hunter RO (1980) Effects of grazing on the quantity and quality of freshwater aufwuchs. Hydrobiology 69:251–259.

Jacobs LL (1992) Aquatic ecology of two subarctic lakes: Big and Little Minto Lakes, Alaska Unpublished M.S. thesis, University of Alaska, Fairbanks.

Kafka DM (1988) Use of beaver-influenced wetlands by waterfowl on the Kanuti National Wildlife Refuge. Unpublished M.S. thesis, University of Alaska, Fairbanks.

Kairesdo T (1980) Diurnal fluctuations within a littoral plankton community in oligotrophic Lake Paajarvi, southern Finland, Freshwater Biol 10:533–537.

Kaminski RM, Prince HH (1981) Dabbling duck activity and foraging responses to aquatic macroinvertebrates. Auk 98:115–126.

Kempka RG, Reid FA, Altop RC (1993) Developing large regional databases for waterfowl habitat in Alaska using satellite inventory techniques: A case study of the Black River. GIS 93 Symp. Vancouver, BC, 581–191.

Kincheloe KL, Stehn RA (1991) Vegetation patterns and environmental gradients in coastal meadows on the Yukon–Kuskokwim delta, Alaska. Can J Bot 69:1616–1627.

King JG (1985) Evaluating Alaska habitats for ducks. Unpublished U.S. Fish and Wildlife Service Report. Juneav, AK.

King JG, Lensink CJ (1971) An evaluation of Alaskan habitat for migratory birds. Unpublished Report, Department of the Interior Bureau of Sport Fisheries and Wildlife, Washington, DC.

King R (1979) Results of aerial surveys of migratory birds on NPR-A in 1977 and 1978. Studies of selected wildlife and fish and their use of habitats on and adjacent to the National Petroleum Reserve in Alaska 1977–1978. Vol 1. U.S. Dept of the Int. National Petroleum Res in Alaska 105 (c) land use study. Anchorage, AK, 187–226.

Kirchoff MD (1978) Distribution and habitat relations of Pintails on the coast of the Yukon delta, Alaska. Unpublished M.S. thesis, University of Maine, Orono, ME.

Krapu GL, Reinecke KJ (1992) Foraging ecology and nutrition. In Batt BDJ, Afton AD, Anderson MG, et al. (Eds) Ecology and management of breeding waterfowl. University of Minnesota Press, St. Paul, MN, 1–29.

Krapu GL, Swanson GA (1975) Foods of juvenile, broodhen and post-breeding Pintails in North Dakota. Condor 79:504–507.

Krull JN (1970) Aquatic plant–macroinvertebrate associations and waterfowl. J Wildl Manage 34:707–718.

LaBaugh JW (1989) Chemical characteristics of water in northern prairie wetlands. In Van der Valk A (Ed) Northern prairie wetlands. Iowa State University Press, Ames, IA, 56–90.

Larsson K, Forslund P (1991) Environmentally induced morphological variation in the Barnacle Goose, Branta leucopsis. J Evol Biol 4:619–636.

Lensink CJ (1965) Waterfowl and waterfowl habitats of the Yukon Flats Unpublished Report. Bureau of Sport Fisheries and Wildlife, Bethel, AK.

Lensink CJ, Derksen DV (1990) Evaluation of Alaskan Wetlands for Waterfowl. Proc Workshop Alaska: Regional wetland functions. Anchorage, AK, 45–84.

Lynch J (1984) Escape from mediocrity: A new approach to American waterfowl hunting regulations. Wildfowl 35:5–13.

Maciolek JA (1989) Tundra ponds of the Yukon delta, Alaska, and their macroinvertebrate communities. Hydrobiology 172:193–206.

MacLean SF Jr (1980) The detritus-based trophic system. In Brown J, Miller PC, Tieszen LL, Bunnell FL (Eds) An arctic ecosystem the coastal tundra at Barrow, Alaska. Dowden, Hutchinson and Ross, Stroudsburg, PA, 411–457.

Mann FE, Sedinger JS (1993) Nutrient reserve dynamics and control of clutch size in Northern Pintails breeding in Alaska. Auk 110:264–278.

McKendrick JD, Ott VJ, Mitchell GA (1978) Effects of nitrogen and phosphorus fertilization on carbohydrate and nutrient levels *Dupontia fisheri* and *Arctagrostis latifolia*. In Tieszen LL (Ed) Vegetation and production ecology of an Alaskan arctic tundra. Springer–Verlag, New York, 509–537.

McNaughton SJ (1966) Ecotype function in the *Typha* community type. Ecol Monogr 36:297–325.

Mickelson PG (1975) Breeding biology of cackling geese and associated species on the Yukon–Kuskokwim delta, Alaska. Wildl Monogr 45.

Miller MC, Prentki RT, Barsdate RJ (1980) Physics. In Hobbie JE (Ed) Limnology of tundra ponds. Dowden, Hutchinson and Ross, Stroudsburg, PA, 51–75.

Miller PC, Webber PJ, Oechel WC, Tieszen LL (1980) Biophysical processes and primary production. In Brown J, Miller PC, Tieszen LL, Bunnell FL (Eds) An arctic ecosystem the coastal tundra at Barrow, Alaska. Dowden, Hutchinson an Ross, Stroudsburg, PA, 66–101.

Mitsch WJ, Gosselink JG (1986) Wetlands. Van Nostrand Reinhold, New York.

Murkin HR (1989) The basis for food chains in prairie wetlands. In van der Valk A (Ed) Northern prairie wetlands. Iowa State University Press, Ames, IA, 316–338.

Murkin HR, Kaminski RM, Titman RD (1982) Responses by dabbling ducks and aquatic invertebrates to an experimentally manipulated cattail marsh, Can J Zool 60:2324–2332.

Murphy SM, Kessel B, Vining LJ (1984) Waterfowl population and limnologic characteristics of taiga ponds. J Wildl Manage 48:1156–1163.

Neely RK, Davis CB (1985) Nitrogen and phosphorus fertilization of *Spargamium eurycarpum* Engelm and *Typha glauca* Godr. stands. I: Emergent plant production. Aquat Bot 22:347–361.

Nelson JW, Kadlec JA (1984) A conceptual approach to relating habitat structure and macroinvertebrate production in freshwater wetlands. Trans N Am Wildl Natur Res Conf 49:262–270.

Norton D, Weller G (1984) The Beaufort Sea: Background, history and perspective. In Barnes PW, Schell DM, Reimnitz E (Eds) The Alaskan Beaufort Sea: Ecosystems and environments. Academic Press, Orlando, FL, 3–9.

Owen M (1980) Wild geese of the world. Bastsford, Ltd., London.

Owen M, Black JM (1989) Factors affecting the survival of Barnacle Geese on migration from the breeding grounds. J Anim Ecol 58:603–617.

Pehrsson O, Nystrom KGK (1988) Growth and movements of Oldsquaw ducklings in relation to food. J Wildl Manage 52:185–191.

Petrula MJ, MacCluskie MC, Sedinger JS (1992) Ecology of ducks nesting in interior Alaska. Unpublished Report. USFWS-ACWRU RWO No. 39. University of Alaska, Fairbanks, AK.

Pospahala RS, Anderson DR, Henny CJ (1974) Population ecology of the Mallard. II. Breeding populations, and production indices. Resource Published 115. U.S. Fish and Wildlife Service. Washington, DC.

Prentki RT, Miller MC, Barsdate RJ, Alexander V, Kelley J, Coyne P (1980) Chemistry. In Hobbie JE (Ed) Limnology of tundra ponds. Dowden, Hutchinson and Ross, Stroudsburg, PA, 76–178.

Prevett JP, Marshall IF, Thomas VG (1979) Fall foods of Lesser Snow Geese in the James Bay region. J Wildl Manage 43:736–742.

Raveling DG (1979a) The annual cycle of body composition of Canada Geese with special reference to control of reproduction. Auk 96:234–252.

Raveling DG (1979b) The annual energy cycle of the Cackling Canada Goose. In Jarvis RL, and Bartonek JC (Eds) Management and biology of Pacific flyway geese. OSU Book Stores, Inc. Corvallis, OR, 81–93.

Rohwer FC (1984) Patterns of egg laying in prairie ducks. Auk 101:603–605.

Sedinger JS (1986) The status and biology of geese nesting at Tutakoke, 1986, a progress report. Unpublished Report to U.S. Fish and Wildlife Service. Fairbanks, AK.

Sedinger JS (1992) Ecology of prefledging waterfowl In Batt BDJ, Afton AD, Anderson MG, et al. (Eds) Ecology and management of breeding waterfowl. University of Minnesota Press, St. Paul, MN, 109–127.

Sedinger JS, Bollinger KS (1987) Autumn staging of Cackling Geese on the Alaska Peninsula. Wildfowl 38:13–18.

Sedinger JS, Flint PL (1991) Growth rate is negatively correlated with hatch date in Black Brant. Ecology 72:496–502.

Sedinger JS, Raveling DG (1984) Dietary selectivity in relation to availability and quality of food for goslings of Cackling Geese. Auk 101:295–306.

Sedinger JS, Raveling DG (1986) Timing of nesting by Canada Geese in relation to the phenology and availability of their food plants. J Anim ecol 55:1083–1102.

Seppi B (1993) Use of wetlands by waterfowl broods in relation to habitat quality in the lower Innoko River area, Alaska. Unpublished M.S. thesis, University of Alaska, Fairbanks.

Shepherd PEK, Mathews MP (1985) Impacts of resource development and use on the Athabaskan people of Minto Flats, Alaska. Bureau of Indians Affairs, Washington, DC.

Sloan CE (1972) Ground-water hydrology of prairie potholes in North Dakota. U.S. Geological Survey Prof. Paper 585–c. Washington, DC.

Smith RH, Dufresne F, Hansen HA (1964) Northern watersheds and deltas. In Linduska JP (Ed) Waterfowl tomorrow. U.S. Dept of the Interior, Washington, DC.

Stross RG, Miller MC, Daley RJ (1980) Zooplankton. In Hobbie JE (Ed) Limnology of tundra ponds. Dowden, Hutchinson and Ross, Stroudsburg, PA, 251–296.

Sugden LG (1973) Feeding ecology of Pintail, Gadwall, American Wigeon and Lesser Scaup ducklings. Can Wildl Service Sci Rep No. 24. Ottawa, Canada.

Swanson GA, Krapu GL, Serie JR (1979) Foods of laying female dabblings ducks on the breeding grounds. In Bookhout TA (Ed) Waterfowl and wetlands an integrated review. The Wildlife Society, Madison, WI, 47–57.

Tande GF, Jennings TW (1986) Classification and mapping of tundra near Hazen Bay, Yukon Delta National Wildlife Refuge, Alaska. Unpublished Report, U.S. Fish and Wildlife Service. Anchorage, AK.

Taylor EJ (1986) Foods and foraging ecology of Oldsquaws (Clangula hyemalis L.) on the arctic coastal plain of Alaska. Unpublished M.S. thesis, University of Alaska, Fairbanks.

Thomas VG (1983) Spring migration: The prelude to goose reproduction and a review of its implications. In Boyd H (Ed) First western hemisphere waterfowl and waterbird symposium. Canadian Wildlife Service, Ottawa, ON, 73–81.

Thomas VG, Prevett JP (1980) The nutritional value of arrowgrasses to geese at James Bay. J Wildl Manage 44:830–836.

Thompson SC, Raveling DG (1987) Incubation behavior of Emperor Geese compared with other geese: Interactions of predation, body size and energetics. Auk 104:707–716.

Tieszen LL (1978) Summary. In Tieszen LL (Ed) Vegetation and production ecology of an Alaskan arctic tundra. Springer–Verlag, New York, 621–645.

Tiner RW Jr (1984) Wetlands of the United States: Current status and recent trends. U.S. Fish and Wildlife Service, Washington, DC.

U.S. Department of Commerce (1987) Tidal current tables. Pacific coast of North America and Asia. National Oceanic and Atmospheric Administration, National Ocean Survey. Washington, DC.

Van der Valk AG, Davis CB (1978) Primary production of prairie glacial marshes. In Good RE, Whigham DF, Simpson RL (Eds) Freshwater wetlands: Ecological processes and management potential. Academic Press, New York, 21–37.

Voigts DK (1976) Aquatic invertebrate abundance in relation to changing marsh vegetation. Am Midland Naturalist 93:313–322.

Ward DH, Stehn RA (1989) Response of Brant and other geese to aircraft disturbance at Izembek Lagoon, Alaska. Final Report No. 14-12-0001-30332. U.S. Fish and Wildlife Service, Anchorage, AK.

Webber PJ (1978) Spatial and temporal variation of the vegetation and its productivity. In Tieszen LL (Ed) Vegetation and production ecology of an Alaskan arctic tundra. Springer–Verlag, New York, 37–112.

Wellein EG, Lumsden HG (1964) Northern forests and tundra. In Linduska JP (Ed) Waterfowl tomorrow. U.S. Dept of the Interior, Washington, DC, 67–76.

Weller MW, Fredrickson LH (1973) Avian ecology of a managed glacial marsh. Living Bird 12:269–291.

White DA, Weiss TE, Trapani JM, Thien LB (1978) Productivity and decomposition of the dominant salt marsh plants in Louisiana. Ecology 3:122–131.

Wypkema PDP, Ankney CD (1979) Nutrient reserve dynamics of Lesser Snow Geese staging at James Bay, Ontario. Can J Zool 57:213–219.

Yokum CF (1970) Weights of ten species of ducks captured at Ohtig Lake, Alaska– Aug 1962. Murrelet 51:21.

Zedler JB (1980) Algae mat productivity: Comparisons in salt marsh. Estuaries 3:122–131.

7. The Effect of Salmon Carcasses on Alaskan Freshwaters

Thomas C. Kline Jr., John J. Goering,
and Robert J. Piorkowski

Many Alaskan freshwaters provide important spawning and nursery habitat for salmonid fishes. Pacific salmon are well known for their anadromous and semelparous natural history of rearing in the marine environment and returning to freshwater as adults to spawn once before dying in their natal habitat. Five species of anadromous Pacific salmon, *Oncorhynchus nerka* (sockeye or red salmon), *O. kisutch* (coho or silver salmon), *O. gorbuscha* (pink or humpback salmon), *O. keta* (chum or dog salmon), and *O. tshawytscha* (chinook or king salmon) spawn in Alaskan freshwaters. The time juvenile salmon reside in freshwater following emergence from the gravel as fry until smoltification (physiological preparation for migration to saltwater) depends on species and location. Because freshwater residence can range from virtually no time to several years, considerable variation in dependence on the freshwater habitat as a nursery environment exists. The sockeye salmon is the only Pacific salmon to have a juvenile stage that is usually dependent on a lacustrine habitat and a forage base of zooplankton. Because lakes used for rearing by juvenile salmon are typically oligotrophic, the productivity of sockeye lakes has been studied as a factor limiting sizes of salmon runs (see Chapter 8; and Burgner et al., 1969; Hyatt and Stockner, 1985; Stockner, 1981, 1987).

Limnological investigations of the freshwater life history stages of sockeye salmon, undertaken primarily in the interest of better fisheries management (e.g., Burgner et al., 1969), have resulted in a body of knowledge relating characteristics of the freshwater ecosystem to sockeye productivity. A characteristic of many sockeye nursery lakes is that abundance of forage for sockeye limits production (Koenings and Burkett, 1987a,b; Chapter 8). The sockeye forage base ultimately depends on nutrient availability as well as physical factors regulating primary productivity (reviewed by Koenings and Burkett, 1987a). It has been surmised that nutrients released during the decomposition of spawned-out salmon carcasses are important for the maintenance of adequate nutrient levels in nursery lakes, thus declining stocks may be partly due to reduced nutrient loading because of low escapements (the portion of the spawning population that is not harvested by humans) (Barnaby, 1944; Mathisen, 1972; Koenings and Burkett, 1987a). An important aspect of the limnology of sockeye lakes has been assessing the significance of the salmon carcasses to nutrient pools by the use of a mass-balance approach that is essentially an economic analysis of gains and losses of nutrients vital to the system. Donaldson (1967) determined, through mass-balance calculations, that >50% of the annual phosphorus (P) budget to Iliamna Lake (the largest freshwater body in Alaska) was biogenic P when the escapement was 25 million sockeye. Mass-balance analysis is limited by its whole-lake spatial and per year temporal scales. The mass-balance sampling unit is the marine contribution for the whole lake for that year. Thus there can be only one datum per year per system investigated this way. Thus the mass-balance approach does not reveal intra-annual and spatial variability of the biogenic (i.e., marine) nutrient contribution (vs. other sources) or how biogenic nutrients are utilized by freshwater food webs. Knowledge of the pathways of biogenic nutrients is important for the effective implementation of large-scale fertilization programs used experimentally as a remedy for nutrient deprivation to increase sockeye production (see Chapter 8; Hyatt and Stockner, 1985). Also, interdependency of other organisms on Pacific salmon cannot be determined by a mass balance approach alone. Other interactions between Pacific salmon and resident freshwater fishes, mammals, and birds include scavenging of wasted eggs and carcasses, predation of adult and juvenile salmon, and competition for resources in the freshwater habitat.

This chapter reviews a new approach first, outlined by Kline (1991), using the natural abundance of stable isotopes to study the biogenic fertilization effects of Pacific salmon on their freshwater habitats. The technique makes it possible to obtain a quantitative estimate of the relative contribution of marine-derived nitrogen (MDN) to total nitrogen for a sample of organic matter. Sampling units can be as small as a single fish and therefore stable isotope data can be used to estimate population statistics on the relative importance of marine-derived nutrients. Investigations on the limitation of the freshwater habitat on Pacific salmon production has mainly addressed sockeye salmon lakes. The effects of run size and the potential role played by biogenic nutrients in these studies are reviewed in the following two sections. The spawning migration also provides a direct food source for consumers, as discussed in the third section. Stable isotope methodology and its application to the question of the significance of biogenic nutrients is covered at the end of this chapter.

Anadromous Salmon and Nutrient Limitation Theory in Freshwaters

The evolution of anadromy by Pacific salmon has been hypothesized as a response to the relatively poor productivity of freshwaters at high latitudes in comparison to the marine environment (Gross, 1987; Gross et al., 1988). The 23 and 13 nursery lakes for Alaskan sockeye salmon surveyed by Burgner et al. (1969) and Koenings and Burkett (1987a), respectively, were classified as oligotrophic. The biotic and abiotic factors limiting sockeye salmon productivity have been synthesized into a classification scheme by Koenings and Burkett (1987a; see also Chapter 8), where freshwater production of sockeye salmon is either recruitment limited (type A, limited by mortality before, accessibility to, or area available for spawning) or rearing limited (type B, limited by forage or environment). Because the major sockeye salmon systems of southwestern Alaska are not restricted in terms of accessibility to and availability of spawning grounds, they are not naturally recruitment limited. The number and age of sockeye that are recruited to the fishery is primarily determined by environmental conditions during the portion of salmon life history from spawning through the juvenile growth phase prior to smoltification (when the young sockeye first enter the marine environment) (Burgner et al., 1969; Eggers and Rogers, 1987). Because mortality during the freshwater stage of the life history is both variable and high, run size (a given cohort of salmon) is largely established before Pacific salmon reach the sea (Burgner et al., 1969).

Recruitment limitation is less likely where spawning runs are large (e.g., Fig. 7.1). Koenings and Burkett (1987a) found that size of sockeye salmon smolts was dependent on the type of limitation imposed by nursery lakes. Lakes with natural runs that were recruitment limited produced large smolts (>2 g, >60 to 65 mm, i.e., >threshold size for smoltification) in one growing season. Lakes that normally did not have sockeye populations (inaccessible, therefore also recruitment limited) produced large smolts when initially stocked at low population density (Koenings and Burkett, 1987a). Increasing the stocking density resulted in smaller and older smolts (i.e., slower growth) with concurrent reduction of zooplankton in the size range important as sockeye forage, suggesting incipient rearing limitation. This condition can be alleviated by artificial fertilization (see Chapter 8), demonstrating that nutrient availability is very important for sustaining a forage base for sockeye salmon production.

Figure 7.1. An aerial photograph showing typical densities of spawning (in upper section) and schooling sockeye salmon on spawning grounds in a tributary of Iliamna Lake (photo © Thomas C. Kline, Jr.).

The significance of smolt size for marine survival was suggested by comparisons of ocean survival rates versus smolt length (Koenings and Burkett, 1987a). Accumulated data from natural systems show increased ocean survival with an increase in smolt size up to 110mm (fork length). Smolts > 110mm, however, have a decreased survival. A concomitant decrease in ocean survival with decreasing smolt size was also seen when stocking density was increased in artificially stocked, naturally recruitment-limited lakes that became nutrient limited through overstocking. By manipulating stocking density and nutrient loading and then monitoring the outcome, an optimum stocking density and fertilization program can be derived empirically (Koenings and Burkett, 1987a; Chapter 8).

Other factors not related to density of salmon affect nursery productivity. Because of the extreme variability in water clarity of Alaskan sockeye nursery lakes, light transmittance was found to be the main factor affecting primary productivity and smolt production (Koenings and Burkett, 1987a; see also Chapter 8). Koenings and Burkett (1987a) found that the relationship of water clarity with sockeye productivity can be made in terms of a "euphotic volume," the product of lake surface area and depth to which 1% of surface light penetrates. Thus, both the more productive nonglacial lakes and the less productive glacial lakes were equated in terms of carrying capacities for smolt production, that is, a capacity of $23,000 \times EV^{-1}$ (EV = euphotic volume unit = $10^6 \, m^3$) or a smolt biomass of $81 \, kg \times EV^{-1}$ (Koenings and Burkett, 1987a; see also Chapter 8).

Marine Biogenic Nutrients in Freshwater Systems

A unique limnological setting is created by the return migration of large numbers of salmon spawners (Fig. 7.2) and the subsequent impact of their carcasses and offspring on Alaskan freshwater ecosystems, in marked contrast to systems supporting food webs solely based on allochthonous (e.g., leaf detritus) and autochthonous (e.g., phytoplankton) production supported by runoff and atmospherically derived nutrient inputs. The freshwater environment supports the early life history stages of the salmon populations, from emergence (following yolk-sac resorption) to smolting. Although adult salmon may reside in freshwater during the completion of gametic development while metabolizing accumulated reserves, most of the somatic growth of anadromous salmon takes place in the marine environment. However, because Pacific salmon cease feeding prior to reentering freshwater and contain insignificant levels (<1%) of residual freshwater-derived constituents from when they were ocean-bound smolts, the adults are composed almost entirely of marine-derived carbon and nitrogen (and other elements that are potential nutrients). The marine-derived nutrients released by salmon can significantly impact their freshwater nurseries because these ecosystems are typically nutrient impoverished or oligotrophic (Burgner et al., 1969; Chapter 8). Returning adults thus have the potential to enhance the productivity of freshwater nurseries, particularly lakes that support planktivorous fry of sockeye salmon. The peak in nutrient abundance observed in streams following spawning by semelparous salmon (Brickell and Goering, 1970; Richey et al., 1975; Sugai and Burrell, 1984) suggests that natural perturbations exist in ecosystems from biogenic nutrient inputs. Also, Mathisen (1972) suggested that increased plankton standing stocks in years following large spawnings resulted from marine biogenic nutrient inputs. Because of their biogenic nature, the relative abundance of elements released should approximate the elemental composition of plankton known as the Redfield ratio (Redfield et al., 1963), and thus constitute an ideal "fertilizer." This is borne out by the similarity of the N:P ratio of 12.2 in adult sockeye salmon upon entry to freshwater (Mathisen et al.,

Figure 7.2. The massive dimensions of Pacific salmon runs seen during the spawning migration in an aerial photograph of the Newhalen River, Iliamna Lake (photo © Thomas C. Kline, Jr.).

1988) and the mean dissolved N:P ratio of 12.3 in Iliamna Lake, an important nursery lake system (Poe, 1980). The continuity of the elemental ratio seemed to justify the use of one element (N) as a proxy for biogenic nutrients in our studies.

The mass-balance analyses of P, using estimates of non-salmon P sources and sinks compared to P inputs contributed by salmon (based on escapement counts) to several lakes, were used by Krohkin (1967) to conclude that the biogenic P input from salmon would be significant (>25%) if a minimum of 500 spawners per $10^6\,m^3$ of lake volume were present. Donaldson (1967) modeled the P budget of Iliamna Lake, which is the largest contributor to the production of the immense Bristol Bay sockeye salmon fishery. Donaldson (1967) concluded that P input from salmon equalled that of the terrestrial input in a peak year of sockeye salmon returns in the Iliamna system. During his investigation, the peak return of 1965, with a total run of over 50 million sockeye salmon, was the largest in the 100 year history of the fishery. Of these, about 25 million spawned and died in the Kvichak River watershed (Iliamna Lake and tributaries, including Lake Clark). However, the density of spawners produced by this escapement was less than the Krohkin criterion for >25% marine P that would be met at escapements of >60 million to lliamna Lake. Thus density of spawners alone probably does not adequately assess the significance of carcass-derived nutrients.

A mass-balance of nitrogen (N) for Iliamna Lake can be constructed, based on three components: the chemical composition of returning salmon (Mathisen et al., 1988), the dissolved inorganic N pool (Poe, 1980), and the flushing rate. Possible processes that result in loss components in the N mass balance of lliamna Lake are sedimentation, denitrification, fish emigration, and outflow loss of dissolved and particulate forms of N via the Kvichak River. Sedimentation losses are probably small because of the oligotrophic nature of Iliamna Lake and the probable tendency to recycle N from the sediments. Denitrification that occurs in low dissolved oxygen environments is not likely because of the well-oxygenated nature of Iliamna Lake and its sediments (Donaldson, 1967). Because the nutrient balance resulting from fish emigration is dependent on relative mass of fish cohorts at immigration versus emigration (Deegan, 1993), nutrient flow from fish emigration is small because of the small biomass of smolts compared to returning adults (Donaldson, 1967; Stockner, 1987). Emigration loss in the Kvichak system is estimated to be less than 10% of the N input by adults (Kline et al., 1993). The remaining loss pathway, flushing, is estimated to be 7×10^8 g yr^{-1}, based on the mean amount of N dissolved in the lake (Poe, 1980), divided by the flushing rate. A steady-state input rate, assumed to be approximated by the flushing of the dissolved N pool, is compared to marine-derived N input from salmon in Fig. 7.3. The mean N input from salmon from 1957 to 1989 of 4.1×10^8 g yr^{-1} is equivalent to 59% of the estimated loss by flushing. If one assumes a long-term steady state in the N balance of lliamna Lake, then input equals output. Assuming that terrestrial N inputs to the lake (ultimately atmospherically derived) are constant on a year to year basis, then terrestrial N input is 2.9×10^8 g yr^{-1} (steady-state N input minus marine N input). Episodes of large marine N input occurring about every 5 years are suggested by Fig. 7.3. Although marine N input is on average greater than terrestrial, >50% of escapements provide N input less than the terrestrial N input; hence total N input is less than steady state for >50% of the years. Thus the N mass-balance analysis demonstrates that cycling in size of escapement results in a cyclic perturbation to the N loading and by extension, nutrient loading in Iliamna Lake.

Extensive growth of benthic algae (periphyton) near the major spawning sites was observed by Donaldson (1967) who suggested algal uptake was a

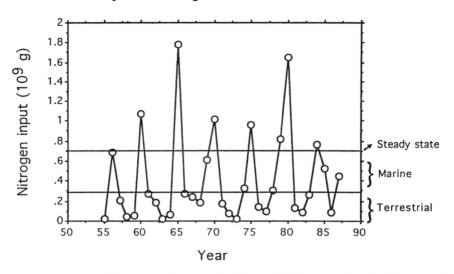

Figure 7.3. Mass balance of nitrogen for Iliamna Lake suggesting significance of marine-derived nitrogen versus terrestrial nitrogen with respect to the cyclic nature of salmon runs. Steady-state input determined as described in text is compared to the mean marine nitrogen input from 1955 to 1987.

potential mechanism for concentration and retention of nutrients subsequent to spawning. Concentration of nutrients in the photic zone counteracts their dilution into the total lake volume (e.g., Iliamna Lake, 117km³; c.f. Durbin et al., 1979) suggesting that the mass balance approach may underestimate the biological significance of salmon inputs. Subsequent investigations (Mathisen, 1972; Poe, 1980) in lliamna Lake show very large concentrations of periphyton (>10g chlorophyll a m⁻²) were stimulated near major spawning sites confirming this viewpoint. Large growths of periphyton can also be found in streams concomitantly with the decomposition of salmon carcasses (Kline et al., 1990) and measured increase of dissolved nutrient abundance (Brickell and Goering, 1970). These levels were markedly higher than the 0.1 to 0.3g chlorophyll a m⁻² found at control (nonspawning) sites (Mathisen, 1972; Poe, 1980). A peak in the nutrient concentration profile just below the ice during March 1976 in Iliamna Lake (P.H. Poe and W.S. Reeburgh, unpublished data) suggests that nutrients initially bound up in periphyton are remineralized during winter. Additional errors in determining the significance of marine-derived nutrients based on mass balance may arise because some loss terms are difficult to determine, for example, they may be based on small subsamples from the ecosystem. Donaldson (1967) used a single core from one bay to estimate sedimentation loss for all of Iliamna Lake. Thus the mass-balance approach primarily provides a point of departure for further studies and is not an end in itself. Its strength is that the magnitude of the marine input can be established, thus providing the premise for more elaborate biogeochemical studies.

Returning Salmon as a Food Source for Freshwater Consumers

Whereas anadromous salmon commonly deliver nutrients directly to terrestrial and aquatic consumers, the extent to which affected ecological communities along upper stream reaches depend upon the yearly flush of marine-derived foodstuffs for support is undetermined. Despite the strongly oligotrophic nature of watersheds in Alaska, systems with salmon runs appear to display higher production of aquatic invertebrates, fish, birds, and mammals. Live and dead spawners are undoubtedly significant, if not essential components of many northern food webs. Investigations by one of us (R.J.P.) appear to show that salmon streams possess a different community structure (functional group composition) of aquatic macroinvertebrates than non-salmon streams and possibly serve as highly productive and diverse oases within an oligotrophic subarctic environment (Fig. 7.4). These ecosystems may be viewed as highly evolved organisms,

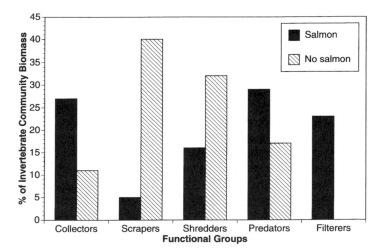

Figure 7.4. Comparison of aquatic macroinvertebrate functional group biomass in salmon carcass enriched and unenriched streams located in south central Alaska. Scrapers and shredders were more common in unenriched streams while collectors and predators were more common in enriched streams. Filter feeders, that require a seston-rich environment, were rare in unenriched streams but made up a significant part of the aquatic macroinvertebrate biomass in salmon carcass enriched streams.

exceptionally efficient at, and dependent on, retaining and recycling marine-derived nutrients.

Overall, the natural history of salmon carcass decomposition has surprisingly received little attention. Bald eagles, brown bears, and occasionally white-tailed deer fed on carcasses of the once numerous kokanee (landlocked sockeye salmon) runs in McDonald Creek, Montana (Spencer et al., 1991). Retention rates of salmon carcasses are high on the Olympic Peninsula, Washington where 22 species of birds and mammals consume them (Cederholm et al., 1989). Bears living near the Karluk River, Kodiak Island prey on adult salmon but switch to berries for food when salmon numbers are low (McIntyre et al., 1988). Scavenging of salmon carcasses by both aquatic and terrestrial macroinvertebrates has been observed in Alaskan streams and lakes (Kline et al., 1990; R.J. Piorkowski, personal observation). Although Minshall et al. (1991) noted no increase in algal biomass downstream from rainbow trout carcasses in an Idaho stream, this may have been a result of tight nutrient spiraling. Richey et al. (1975) found that periphyton production, heterotrophic microbe activity, and nutrient concentration were greater downstream of the spawning area of kokanee in a California stream. Figure 7.5 summarizes the known and potential effects of salmon carcass decomposition on watersheds with spawning salmon. Much work remains to clarify the impact of salmon carcasses on lotic food webs and terrestrial omnivore populations.

J. Salmon carcass decomposition has been investigated by one of us (R.J.P.) through examining hundreds of dead salmon (*Oncorhynchus* spp.) and numerous salmon body parts from 1988 through 1991 in two interior and six southcentral Alaska spawning streams (Clearwater River, Kaltag River, Troublesome Creek, Byers Creek, Pass Creek, Horseshoe Creek, Honolulu Creek, and the Middle Fork Chulitna River). Initial breakdown of immersed carcasses appeared to be predominantly microbial with carcasses covered by varying amounts of sewage fungus complex at time of death. Unlike terrestrial insect-assisted decomposition of vertebrates in woodlands (Payne, 1965), aquatic macroinvertebrates neither facilitate decomposition by burrowing into salmon flesh nor display an evident succession of taxa utilizing the carcasses.

Within hours of the known time of death of individual fish, aquatic macroinvertebrates invaded mouth and gill areas of fish in fast currents and abraded body parts (tail, fins, nose) of fish in slower currents. These macroinvertebrates appeared to feed on the gill membranes in the oral cavity but choose exposed muscle in the abraded areas, or fungal patches on exposed parts of fish (Figs. 7.6). Most insects positioned themselves beneath the carcasses, possibly for predator protection or avoidance of current. In still waters where current was negligible, carcasses or part of carcasses were nearly covered by insects, with over 1000 caddisfly larvae (Trichoptera: Limnephilidae *Ecclisomyia*) observed on one fish head (Fig. 7.7). Feeding

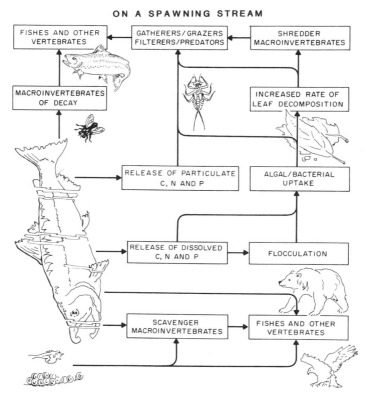

Figure 7.5. Conceptual model of salmon carcass utilization by aquatic and terrestrial watershed fauna.

and residence times vary depending on the taxon and feeding conditions. Limnephilid caddisfly larvae were observed moving to salmon carcasses and feeding for up to 15 minutes before leaving whereas chironomid midges colonized the carcass fungal mat for indeterminate longer periods.

Within minutes of exposure to air, carcasses (either on riffles or near the shoreline) supported blowfly (Diptera: Family Calliphoridae) (Fig. 7.8) whose burrowing larva subsequently aided in rapid carcass deterioration. Rising water levels were observed to float salmon carcasses previously stranded during water lowering. Larvae spilled out of these moving carcasses and became available as food for fishes and predatory aquatic insects. Johnson and Ringler (1979) reported that blowfly larvae, apparently from salmon carcasses, were consumed by young salmon and trout. Brusven and Scoggan (1969) observed blowfly larvae eating dead squawfish in Idaho.

Figure 7.6. Abraded tail of salmon carcass being fed on by caddis flies (Limnephilidae *Psychoglypha*) (photo © Robert Piorkowski).

Figure 7.7. This salmon head was photographed after 1 day in a stream. The large number of caddis flies (Limnephilidae *Ecclisomyia*) feeding on it demonstrate the importance of salmonid protein to aquatic macroinvertebrates (photo © Robert Piorkowski).

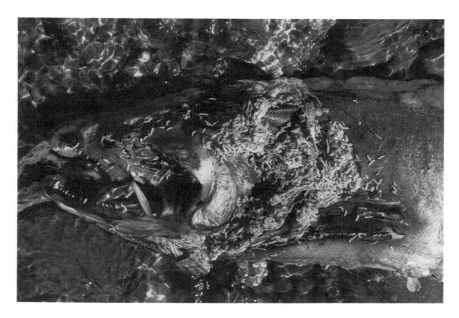

Figure 7.8. Salmon carcass partially exposed to air in stream riffle. Diptera larvae (Calliphoridae) untilizing gill and head regions of carcass (photo © Robert Piorkowski).

Carcasses in advanced states of decomposition commonly became fragmented in turbulent streams (Fig. 7.9). Rapid colonization of these small body fragments by numerous taxa resulted in high insect densities. Long nose suckers (*Catastoma catastoma*), generally regarded as herbivores, fed on these insects, as evidenced by stomachs containing both a wide range of aquatic insect taxa and decomposing salmon flesh. Sport fishermen in Alaska have designed a successful artificial fly for rainbow trout that mimics a strip of decomposing salmon flesh.

In addition to *Ecclisomyia*, another Limnephilid, *Psychoglypha* has been commonly observed on both carcasses and scattered body parts. Other stream macroinvertebrates (Trichoptera: Limnephilidae *Nemotaulius*, Brachycentridae *Brachycentrus*, Glossosomatidae *Glossosoma*; Plecoptera: Nemouridae *Zapada*; Ephemeroptera: Baetidae *Baetis* and Ephemerellidae *Ephemerella*, Heptageniidae *Ironodes*; and Tricladida: Planariidae *Phagocata*) were less commonly observed. Certain taxa may have a proclivity for salmon flesh or they may simply feed opportunistically on decaying fish during the spawning and decomposition period. Anderson (1976) suggested that ingestion of animal material (especially late in the larval or nymphal growth cycle) may be critical to growth of many stream invertebrates including detritivores and herbivores. In a laboratory experiment, stonefly nymphs (Plecoptera) consumed steelhead (nonanadromous

Figure 7.9. A salmon carcass disintegrates as it moves downstream with the current. These particles are entrained by the stream bed for utilization by various biota (photo © Robert Piorkowski).

Oncorhynchus mykiss) eggs and alevins (Claire and Philips, 1968). A large net-building caddisfly (Trichoptera: Hydropsychidae *Arctopsyche*) was never observed on carcasses but was seen on two occasions pulling pieces of salmon, many times the size of its normal prey, into its net.

Depending on stream substrate size, carcasses can be entrained behind rocks or logs and rapidly covered by shifting gravels and sands. These carcasses remain in the aquatic system longer than those exposed to the current and may serve an important function as "time-release capsules" of energy and nutrients in the hyporheic zone. The oligochaete *Eclipidrilus* (Lumbriculidae) was common in carcass deposition areas in Byers Creek (Chulitna River watershed, southcentral Alaska) but rarely found in non-salmon areas. Their stable carbon isotope ratio was similar to that of salmon flesh (R.J. Piorkowski, unpublished data) indicating probable direct utilization of carcasses. The ability of the hyporheic zone to provide significant habitat for invertebrates and storage for marine-derived nutrients in either bacterial, fungal, or invertebrate biomass may be of key importance in determining the ability of a system to consistently produce large salmon runs.

In summary, utilization of decomposing salmon as food by a wide range of aquatic invertebrate taxa appears to be an important mechanism for retaining marine nutrients in lotic systems. Moreover, present studies show that invertebrate communities in salmon spawning streams appear to respond to increased nutrient levels as indicated by increases in taxonomic and functional group richness (especially filter feeders, collectors, and predators), compared to otherwise comparable non-salmon streams.

Although no systematic observations of salmon decomposition have been made in other regions, incidental observations in the Kvichak system suggest that decomposition processes may have marked differences depending on local conditions. Years of observing the effects of precipitation patterns on the distribution of carcasses on shorelines of the Kvichak system during otolith sampling surveys suggest that there can be significant interannual variability in the location of carcass accumulation (P.H. Poe, personal communication). Generally, low precipitation during the late summer results in the deposition on gravel bars and banks of carcasses that would otherwise be transported downstream into Iliamna Lake, or be buried in (trapped between branches) and under snags in the streams. Figures 7.10 and 7.11 illustrate the effects of water lowering during the decomposition process. The relative stability of the carcass location in a stream and aerial exposure appear to be important factors regulating their direct use by terrestrial insect larvae. The carcass in Figure 7.11 was among several others in the area that appeared to have been deposited by currents during high water. The head end of the carcass appears to have first emerged as

Figure 7.10. Spawned-out sockeye salmon carcasses on left on the banks of the Newhalen River, Iliamna Lake, 1983, following a freshet (photo © Thomas C. Kline, Jr.).

Figure 7.11. Sockeye salmon carcass undergoing decomposition by dipteran larvae, Chinkleyes Creek, Iliamna Lake, 1979. Decomposition seen at different stages due to lowering of the water level that has gradually exposed this carcass to terrestrial insects. Viewpoint is from the shore (photo © Thomas C. Kline, Jr.).

water level subsided because it is decomposed the most. The portion closest to the water's edge is undergoing scavenging by dipteran larvae whereas the submerged tail portion has a layer of fungus suggesting microbial decomposition. During surveys in this same location in other years, the stream was found to run up to its cut banks at the point where this photograph was taken (T.C. Kline, unpublished data). Decomposing carcasses found floating can be described as "bags of bones" because the muscle tissue will have completely decayed, leaving only the skeleton and skin. Many such carcasses have been collected in both streams and Iliamna Lake starting about 3 weeks from commencement of spawning, and therefore suggesting relatively rapid decomposition (P.H. Poe and T.C. Kline, unpublished data). In contrast, the remains of decomposing salmon have also been noted during spring. Bones can be found lining the high water line at certain sites.

Incompletely decomposed salmon have been noted during a remotely operated vehicle survey of Iliamna Lake in July 1988 (Kline et al., 1993). Thus, the process of salmon decomposition is influenced by hydrology, and results in variable availability of carcasses to aquatic consumers. Reduced transport away from spawning sites, as occurs in smaller interior streams, may improve in situ utilization by consumers.

Stable Isotopes

Measurement of the natural abundance of stable isotopes in a wide range of materials make it possible to trace the flow of selected chemical elements (i.e., C, N, S) in ecosystems (Fry and Sherr, 1984; Owens, 1987; Peterson and Howarth, 1987; Wada et al., 1991). This has application in anadromous Pacific salmon systems because of the dichotomous nature of N sources (oceanic N in Pacific salmon vs. atmospheric N_2 in terrestrial and freshwater food chains) and correspondingly distinctive isotope ratios (Kline, 1991).

The premise of using stable isotope abundance is the relative enrichment of ^{15}N in Pacific salmon N in comparison with atmospheric N_2. Food webs based on N_2 fixation tend to be low in ^{15}N (Minagawa and Wada, 1984; Owens, 1987; Wada and Hattori, 1991). These include freshwater systems and some tropical marine systems where N_2 fixation is important (Minagawa and Wada, 1984). The relative enrichment of ^{15}N in certain marine organisms has made it possible to determine the significance of marine versus terrestrial diets in prehistoric human diets (Schoeninger et al., 1983). An application more analogous to salmon-derived nutrients is the traceability of guano-derived nutrients by elevated ^{15}N in biota near seabird rookeries (Mizutani and Wada, 1988).

The difference in ^{15}N abundance between Pacific salmon and atmospheric N_2 is the "source effect" affecting biota $\delta^{15}N$ (the recognized expression of stable isotope abundance; the per mil, ‰, deviation of $^{15}N/^{14}N$ from air N_2) values. Other factors that affect ^{15}N abundance are: N pool isotopic enrichment in ^{15}N resulting from ^{14}N depletion because of previous algal uptake; isotopic fractionation of N pools due to N cycling (denitrification can significantly increase the ^{15}N level in an N pool); and enrichment of ^{15}N at higher trophic levels. These factors are assessed as potential problems in the interpretation of $\delta^{15}N$ data in the following.

The problem of N pool depletion is most likely in an extremely oligotrophic system. Depletion changes the fractionation effect during uptake of dissolved fixed N and is dependent on whether the system is "open" or "closed" (Fritz and Fontes, 1980). In the case of an open system with constant removal of product (i.e., organic matter), the primary producers are increasingly enriched in ^{15}N as the N pool is depleted. The preference shown by plants for $^{14}NO_3^-$ or $^{14}NH_4^+$ results in the enrichment of ^{15}N in the remaining pool. Thus plants that utilize the remaining pool have a relatively ^{15}N enriched dissolved N source and so have a concomitant increase in $\delta^{15}N$ value as the N supply is depleted. In the case of a closed system the isotopic ratio of the biota approaches that of the initial dissolved N pool as N becomes depleted. If there is a 100% conversion of the dissolved N pool into biota, then by conservation of matter, the $\delta^{15}N$ value of the total biota will equal the initial $\delta^{15}N$ value of the dissolved N pool. For this reason, the extremely oligotrophic systems based on N_2 fixation tend to have primary producer $\delta^{15}N$ values near 0‰. This appears to be the case in non-anadromous salmon control sites (three stream sites, two control sites within Iliamna Lake, and four lakes) that have been used for comparison with salmon spawning systems (Kline et al., 1986, 1990, 1993; Kline, 1991). Here periphytic algae had $\delta^{15}N$ of 0 ± 2‰ (range). Assessment of change in discrimination is predictable and dependent on quantification of nutrient depletion (Fritz and Fontes, 1980; Owens, 1987). If less than 20% of the N pool is depleted, the change in discrimination is <1‰ and so for practical purposes can be ignored. This was the case in Iliamna Lake because NO_3^- levels were consistently 4 to $5 \mu M$ NO_3^- (Mathisen et al., 1988).

Denitrification has been shown to enrich ^{15}N in the dissolved N pool. The denitrification that occurs below areas of upwelling may be a significant cause of ^{15}N enrichment in marine NO_3^- (Cline and Kaplan, 1975). Denitrification may also significantly enrich the $\delta^{15}N$ value of dissolved N in lakes with long residence times. In Pyramid Lake, Nevada, a closed basin or terminal lake (no outlet), the only possible losses of N are sedimentation and denitrification. Although Pyramid Lake has a very high percentage of N input from N_2 fixation (Horne and Galat, 1985), food webs based on phytoplankton utilizing the dissolved fixed N pool have $\delta^{15}N$ values greater than that expected of N derived from N_2 fixation (Estep and Vigg, 1985). A possible explanation could very well be ^{15}N enrichment of the dissolved N pool by denitrification. This situation is unlikely to occur in anadromous salmon lakes because of the outlet to the ocean unlike terminal lakes. The coastal environment of the northeast Pacific Ocean where salmon habitats exist is well known for copious precipitation. Thus lakes and streams tend to have rapid flushing (mean residence time of 13 sockeye lakes is 5 years, Koenings and Burkett, 1987a) and little likelihood of isotopic alteration due to long-term N cycling. Furthermore, low dissolved O_2 conditions that favor denitrification are incompatible with salmonid oxygen requirements as borne out by the low $\delta^{15}N$ values found for all organisms collected at anadromous salmon-free sites (Kline et al., 1990, 1993; Kline, 1991).

Enrichment of $\delta^{15}N$ in animals relative to their diet is well documented (DeNiro and Epstein, 1981; Minagawa and Wada, 1984; Fry, 1988), and trophic enrichment was taken into account in the development of a model to estimate MDN from $\delta^{15}N$ of different trophic levels by establishing mixing lines (correlations of $\delta^{15}N$ with 0 to 100% MDN) for each trophic level (Kline et al., 1990). Pacific salmon nursery systems at high latitudes are ideal for the application of the stable isotope natural abundance technique because they are relatively "simple" ecosystems with only a few trophic interactions. The complication provided by many intermediate trophic

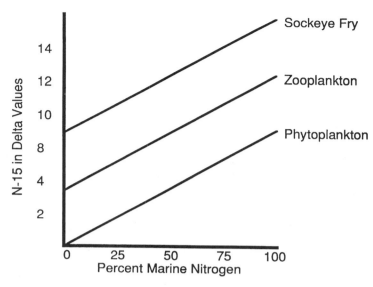

Figure 7.12. The nitrogen stable isotope mixing model (Kline et al., 1990) adapted to the limnetic food web of juvenile sockeye salmon.

levels in other systems is minimized. ^{15}N data expressed as $\delta^{15}N$ are converted to percent MDN using a mixing model (Kline et al., 1990, 1993; Kline, 1991) shown in graphic form in Fig. 7.12. Because juvenile sockeye salmon do not occur naturally in Alaskan nonanadromous systems, it was not possible to make the salmon/non-salmon comparisons as in Table 7.1. Instead, because of their zooplanktivorous nature they were assumed to be

Table 7.1. Difference in $\delta^{15}N$ for Various Carnivorous Fishes Compared to Controls

Species	Control $\delta^{15}N$ (‰)	Sashin Creek $\Delta\delta^{15}N$[a]	Iliamna Lake $\Delta\delta^{15}N$[a]
Rainbow trout			
Sashin Creek	8.4	4.1	—
Iliamna watershed	8.6	—	4.9
Dolly Varden	9.2	4.6	4.0
Coho fry	(6.8)[b]	4.5	—
Coast range sculpin			
≥60 mm	7.7	5.4	4.6
<60 mm	6.8	—	4.4

Control values were obtained from fishes in the Iliamna Lake watershed except for rainbow trout where controls were obtained from both Iliamna Lake and Sashin Creek watersheds. Control values were found to be significantly different from spawning system values (*t* test; SD of fish subpopulations shown were about 2‰. Data from Kline et al. (1990, 1993) and Kline (1991).
[a] $\delta^{15}N$ enrichment relative to control.
[b] Estimated based on an assumption that they were feeding at trophic level 3.

trophic level (TL) 3 organisms (phytoplankton = TL_1 and zooplankton = TL_2), see Fig. 7.12, to calculate MDN.

Nonspawning areas free of any salmon biogenic nutrient input (controls for comparison with salmon-influenced habitats) were used to establish the stable isotope ratios of freshwater systems unaffected by salmon and thus based solely on terrestrial and atmospheric nutrient inputs (Kline et al., 1990, 1993; Kline, 1991). Stable isotope values of consumers and algae verified the hypothesis that N in these systems was atmospherically derived. Variation in escapement and use of C isotopes were also used to examine the hypothesis that marine-derived nutrients play a significant role in anadromous Pacific salmon habitats (Kline, 1991). A consistent pattern of $\delta^{15}N$ enrichment of +4 to 5 was found in predatory fishes from both Iliamna Lake and Sashin Creek when compared with portions of these systems not accessible to anadromous salmon (Table 7.1). The $\delta^{15}N$ enrichment divided by 7 (7 was the difference in $\delta^{15}N$ between 0 and 100% MDN in the mixing model, Fig. 7.4) gives an estimate of percent MDN. These enrichments thus indicate that MDN ranges from 57 to 77% (Table 7.1). These data provide direct evidence that MDN plays a significant role in both Sashin Creek, a rapidly flushing stream (dissolved nutrient peak of about 1 month duration following spawning; Brickell and Goering, 1970) and Iliamna Lake, a lake with a flushing rate of 7.2 years (Kline, 1991).

The analysis of food webs using stable isotopes, in addition to tracing MDN, can be applied to other factors affecting Pacific salmon populations not directly related to freshwater productivity: predation and fishing. Predation and fishing may have roles in the maintenance of cyclic dominance (Collie and Walters, 1987) or diminishment of run size (McIntyre et al., 1988). Predation effects on juvenile salmon become significant when run size is small because predators are satiated when runs are large (Mathisen, 1972) and when exploitation exceeds a threshold amount (McIntyre et al., 1988). Fishing, essentially predation by humans, also affects recruitment and may be either constant or proportional to run size as determined by fishing quotas. The effects of predation and fishing on transport of marine nutrients may be both quantitative (change in potential input to freshwater systems) as evidenced by the change in MDN with respect to escapement (Table 7.2) and qualitative (change in the structure and complexity of food webs) as evidenced by stream invertebrate composition (Fig. 7.4). Salmon, and therefore marine-derived allochthonous matter, may be incorporated into freshwater food webs directly through predation (as well as by scavenging). Autochthonous production is based on, at least in part, nutrients released from adult salmon by excretion or decomposition. These two alternative mechanisms for incorporation of marine-derived nutrients into freshwater food webs have been distinguished by isotopic analyses of food webs (Kline, 1991) because distinctive C and N isotopic signatures can be established for allochthonous marine-derived production (adult salmon, eggs, and fry) and autochthonous production using marine-derived nutri-

Table 7.2. Percent Marine Nitrogen Derived From Adult Salmon in Sockeye Fry Determined by ^{15}N Content

Brood Year (Escapement)	Sample Date	Age of Fry	Marine Nitrogen (%)
Iliamna Lake			
1984 (10.2 million)	Spring 1986	1+	71
	Summer 1986	1+	57
1985 (7.2 million)	Summer 1986	0+	39
	Summer 1987	1+	39
1986 (1.2 million)	Summer 1987	0+	27
Karluk Lake			
1983–1985 (<0.5 million)	Summer 1986	0+, 1+, 2+	71
1985–1987 (0.5 to 1 million)	Summer 1988	0+, 1+, 2+	91
1986–1988 (0.5 to 1 million)	Summer 1989	0+, 1+, 2+	91

Offspring from three different cohorts show relation of marine nitrogen to escapement size in Iliamna Lake. Karluk Lake fry distinguished by year of sampling show relation of marine nitrogen to recent escapements. Data from Kline et al. (1993) and T.C. Kline unpublished data.

ents. Autochthonous production is dependent on primary producer uptake of remineralized nutrients delivered into freshwater by anadromous salmon. Salmon eggs may be an important food item because of: (1) their availability when not buried correctly in the gravel nest made by the females, accidently released by another female subsequently digging a nest at the same spawning site, or washed out by stream errosion; and (2) their high nutritional value from energy stored for use during the long nonfeeding salmonid embryonic development (that generally lasts until the following spring). Furthermore, drifting salmon eggs are frequently seen downstream of spawning areas (T.C. Kline, unpublished data) and thus are plentiful. Consumption of adult salmon, salmon eggs, and salmon juveniles results in the acquisition of the marine isotopic signature. The isotopic comparison of dietary alternatives with the isotopic signature of the consumers indicated that the dietary significance of predation on salmon eggs and fry was small except for coast range sculpin that were up to 34% dependent on salmon (Kline, 1991). The technique of using multiple stable isotope ratios in food web studies has been reviewed recently by Wada et al. (1991).

The effect of run size and cycling are quantifiable through changes in $\delta^{15}N$ when escapement or food webs shift in response. The sockeye salmon returns to Bristol Bay, Alaska (and especially the Kvichak watershed) occur in a 5 year cycle. The corresponding cyclic fluctuation in the biogenic nutrient inflow (Fig. 7.1) led to a notable shift in $\delta^{15}N$ values of juvenile sockeye (Table 7.2). Because of the specialized nature of juvenile sockeye diet (e.g. Kyle et al., 1988), they were a good indicator of MDN shifts in response to fluctuations in escapement. The analysis of sockeye fry $\delta^{15}N$ paralleled the expected MDN based on the mass balance approach (Fig. 7.3). The Kvichak system was therefore an ideal area in which to study the effects of variable input of MDN using the stable isotope technique. Recent extension of this technique into the Karluk Lake system on Kodiak Island suggests a greater dependence on MDN (T.C. Kline, unpublished data). Preliminary analysis suggests that there was a $\delta^{15}N$ response to changes in size of escapement (Table 7.2).

Although important (see Chapter 8), nutrients per se do not result in fish production. Nutrients must first be assimilated by primary producers, and through trophic interactions be incorporated into the planktonic forage of juvenile sockeye salmon. It is now possible to trace MDN through the food chain and thus follow the fate of nutrients released from decomposition. Lack of concordance of forage availability with demand by fry has been cited as a limiting factor in the decline of the Karluk Lake sockeye salmon production (Koenings and Burkett, 1987b). By measuring the stable isotope ratios of forage, it is possible to measure response to increased nutrients following release by carcasses. Thus the structure of the food webs yielding sockeye salmon forage can be considered with respect to biogenic nutrients. Forage availability at time of fry emergence may be the most critical factor affecting survival of a cohort (Koenings and Burkett, 1987b); thus future studies should include early season measurements. Spawning is timed so that emergence occurs in the spring at an optimal time for forage availability (Foerster, 1968) and the timing is site specific because of local temperature regime. Consequently, several subpopulations may exist in any one system, each with different spawning times that correspond to specific sites within the system (Gard et al., 1987; P.H. Poe, personal communication). The mistiming of fry occurrence relative to forage availability (Koenings and Burkett, 1987b) may be a consequence of fishing activity targeted against that portion of the run (the peak) most likely to produce offspring with the best chance of forage availability. Salmon fry can have an impact on population characteristics of limnetic forage (e.g., size distribution of zooplankters) as shown by the planting of fry in previously salmon-free lakes (Koenings and Burkett, 1987a; Kyle et al., 1988). Thus the nature of fry demand on forage, in addition to forage quantity, may possibly be related indirectly to escapement. By carrying out carefully conducted fertilization experiments, it will be possible to test if timing of nutrient input is important. This can be accomplished in systems with a high MDN because of the similarity of fertilizer $^{15}N/^{14}N$ ratios to air N_2. Thus (fertilizer + atmospheric) N takes the place of atmospheric N in the mixing model (Fig. 7.4). As atmospheric N input approaches 0%, then the low $^{15}N/^{14}N$ end member in the model becomes representative of fertilizer N, hence making it possible to measure effectiveness of a fertilization experiment.

Our principal reason for measuring the natural abundance of stable isotopes has been to determine the significance of MDN in the biogeochemistry and production of anadromous Pacific salmon freshwater habitats (Kline et al., 1986, 1990, 1993; Mathisen et al., 1988; Kline, 1991). This was accomplished by the development of an empirical model that related the

$\delta^{15}N$ values of the biota to MDN based on comparisons of isotope ratios of primary producers from salmon spawning habitats and nonspawning habitats. The high $\delta^{15}N$ found in anadromous Pacific salmon habitats compared with controls suggests that there is a very strong effect of salmon runs on production in their freshwater aquatic ecosystems. Furthermore, isotope ratio measurements in conjunction with natural history observations suggest that the organismal make-up and flow of matter in anadromous salmon spawning systems is fundamentally altered when compared with similar systems without salmon.

Acknowledgments. The stable isotope studies reviewed in this chapter were supported by grants from the National Science Foundation Division of Polar Programs (DPP 844112285), and by Alaska Sea Grant with funds from the national Oceanic and Atmospheric Administration, office of Sea Grant, Department of Commerce, under grant no. NA86AA-D-SG041, projects R/06-30A and m/81-01, and from the University of Alaska with funds appropriated by the state, and a University of Alaska Water Research Center research fellowship. Our colleagues Ole Mathisen, Patrick Parker, Patrick Poe, and Jeff Koenings have stimulated many of the ideas presented here. Extensive technical support in the preparation of samples and mass spectrometry came from Brenda Holladay, Norma Haubenstock, Brenda Anderson, Richard Anderson, and Della Scalan. Extensive field support was provided by the University of Washington Fisheries Research Institute at Iliamna; the Alaska Department of Fish and Game Division of Fisheries Rehabilitation, Enhancement, and Development at Karluk Lake; and the National Marine Fisheries Service Little Port Walter field station at Sashin Creek. We also thank the numerous local residents, students, and field technicians at these sites for their generous help in our studies there. The natural history of salmon carcasses study reviewed in this chapter was funded by fellowships from the University of Alaska Foundation and the Territorial Sportsmens' Association. Further funding as research assistantships came from the Alaska Department of Fish an Game through the Alaska Cooperative Fish and Wildlife Unit and the National Science Foundation Taiga Forest Long Term Ecological Research Grant. This manuscript was aided by two anonymous reviewers who provided beneficial suggestions for the final draft.

References

Anderson NH (1976) Carnivory by an aquatic detritivore, *Clistoronia magnific* (Trichoptera: Limnephilidae). Ecology 57:1081–1085.

Barnaby JT (1944) Fluctuations in abundance of red salmon, *Oncorhynchus nerka*, (Walbaum), of the Karluk River, Alaska. Fish Bull 50:237–295.

Brickell DC, Goering JJ (1970) Chemical effects of salmon decomposition on aquatic ecosystems. In Murphy RS (Ed) First international symposium on water pollution control in cold climates. U.S. Government Printing Office, Washington, DC, 125–138.

Brusven MA, Scoggan AC (1969) Sarcophagus habits of Tricopteral larvae on dead fish. Entomol News 80:103–105.

Burgner RL, DiCostanzo CJ, Ellis RJ, et al. (1969) Biological studies and estimates of optimum escapements of sockeye salmon in the major river systems in southwestern Alaska. Fish Bull 67:405–459.

Cederholm CJ, Houston DB, Cole DL, Scarlett WJ (1989) Fate of coho salmon (*Oncorhynchus kisutch*) carcasses in spawning streams. Can J Fish Aquat Sci 46:1347–1355.

Claire EW, Phillips RW (1968) The stonefly *Acroneuria pacifica* as a potential predator on salmonid embryos. Trans Am Fish Soc 97:50–52.

Cline JD, Kaplan IR (1975) Isotopic fractionation of dissolved nitrate during denitrification in the eastern tropical North Pacific Ocean. Mar Chem 3:271–299.

Collie JS, Walters CJ (1987) Alternative recruitment models of Adams River sockeye salmon, *Oncorhynchus nerka*. Can J Fish Aquat Sci 44:1551–1561.

DeNiro MJ, Epstein S (1981) Influence of diet on the distribution of nitrogen isotopes in animals. Geochim Cosmochim Acta 45:341–353.

Donaldson JR (1967) The phosphorus budget of Iliamna Lake, Alaska, as related to the cyclic abundance of sockeye salmon. Ph.D. thesis, University of Washington, Seattle.

Durbin AG, Nixon SW, Oviatt CA (1979) Effects of the spawning migration of the alewife, *Alosa pseudoharengus*, on freshwater ecosystems. Ecology 60:8–17.

Eggers DM, Rogers DE (1987) The cycle of runs of sockeye salmon (*Oncorhynchus nerka*) to the Kvichak River, Bristol Bay: Cyclic dominance or depensatory fishing? In Smith HD, Margolis L, Wood CC (Eds) Sockeye salmon (*Oncorhynchus nerka*) population and future management. Can Spec Publ Fish Aquat Sci 96:343–366.

Estep MLF, Vigg S (1985) Stable carbon and nitrogen isotope tracers of trophic dynamics in natural populations and fisheries of the Lahontan Lake system, Nevada. Can J Fish Aquat Sci 42:1712–1719.

Foerster RE (1968) The sockeye salmon, *Oncorhynchus nerka*, Fish Res Board Can Bull 162.

Fritz P, Fontes JC (1980) Introduction. In Fritz P, Fontes JC (Eds) Handbook of Isotope Geochemistry, Vol 1, The Terrestrial Environment, A. Elsevier, Amsterdam, 1–19.

Fry B (1988) Food web structure on Georges Bank from stable C, N, and S isotopic compositions. Limnol Oceanogr 33:1182–1190.

Fry B, Sherr EB (1984) $\delta^{13}C$ measurements as indicators of carbon flow in marine and freshwater ecosystems. Contrib Mar Sci 27:13–47.

Gard R, Drucker B, Fagen R (1987) Differentiation of subpopulations of sockeye salmon (*Oncorhynchus nerka*), Karluk River system, Alaska. In Smith HD, Margolis L, Wood CC (Eds) Sockeye salmon (*Oncorhynchus nerka*) population and future management. Can Spec Publ Fish Aquat Sci 96:408–418.

Gross MR (1987) Evolution of diadromy in fishes. Am Fish Soc 1:14–25.

Gross MR, Coleman RM, McDowall RM (1988) Aquatic productivity and the evolution of diadromous fish migration. Science 239:1291–1293.

Horne AJ, Galat DL (1985) Nitrogen fixation in an oligotrophic, saline desert lake: Pyramid Lake, Nevada. Limnol Oceanogr 30:1229–1239.

Hyatt KD, Stockner JG (1985) Responses of sockeye salmon (*Oncorhynchus nerka*) to fertilization of British Columbia coastal lakes. Can J Fish Aquat Sci 42:320–331.

Johnson JH, Ringler NH (1979) The occurrence of blowfly larvae (Diptera: Calliphoridae) on salmon carcasses and their utilization as food by juvenile salmon and trout. Great Lakes Entomol 12:131–140.

Kline TC Jr (1991) The significance of marine-derived biogenic nitrogen in anadromous Pacific salmon freshwater food webs. Ph.D. thesis, University of Alaska Fairbanks.

Kline TC, Goering JJ, Mathisen OA, Poe PH, Parker PL, Scalan RS (1986) $\delta^{15}N$ evidence for the transport of marine nitrogen into freshwater Pacific salmon habitats. Eos Trans AGU 67:989–990.

Kline TC Jr, Goering JJ, Mathisen OA, Poe PH, Parker PL (1990) Recycling of elements transported upstream by runs of Pacific salmon: I. $\delta^{15}N$ and $\delta^{13}C$ evidence in Sashin Creek, southeastern Alaska. Can J Fish Aquat Sci 47: 136–144.

Kline TC Jr, Goering JJ, Mathisen OA, Poe PH, Parker PL, Scalan RS (1993) Recycling of elements transported upstream by runs of Pacific salmon: II. $\delta^{15}N$ and $\delta^{13}C$ evidence in the Kvichak River watershed, Bristol Bay, southwestern Alaska. Can J Fish Aquat Sci 50:2350–2365.

Koenings JP, Burkett RD (1987a) Population characteristics of sockeye salmon (*Oncorhynchus nerka*) smolts relative to temperature regimes, euphotic volume, fry density, and forage base within Alaskan lakes. In Smith HD, Margolis L, Wood CC (Eds) Sockeye salmon (*Oncorhynchus nerka*) population and future management. Can Spec Publ Fish Aquat Sci 96:216–234.

Koenings JP, Burkett RD (1987b) An aquatic Rubic's cube: Restoration of the Karluk Lake sockeye salmon (*Oncorhynchus nerka*). In Smith HD, Margolis L, Wood CC (Eds) Sockeye salmon (*Oncorhynchus nerka*) population and future management. Can Spec Publ Fish Aquat Sci 96:419–434.

Krohkin EM (1967) Influence of the intensity of passage of the sockeye salmon *Oncorhynchus nerka* (Wald.) on the phosphate content of spawning lakes. Izdanija "Nauka," 15:26–31 [Trans from Russian (1968) by Fish Res Board Can Trans Ser 1273].

Kyle GB, Koenings JP, Barrett BM (1988) Density-dependent, trophic level responses to an introduced run of sockeye salmon (*Oncorhynchus nerka*) at Frazer Lake, Kodiak Island, Alaska. Can J Fish Aquat Sci 45:856–867.

Mathisen OA (1972) Biogenic enrichment of sockeye salmon lakes and stock productivity. Verh Int Verein Limnol 18:1089–1095.

Mathisen OA, Parker PL, Goering JJ, Kline TC, Poe PH, Scalan RS (1988) Recycling of marine elements transported into freshwater by anadromous salmon. Verh Int Verein Limnol 23:2249–2258.

McIntyre JD, Reisenbichler RR, Emlen JM, Wilmot RL, Finn JE (1988) Predation of Karluk River sockeye salmon by coho salmon and char. Fish Bull 86:611–616.

Minagawa M, Wada E (1984) Stepwise enrichment of ^{15}N along food chains: Further evidence and the relation between $\delta^{15}N$ and animal age. Geochim Cosmochim Acta 48:1135–1140.

Minshall GW, Hitchcock E, Barnes JR (1991) Decomposition of rainbow trout (*Oncorhynchus mykiss*) carcasses in a forest stream ecosystem inhabited only by nonanadromous fish populations. Can J Fish Aquat Sci 48:191–195.

Mizutani H, Wada E (1988) Nitrogen and carbon isotope ratios in seabird rookeries and their ecological implications. Ecology 69:340–349.

Owens NJP (1987) Natural variations in ^{15}N in the marine environment. Adv Mar Biol 24:389–451.

Payne JA (1965) A summer carion study of the baby pig *Sus scrofa* linnaeus. Ecology 46:592–602.

Peterson BJ, Howarth RW (1987) Sulfur, carbon and nitrogen isotopes used to trace organic matter flow in the salt-marsh estuaries of Sapelo Island, Georgia. Limnol Oceanogr 32:1195–1213.

Poe PH (1980) Effects of the 1976 volcanic ash fall on primary productivity in Iliamna Lake, Alaska, 1976–1978. M.S. thesis, University of Washington, Seattle.

Redfield AC, Ketchum BH, Richards FA (1963) The influence of organisms on the composition of sea-water. In Hill MN (Ed) The sea Vol 2. Interscience, New York, 26–77.

Richey JE, Perkins MA, Goldman CR (1975) Effects of kokanee salmon (*Oncorhynchus nerka*) decomposition on the ecology of a subalpine stream. J Fish Res Board Can 32:817–820.

Schoeninger MJ, DeNiro MJ, Tauber H (1983) Stable nitrogen isotope ratios of bone collagen reflect marine and terrestrial components of prehistoric human diet. Science 220:1381–1383.

Spencer CN, McClelland BR, Stanford JA (1991) Shrimp stocking, salmon collapse and eagle displacement. BioScience 41:11–21.

Stockner JG (1981) Whole-lake fertilization for the enhancement of sockeye salmon (*Oncorhynchus nerka*) in British Columbia, Canada. Verh Int Verein Limnol 21:293–299.

Stockner JG (1987) Lake fertilization: The enrichment cycle and lake sockeye salmon (*Oncorhynchus nerka*) production. Can Spec Publ Fish Aquat Sci 96:198–215.

Sugai SF, Burrell DC (1984) Transport of dissolved organic carbon, nutriens, and trace metals from the Wilson and Blossom rivers to Smeaton Bay, southeast Alaska. Can J Fish Aquat Sci 41:180–190.

Wada E, Hattori A (1991) Nitrogen in the sea: Forms, abundances, and rate processes. CRC Press, Boca Raton FL.

Wada E, Mizutani H, Minagawa M (1991) The use of stable isotopes for food web analysis. Crit Rev Food Sci Nutr 30:361–371.

8. An Overview of Alaska Lake-Rearing Salmon Enhancement Strategy: Nutrient Enrichment and Juvenile Stocking

Gary B. Kyle, Jeffrey P. Koenings, and Jim A. Edmundson

Among the fundamental factors limiting the productive (rearing area) capacity of salmon nursery lakes is the quantity of nutrients needed for protoplasmic growth. Other essential factors include light and heat; however, without nutrients primary production is limited, which in turn, regulates zooplankton production. In addition, differences in the capacity of a lake to produce food (i.e. zooplankton) results in distinctive capacities of a lake to support planktivores such as juvenile sockeye salmon (*Oncorhynchus nerka*).

Relationships exist that show growth rate and survival of juvenile salmon rearing in lakes are functions of preceding trophic levels that can be affected by the physical environment or natural events. For example, the biogenic input of nutrients from decomposing salmon carcasses has been strongly correlated to lake productivity (Donaldson, 1967; Gilbert and Rich, 1927; Juday et al., 1932; Krohkin, 1967). In cold regions where sockeye salmon thrive, solar penetration can alter trophic linkages leading to fish production (Koenings et al., 1986; Lloyd et al., 1987), and temperature can affect fish growth (Brett, 1969; Peltz and Koenings, 1989). Similarly, outplanting of fry into a lake (Koenings and Kyle, 1991) or increased natural fry recruitment (Kyle et al., 1988) can have a profound effect on subsequent brood years through density-dependent effects on the secondary trophic structure. Moreover, in lakes with a high or fluctuating population of planktivores, changes in solar penetration due to the loss of large zooplankton (by size-selective fish predation) that feed on small algae can ultimately alter the water clarity (Hrbacek et al., 1961), phytoplankton abundance and nutrient concentrations (Carpenter et al., 1985), and the thermal structure (Mazumder, 1990).

Because of the interacting links between fish and lake ecology, salmon enhancement or restoration programs such as lake stocking and nutrient enrichment require a systematic approach to evaluation. In 1979, the Division of Fisheries Rehabilitation, Enhancement, and Development (FRED) of the Alaska Department of Fish and Game (ADF&G) launched a statewide limnology program aimed at rehabilitation and enhancement of in-lake salmon production through nutrient enrichment and/or fry outplants. Since 1980, 16 Alaskan lakes have undergone nutrient enrichment for either restoration or enhancement of sockeye or coho salmon (*O. kisutch*) stocks (Fig. 8.1). In addition, since 1974, 43 lakes have been stocked with nearly 575 million sockeye salmon fry, fingerlings, or smolts from 14 hatcheries (Table 8.1). Evaluation has focused on monitoring responses in algal biomass, zooplankton community structure and biomass, and age and size of salmon juveniles produced in the treated lakes. Herein, we present an overview of the lake nutrient enrichment technique, the prevailing strategy used in Alaska for determining potential sockeye production in lakes, and results of nutrient enrichment and stocking projects for three sockeye and three coho salmon lakes located within different geographical areas of Alaska.

Figure 8.1. Location of the 16 Alaskan lakes that were treated with nutrients to enhance salmon production.

Table 8.1. Summary of Sockeye Salmon Juveniles Released From Alaskan Hatcheries, 1974–1990

Hatchery	Geographical Region	Release Years	No. Juveniles Released[a]
Beaver Falls	Southern southeast	1985–1990	18,015,700
Klawock	Southern southeast	1987–1990	4,162,300
Snettisham	Northern southeast	1989–1990	3,949,800
Auke Greek	Northern southeast	1987–1989	132,500
Gulkana	Cooper River	1974–1990	178,892,700
Noerenberg	Prince William Sound	1987–1989	3,195,800
Main Bay	Prince William Sound	1988–1989	2,162,400
Crooked Creek	Cook Inlet	1976–1990	182,145,600
Trail Lakes	Cook Inlet	1986–1990	30,494,400
Big Lake	Cook Inlet	1977–1990	114,672,200
Tutka	Cook Inlet	1990	350,000
Kitoi Bay	Kodiak	1990	828,000
Thumb River	Kodiak	1979–1986	34,463,000
Russell Creek	Alaska Peninsula	1989	930,000
		Total	574,394,400

[a] Includes fry, fingerlings, and smolts.

Background and Fundamentals of Nutrient Enrichment

Artificial aquatic enrichment dates back over 2000 years to when the Chinese fertilized carp (*Cyprinus* sp.) ponds. It was not until the late 19th century that scientific literature on pond-culture fertilization in Europe first appeared. Review of these early nutrient enrichment ventures (Davis and Wiebe, 1930; Ness, 1949; Smith, 1934) revealed that initial experiments were directed toward the production of plankton and later were applied to carp culture. Nutrient enrichment work began in a similar fashion in America in the early 1920s, focusing mainly on pond-culture fertilization experiments (Embody, 1921; Weibe, 1930; Wiebe et al., 1929).

It was not until the early 1950s, that the first nutrient enrichment project in a sockeye salmon nursery lake was conducted. Bare Lake on Kodiak Island was treated with nutrients from 1950 to 1953 (Nelson, 1958; Nelson and Edmondson, 1955); however, nutrient enrichment was not conducted again in Alaska until 1970 at Little Togiak Lake (Rogers, 1979). Results from these two projects indicated some positive change in primary production; however, the affect of fertilization on salmon production was largely unknown. In 1970, the lake fertilization program of Canada began with nutrient additions to Great Central Lake. Increased production of lower trophic levels resulted in a higher in production of sockeye salmon in Great Central Lake; in particular, increased smolt growth and survival yielded higher adult returns (LeBrasseur et al., 1978).

In Alaska during the early 1970s, record low returns of adult sockeye salmon occurred and as a result nutrients supplied through salmon carcasses in nursery lakes began to decrease. Over years of depressed returns, the presumption was that nutrient levels in nursery lakes were progressively decreasing and lowering the rearing capacity for the new recruitment of fry. As part of the ADF&G efforts to rehabilitate and/or enhance depressed salmon stocks, in 1980 the Limnology Section of the FRED Division began conducting nutrient enrichment research projects on lakes, as well as evaluation of fry outplants to assess changes in production at all trophic levels.

The addition of nutrients to salmon nursery lakes is designed to elevate concentrations of nitrogen and phosphorus such that the atomic N:P ratio approximates 20:1 for optimum phytoplankton production. The goal is to produce a sustained increase in primary and secondary production throughout the summer growing season without negatively altering the plankton community structure or changing the lake's oligotrophic state. Phosphorus is added in the form of inorganic phosphate and nitrogen is applied in a mixture of ammonium, nitrate, and urea. The supplemental loading of phosphorus is based on 90% of the critical annual loading after Vollenweider (1976).

Surface specific loading:

$$L_p\left(\mathrm{mg\,P\,m^{-2}\,yr^{-1}}\right) = \frac{[P]^{sp} \times \overline{z}\left(1 + \sqrt{T_w}\right)}{T_w}.$$

Surface critical loading:

$$L_c\left(\mathrm{mg\,P\,m^{-2}\,yr^{-1}}\right) = \frac{10\,\mathrm{mg\,P\,m^{-2}} \times \overline{z}\left(1 + \sqrt{T_w}\right)}{T_w}.$$

Permissible supplemental P ($\mathrm{mg\,m^{-2}\,yr^{-1}}$) loading = $L_c \cdot 90\% - L_p$ where $[P]^{sp}$ = spring overturn period total P ($\mathrm{mg\,m^{-2}}$), $\overline{z}$ = mean depth (m), T_w = water resident time (years), and $10\,\mathrm{mg\,P\,m^{-2}}$ = lower critical phosphorus concentration.

Nutrient enrichment is by no means a simple technique to implement or a simple process to evaluate. For example, in Alaska 2 years of limnological and fisheries sampling is conducted to assess the lake's suitability for enrichment (Table 8.2). In addition, the influence of environmental parameters and inherent biological variation must be delineated to determine the responses and effects of nutrient enrichment. To address the above concerns, ADF&G approached lake enrichment in a systematic manner through developing selection criteria (Koenings et al., 1979), and through standardized data collection and analytical procedures (Koenings et al., 1987). Moreover, evaluation and implementation of nutrient enrichment (and stocking) projects in Alaska were initially conducted in a research manner to assess parameters from all trophic levels within a lake so that pathways

Table 8.2. Morphometric Information and Preenrichment Nutrient and Algal Concentrations (Seasonal Mean Epilimnetic Values) for Treated Alaskan Lakes

Lake	Elevation (m)	Surface Area (km²)	Mean Depth (m)	Water Residence Time (yr)	Application Area (km²)	Euphotic Volume (× mill m³)	Total P (µg L⁻¹)	Filterable Reactive P (µg L⁻¹)	Inorganic N (µg L⁻¹)	Organic N (µg L⁻¹)	Chlorophyll a (µg L⁻¹)
Hugh Smith	12	3.24	61.0	1.13	0.81	17.8	5.8	2.1	17.3	181.1	NA
McDonald	15	4.30	45.6	0.83	1.42	27.7	5.7	1.9	18.1	104.1	NA
Falls	6	0.95	31.5	0.47	0.31	9.3	3.7	1.3	37.5	18.5	0.62
Redoubt	4	12.80	95.2	3.43	1.55	130.6	2.9	1.2	38.0	33.2	0.76
Sealion Cove	67	0.08	4.3	0.31	0.08	0.5	5.9	2.4	28.3	80.2	0.78
Deer	114	4.02	189.0	3.51	1.34	114.2	1.8	0.8	31.4	27.9	0.23
Tokun	54	1.80	20.9	0.49	0.72	27.5	8.1	1.5	16.7	28.6	0.82
Pass	24	0.50	12.0	0.37	0.25	6.0	2.7	1.5	17.1	59.9	0.31
Bear	10	1.80	10.5	0.99	0.45	18.9	6.8	1.0	11.6	105.0	2.65
Larson	186	1.80	16.4	1.56	0.53	17.2	6.9	1.5	587.0	170.2	0.67
Packers	16	2.10	13.3	2.84	0.68	8.1	7.6	5.0	78.7	210.1	2.56
Chenik	46	1.20	28.7	4.83	0.40	33.5	3.5	1.7	438.0	46.8	0.49
Leisure	46	1.05	16.4	0.57	0.35	15.4	4.3	2.4	417.7	63.1	0.27
Afognak	21	5.30	8.6	0.40	1.75	58.3	8.0	2.5	81.1	129.1	1.10
Frazer	108	16.60	33.2	2.10	9.96	277.8	5.8	2.3	70.4	80.5	0.95
Karluk	106	39.40	48.6	4.37	13.00	949.2	6.4	0.9	49.4	77.7	1.29

leading to increased production could be ascertained, and rearing-capacity models could be developed and applied to other lakes.

Lake Classification, Rearing Models, and Enhancement Strategy

Synthesis of research results obtained from Alaskan lakes formed the basis for the development of predictive models of the carrying capacity and production of sockeye salmon. By using fish growth and production data from a variety of lake types, lakes were classified to reflect rearing quality. Lakes limited by a low recruitment of fry entering the rearing area were classified as recruitment limited; lakes limited by the quality and quantity of forage production in the rearing area were classified as rearing limited. Each type of limitation is linked to specific characteristics of the smolt populations (age, size, number) and to the rearing area (forage, temperature, light). In particular, the rearing-limited lakes were further classified into forage-limited and environment-limited systems (Koenings and Burkett, 1987).

Furthermore, lakes were subclassified according to numbers of sockeye salmon fry per lake unit (population density) and lake fertility (capacity to produce suitable forage):

A. Recruitment limited (low initial input and density of fry)
  1. Escapement limited (density independent)
  2. Spawning-area limited (density independent)
B. Rearing limited (poor lacustrine conditions or fry–forage interaction)
  1. Forage limited
    a. Poor quantity and quality of forage base (density dependent)
    b. Poor spatial/temporal concurrence of fry and forage (density independent)
  2. Environment limited
    a. Unfavorable temperature regime (density independent)
    b. Short growing season (density independent).

Thus, the above classification reveals that rearing limitation can be either forage or environment based, or both; that forage limitation can be density dependent or independent; and that density-independent growth can be either recruitment or rearing limited.

The development of a model for estimating the sockeye salmon rearing capacity and production within nursery lakes utilized the number and biomass of threshold-sized sockeye smolts from several lakes throughout Alaska and Canada. We included only those lakes that produced smolt populations consisting of 60 to 65 mm age^{-1} smolts, that is, the sockeye juveniles grew only to the minimal smolt size in one growing season. Threshold-sized smolts indicate, under density-dependent rearing regimes, that sufficient juveniles are being produced by the number of spawners to cause the amount of rearing capacity to limit numbers and biomass of smolts. The missing link was to relate the number and biomass of smolts to some measure of the capacity of a lake's rearing environment to predict potential fish production based on a normalized but lake-specific unit. This unit turned out to be euphotic volume (EV) (Koenings and Burkett, 1987), which is the volume of water within a lake within which photosynthesis (the basis of the aquatic food chain leading to fish production) can occur (Fig. 8.2). One million cubic meters of lake water represents one EV unit, and the mean number of threshold-sized smolts produced per unit of EV was found to be 23,000.

In addition, the relationship between the size of smolts and their survival at sea was needed for the EV model to estimate the number of threshold-size smolts required to produce one adult sockeye returning to the lake. On average, it takes eight to nine threshold-size smolts (60 to 65 mm, 1.8 to 2.2 g) to produce one adult sockeye (ocean survival of about 12%). Thus, if a lake produces an average of 23,000 threshold-sized smolts per EV unit, about 2,800 adults per EV unit (23,000 × 0.12) would result. In comparison, empirical data from a variety of lakes revealed a production between 2,400 and 2,500 adults per EV unit (Koenings and Burkett, 1987).

Furthermore, Koenings and Burkett (1987) found that one threshold-size smolt was produced from on average four to five sockeye salmon fry planted. If a lake produces on average 23,000 threshold-sized smolts per EV

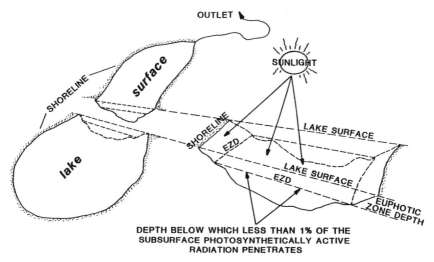

EUPHOTIC VOLUME = EUPHOTIC ZONE DEPTH (EZD) x LAKE SURFACE AREA

Figure 8.2. Schematic diagram of euphotic zone depth and volume used to model sockeye salmon production.

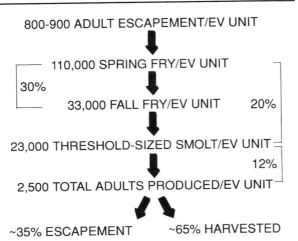

Figure 8.3. Sockeye salmon life-stage survivals at rearing limitation based on euphotic volume.

unit, between 92,000 and 115,000 spring juveniles per unit of EV would be required, resulting in a freshwater survival of about 22%. In addition, through planting increasing numbers of sockeye salmon fry in previously barren Leisure Lake (located in lower Cook Inlet of Alaska), a total of 110,000 juveniles per EV unit produced threshold-sized smolts, and 54,000 juveniles per EV unit produced optimum size (~80 to 90 mm, 5 g) smolts relative to ocean survival (Koenings et al., 1993). As empirical evidence agreed with experimental results, a functional production model became available that is based on lake productivity. Thus, the EV life-stage survival model (Fig. 8.3) can be used to approximate sockeye salmon smolt and adult production from stocking or natural fry recruitment based on euphotic volume and the knowledge of density-dependent versus density-independent rearing.

In addition to the EV rearing capacity/production model for sockeye salmon, zooplankton biomass can be used to fine tune the rearing capacity of lakes and collaborate the EV model when such information is available. That is, in a variety of Alaskan lakes it was found that when competition for food is severe enough to limit juvenile growth (threshold-size age 1 smolts), smolt biomass production is a function of zooplankton biomass (Koenings and Kyle, 1991). Thus, from this model, an estimate can be derived of the expected sockeye smolt biomass from a measured standing crop of zooplankton when the rearing area is fully utilized. In addition, by applying appropriate freshwater survivals (Fig. 8.3), the number of rearing juveniles, either supplied through natural recruitment or from hatchery outplants, that a lake can support and produce can be determined. This knowledge prevents exceeding the lake's rearing capacity and the establishment of top-down control of the zooplankton community that may lead to indirect brood year interactions caused by intraspecific competition (Koenings and Kyle, 1991).

Nutrient Enrichment and Stocking Results

Nutrients and Chlorophyll *a*

In the 15 lakes that received fertilizer containing inorganic phosphate, the phosphorus load increased by 9 to 366%, and in most lakes, represented a substantial amount of the phosphorus entering the lake during the summer (Table 8.3). In addition, the nitrogen added to the lakes increased the inorganic fraction (compared to preenrichment concentrations) by 2 to over 1,200%. In all the lakes before treatment (except Bear Lake), phosphorus concentrations were low, and during the summer a nitrogen depletion existed. Thus, nitrogen was added in all treated lakes to either ameliorate the depression of nitrogen during peak photosynthetic activity, or to ensure maintenance of the desired N/P ratio (~20:1) to avoid production of nuisance blooms of unusable phytoplankton.

The addition of nutrients resulted in higher concentrations of chlorophyll *a* (chl *a*) than during preenrichment in the three sockeye and coho nursery lakes (Figs. 8.4, 8.5). Seasonal and annual fluctuations occurred, and in some years mean chl *a* was less than preenrichment years as in Bear Lake. This phenomena in Bear Lake could be due to an inconsistent algae layer observed at the mean depth (~10m) that, dependent upon density, could act as a partial nutrient trap, that in turn would limit primary production (Woods, 1986). In addition, other processes such as environmental variables (Harris, 1980) and consumer regulation (Carpenter, 1985; Elser and Goldman, 1990; Kitchell et al., 1979) that operate over a wide temporal and spatial scale may explain the variation in primary productivity independent of nutrient supply.

Zooplankton Response

The principal food of underyearling sockeye salmon is macrozooplankton; and because nutrients are important in determining the phytoplankton community, which in turn serve as the food resource for zooplankton, there should be an increase in the zooplankton and/or fish biomass with nutrient additions. Furthermore, prey density and availability limit predator biomass or size, and prey density can be controlled by predator density (Brocksen et al., 1970; Johnson, 1965; Koenings and Burkett, 1987; Kyle et al., 1988). Thus, a large natural recruitment of fry through high escapements or outplanting of fry in nursery lakes can reduce zooplankton biomass.

In the three sockeye nursery lakes that were nutrient enriched and where the number of rearing fry was increased either through sockeye fry

Table 8.3. Mean Phosphorus and Nitrogen Loading Rates and Percent Increases for Nutrient Enriched Lakes in Alaska

Lake	P Load (mg m⁻² yr⁻¹)	P Critical Load (mg m⁻² yr⁻¹)	Permissible P Fertilizer Load (mg m⁻² yr⁻¹)	P Increase From Fertilizer (%)	N Fertilizer Load (mg m⁻² yr⁻¹)	Inorganic N Increase From Fertilizer (%)
Hugh Smith	610	1114	393	64	2200	104
McDonald	731	1050	214	29	3900	236
Falls	283	1130	734	259	4250	180
Redoubt	153	792	560	366	1560	22
Sealion Cove	67	216	127	190	2210	921
Deer	403	1547	989	245	8240	69
Tokun	414	725	239	58	1860	266
Pass	128	522	342	267	2620	639
Bear	240	212	0	0	2980	1221
Larson	190	236	22	12	820	4
Packers	90	126	36	40	2270	109
Chenik	60	190	111	185	590	2
Leisure	273	505	182	67	3590	26
Afognak	225	351	91	40	1730	124
Frazer	261	387	87	33	950	20
Karluk	284	344	26	9	620	13

In the table headers:
P Load $(\text{mg m}^{-2}\,\text{yr}^{-1})$, P Critical Load $(\text{mg m}^{-2}\,\text{yr}^{-1})$, Permissible P Fertilizer Load $(\text{mg m}^{-2}\,\text{yr}^{-1})$, N Fertilizer Load $(\text{mg m}^{-2}\,\text{yr}^{-1})$

outplants (Leisure Lake) or increased escapements (Frazer and Redoubt lakes), zooplankton biomass dramatically increased (Fig. 8.6). As part of an experimental lake manipulation program, Leisure Lake was annually stocked with 1.5 to 2.1 million sockeye fry during 1982 to 1984; and during 1985 to 1990 when the lake was enriched with nutrients to increase forage production, stocking levels ranged from 2.0 to 2.4 million. During the first year of nutrient enrichment (1985) in Leisure Lake, zooplankton biomass increased by nearly 700% to 414 mg m^{-2} (from a mean of 54 mg m^{-2} before treatment) and in 1990 exceeded 900 mg m^{-2} (Fig. 8.6**A**). The primary population response to the increase in primary productivity in Leisure Lake came first from *Bosmina* and then a year later from *Cyclops*; and zooplankton densities as well as size increased (Koenings and Kyle, 1991). For example, the mean length of *Cyclops* increased from 0.63 mm in 1984 to 0.89 mm in 1985 to 0.90 mm in 1986; *Bosmina* sizes increased from 0.35 mm in 1984 to 0.41 mm in 1985 to 0.38 min in 1986. These data quite clearly indicate that during nutrient enrichment, zooplankters were able to rebound from 3 years of top-down control by juvenile sockeye.

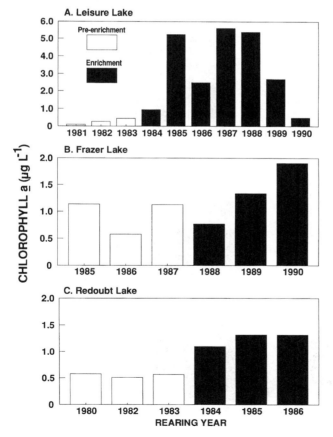

Figure 8.4. Chlorophyll *a* concentrations in Leisure, Frazer, and Redoubt lakes before and during nutrient enrichment.

Kyle et al. (1988) detailed the changes in Frazer Lake and the sockeye population following several years of high adult escapements. Since 1965, the mean adult escapements increased from about 15,000 adults to over 250,000 adults. With increasing numbers of spawners, macrozooplankton densities dropped by 84% from 10,620 m^{-3} to 1450 m^{-3} and smolt sizes decreased from 148 to 89 mm (the system was approaching rearing limitation). For the 1984 to 1986 brood years, escapements equalled 53,500, 485,800, and 126,500 spawners, respectively, and varied from about 25% to almost 2.5-fold the escapement goal (Fig. 8.6**B**). During the next 3 brood years of 1987 to 1989, escapements equalled 40,500, 246,700, and 360,400, respectively, or from 20% to 1.8-fold the goal. Juvenile sockeye from the 1984 to 1986 brood years were reared under natural conditions; during 1987 to 1989 fry were reared while the lake was treated with nutrients to enlarge the forage base. The dramatic differences in the two periods is not just that zooplankton biomass was greater during brood years (1987–1989) in which

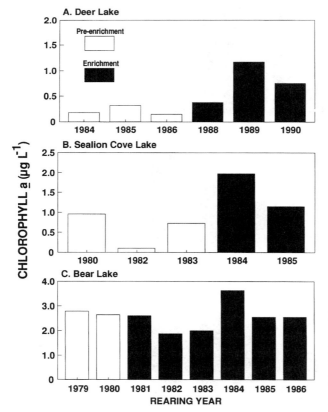

Figure 8.5. Chlorophyll *a* concentrations in Deer. Sealion Cove, and Bear lakes before and during nutrient enrichment.

the lake was enriched; but the record 1985 escapement, well above the escapement goal and before enrichment, had a carryover effect on the zooplankton biomass in the subsequent year (1986), which had a much smaller brood year escapement. In contrast, the large numbers of foraging juveniles from the escapement in 1989 when the lake was enriched were still able to depress zooplankton biomass; however, the decline was muted and not much lower than during the preenrichment 1984 escapement of 53,500.

In Redoubt Lake during preenrichment, sockeye escapements were relatively low (<1000) and zooplankton biomass averaged 90 mg m^{-2} (Fig. 8.6C). During the years when the lake was treated with nutrients, zooplankton biomass ranged between 160 and 200 mg m^{-2} (except for 1986), despite much larger recruitment of fry from adult escapements (2500 to 28,700). Thus, similar to Leisure and Frazer lakes, nutrient enrichment of Redoubt Lake produced higher and relatively consistent zooplankton biomass concurrently while supporting a greater number of rearing fry.

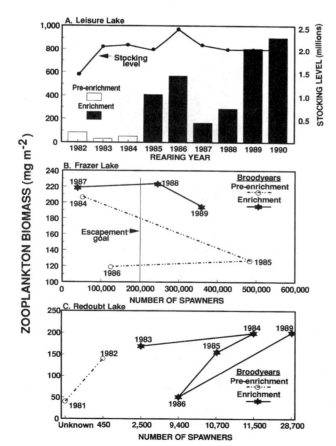

Figure 8.6. Seasonal mean zooplankton biomass as a function of sockeye salmon abundance in Leisure, Frazer, and Redoubt lakes before and during nutrient enrichment.

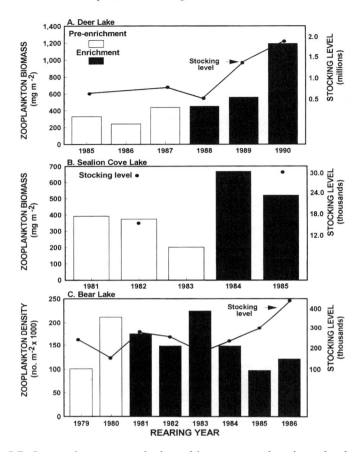

Figure 8.7. Seasonal mean zooplankton biomass as a function of coho salmon stocking level in Deer, Sealion Cove, and Bear lakes before and during nutrient enrichment.

Increased zooplankton biomass also occurred in coho salmon lakes that were enhanced through nutrient enrichment (Fig. 8.7). In Deer Lake during the initial year of nutrient enrichment (1988), zooplankton biomass exceeded the highest level observed during preenrichment, despite annual stocking levels 1.7 to 3.7 times greater than during preenrichment (Fig. 8.7**A**). More importantly, after two consecutive years of high stocking and nutrient enrichment, the highest zooplankton biomass (a twofold increase over preenrichment conditions) was observed in 1990. For the nutrient enrichment experiment in Sealion Cove Lake, Cameron (1990) followed trophic interactions and found zooplankton biomass increased significantly under enriched conditions and high stocking 4000 fry ha^{-1} (30,000 fry) compared to unenriched conditions at a lower stocking density of 2000 fry ha^{-1} (15,000 fry) (Fig. 8.7**B**). In Bear Lake, where Kyle (1990) found that underyearling coho prey on zooplankton as a primary or secondary food source dependent upon fish density and/or environmental conditions, the zooplankton density during preenrichment years 1979 and 1980 averaged $\sim$160,000 m^{-2} when the lake was stocked at an average of 190,000 coho fry yr^{-1} (Fig. 8.7C). During enrichment (1981 to 1986), the zooplankton density also averaged $\sim$160,000 m^{-2} and the annual stocking level was similar (225,000). However, comparison of mean zooplankton densities before and after nutrient enrichment during the period when juvenile coho salmon were virtually absent (June and July) from the limnetic area indicated a significant increase in zooplankton density following nutrient enrichment (Kyle, 1994). The significant increase in zooplankton density during June and July after Bear Lake was fertilized suggests that nutrient enrichment positively affected seasonal zooplankton abundance. The finding of no significant change in seasonal zooplankton density after nutrient enrichment during August to October (when coho are limnetically feeding) is believed to be a result of expanded predation by juvenile coho salmon (Kyle, 1994), as similarly suggested by Olsson et al. (1992) for fertilized Lake Hecklan.

Juvenile Salmon Production

Our empirical knowledge of lake ecosystems and early life history dynamics of limnetic-rearing salmon is that: nutrient additions will increase primary and secondary productivity; and presented with a larger forage base juvenile salmon will experience increased growth and/or survival. In addition, the detection of increases in production at each trophic level depends upon the dynamics and structure of each individual lake, such as the assimilation of nutrients, the uptake of carbon, phytoplankton and zooplankton compositions, and in particular, the abundance of planktivores (juvenile salmon). Intense predation by planktivores causes a restructuring of the zooplankton community to smaller size individuals and to individuals more adapted to predator avoidance. Thus, it is important to note that the degree of response to nutrient enrichment in salmon nursery lakes depends much on the preexisting interactions between the forage base and the population of rearing salmon.

In Leisure Lake, total (ages 1 and 2) smolt biomass during 1986 to 1988 (stocking level averaged 2.2 million) when the lake was enriched with nutrients increased 3.5-fold compared to preenrichment years 1983 to 1985 (Fig. 8.8A) when the stocking level averaged 2.1 million. In addition, age-1 smolt composition increased by 23% and the weight doubled during the enrichment period. The age structure shift and production of much larger smolts after enrichment demonstrated that when food supply is not limiting to juvenile growth and survival, brood years do not interact through effects on the forage base. Moreover, the beneficial shift to more numerous and larger age-1 smolts at similar stocking levels before and after nutrient

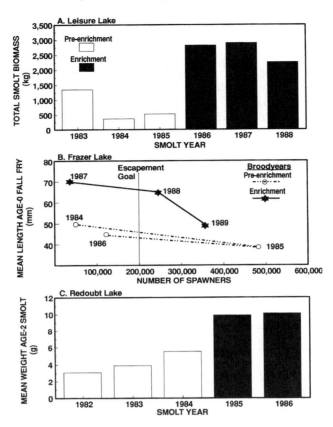

Figure 8.8. Juvenile sockeye salmon production features in Leisure, Frazer, and Redoubt lakes before and during nutrient enrichment.

enrichment provided clear evidence that heavy foraging effects can be reversed.

In Frazer Lake, the large escapement of sockeye salmon in 1985 produced subsequent fall age-0 juveniles of a small size (~40mm) (Fig. 8.8**B**), and below a weight of 1.4g (Koenings and Kyle, 1991), which is typical for the production of threshold-sized smolts (Koenings and Burkett, 1987). Even though before and following the 1985 brood year spawners were below the escapement goal, fall fry size was still relatively small. The dramatic reduction in the size of age-0 fall fry of the 1985 brood year compared to the 1984 brood year fry resulted from heavy foraging by juvenile sockeye that lowered zooplankton biomass by 61% (Fig. 8.6**B**). In comparison, the 1987 brood year escapement, in which the subsequent fry were reared under a nutrient enriched environment, was similar to the number of spawners in the 1984 escapement and both were below the escapement goal; however, the age-0 fall fry size substantially increased to 70mm (Fig.

8.8**B**). Furthermore, the 1988 and 1989 brood year escapements exceeded the goal, but the size of age-0 fall fry continued to be much higher than during preenrichment. Thus, as a result of nutrient enrichment zooplankton biomass and age-0 fall fry size increased, and the numbers of foraging juveniles from the large escapement in 1989 had a modest impact on the standing stock of zooplankton (Fig. 8.6**B**). Thus, similar to the response of juvenile production in Leisure Lake, it is clearly evident that top-down effects can be ameliorated and even reversed by bottom-up control of trophic dynamics through nutrient treatment.

Juvenile biomass is not available for Redoubt Lake; however, the weight of the dominant age-2 sockeye smolts reveal the effect of nutrient enrichment (Fig. 8.8**C**). That is, during preenrichment the mean weight ranged from ~2.5 to 5.0 g (1982 to 1984), whereas during enrichment (1985 and 1986) age-2 smolts were consistently 10 g.

Increased seasonal growth and/or juvenile biomass during nutrient enrichment of the three coho lakes was also evident (Fig. 8.9). As coho stocking levels were considerably greater during nutrient enrichment of

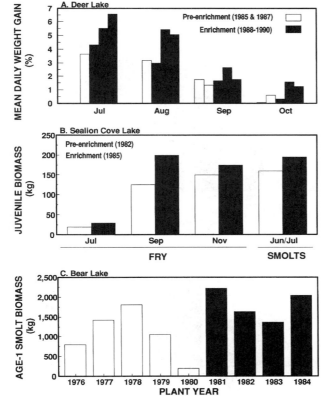

Figure 8.9. Juvenile coho salmon production features in Deer, Sealion Cove, and Bear lakes before and during nutrient enrichment.

Deer Lake (Fig. 8.7**A**), seasonal growth could theoretically be either reduced if enrichment was ineffective or improved if enrichment was effective. The mean daily weight gain of coho rearing in Deer Lake throughout
July to October was dramatically higher during enrichment (Fig. 8.9**A**),
especially during the early segment of the growing season (July and August) when fingerlings are vigorously growing. In Sealion Cove Lake, fry
and smolts accrued greater biomass during nutrient enrichment (Fig. 8.9**B**).
Because this lake is small and without a vegetated littoral zone (limited
habitat for insect populations), nutrient enrichment provided an expanded
food source (mainly large zooplankters) that resulted in a twofold increase
in the capacity of the lake to rear coho fry.

Finally, although there was no measurable change in the mean standing
stock of zooplankton during August to October after nutrient enrichment
of Bear Lake (Fig. 8.7**C**), zooplankton biomass was apparently transformed
into juvenile coho biomass. Underyearling coho in Bear Lake are known to
prey on zooplankton (Kyle, 1990); therefore, favorable changes in smolt
yield in the form of increased growth/survival (i.e. biomass) and production
efficiencies would be expected during the enrichment period. Before enrichment the fingerling to smolt survival was 37% which increased to 50%
after nutrient enrichment, and age-1 smolts comprised 63% of the annual

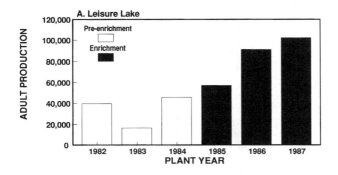

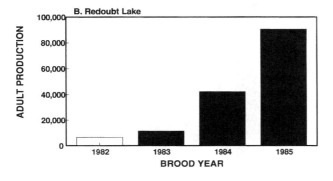

Figure 8.10. Adult
sockeye salmon
production (number
of individuals) in
Leisure and Redoubt
lakes before and
during nutrient
enrichment.

emigration before enrichment compared to 92% afterward (Kyle, 1994). Because age-1 smolts before and after enrichment were almost identical in weight (17.8 vs. 17.5 g), the increase in numbers of age-1 smolts was exclusively responsible for the increase in biomass from an average of 1,055 to 1,819 kg (72% increase in gross yield) after enrichment (Fig. 8.9**C**). Furthermore, the production efficiency defined as the ratio of age-1 smolt yield to production occurring during lake residence (Gillespie and Benke, 1979) was 0.67:1 during enrichment compared to 0.42:1 before (Kyle, 1994). Thus, during enrichment, 67% of the coho biomass produced in Bear Lake became yield as age-1 smolts compared to 42% before. The 25% increase in the lake's ability to produce age-1 smolt biomass during nutrient enrichment indicate that more zooplankton were indeed available and consumed by rearing coho salmon.

Adult Salmon Production

Although fry growth and survival, and smolt biomass are true benchmarks for the effectiveness of lake enhancement projects, the foremost benefit is the number of adult salmon available for harvest and spawning. Of the six study lakes, adult return information is available for four (Figs. 8.10, 8.11).

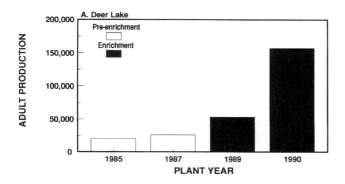

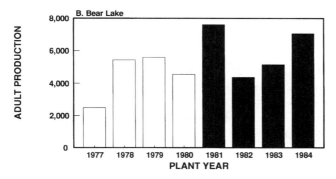

Figure 8.11. Adult coho salmon production (number of individuals) in Deer and Bear lakes before and during nutrient enrichment.

In each of these lakes, production of adult sockeye or coho salmon either from stocking or natural propagation was higher during enrichment. Specifically, the return of adult sockeye salmon to Leisure Lake from near-equivalent stocking levels before and during enrichment increased 2.5-fold (Fig. 8.10**A**). During plant years 1982 to 1984 before enrichment, adult production in Leisure Lake averaged 33,873; in plant years 1985 to 1987 that underwent enrichment adult production progressively increased and averaged 84,392 sockeye. For Redoubt Lake, only one (complete) brood year production is available before nutrient enrichment; however, as with Leisure Lake, adult production progressively increased and was substantially higher (sevenfold increase) than before enrichment (Fig. 8.10**B**).

Coho salmon production in Deer Lake averaged ~23,000 before enrichment (Fig. 8.11**A**) when an average of ~810,000 fingerlings were released. During enrichment (1988 to 1989), adult coho returns averaged 105,000 (550% increase) when the stocking level averaged ~960,000 fingerlings (18.5% increase). Finally, the production of coho salmon from fingerling plants in Bear Lake before enrichment ranged from 2493 to 5588 and averaged 4,500 (Fig. 8.11**B**). During the years of nutrient enrichment, adult coho production averaged 6,050; thus a 34% increase in adult production occurred from the 72% increase in biomass of age-1 smolts.

Conclusions

Sockeye salmon production for a variety of lakes is linked to euphotic volume (EV), which is the volume of water within a lake that photosynthesis (the basis of the aquatic food chain leading to fish production) can occur. EV, in part, determines primary and secondary production. By normalizing freshwater production based on EV, a life-stage relationship can be developed to model the approximate sockeye production in nursery lakes. The estimation of the number of spring rearing juveniles or smolts necessary to match a lake's rearing capacity can be used to develop adult escapement and/or stocking goals. Moreover, use of the lake classification scheme presented provides a strategic approach for matching the appropriate enhancement strategy (e.g., fry stocking or nutrient enrichment) to the limiting feature of a lake. By comparing the existing sockeye production in a lake with rearing capacity models one can:

1. determine the potential for increasing sockeye production;
2. define the enhancement strategy most likely to result in a benefit; and
3. provide a realistic appraisal of expected adult production and the number of fry required to match the lake's rearing capability, both of which are especially useful in planning and benefit/cost projections.

Finally, considering the trophic-level responses presented for the three sockeye and coho lakes that were nutrient enriched, it is evident that both the commonality of results and the agreement between observed and antici-pated results suggest a cause and effect relationship between nutrient en-richment and salmon production. Thus, enhancement of lower trophic levels through nutrient enrichment has increased juvenile and adult salmon production in these lakes as well as other lakes enriched in Alaska. How-ever, from varying responses to nutrient enrichment of Alaskan lakes, it is evident that extrapolation of results from one encircheded lake to another of similar size and morphometry cannot and should not be done. We concur with Stockner (1987) that the efficacy of nutrient enrichment is lake specific and dependent upon biological factors such as food-web processes, fish densities and behavior, predators, competitors, and plankton community structure. However, through a systematic fisheries and limnological preassessment, lakes that offer the most potential relative to existing pro-ductivity and selection criteria can be delineated.

References

Brett JR, Shelbourn JE, Shoop CT (1969) Growth rate and body composition of fingerling sockeye salmon, *Oncorhynchus nerka*, in relation to temperature and ration size. J Fish Res Board Can 26:2363–2394.

Brocksen RW, Davis GE, Warren CE (1970) Analysis of trophic processes on the basis of density-dependent function. In Steel JA (Ed) Marine food chains. Oliver and Boyd, Edinburgh, 468–499.

Cameron WA (1990) Responses to fertilization and fish stocking in the pelagic ecosystem of a naturally-fishless lake. M.S. thesis, Oregon State University, Corvallis, OR.

Carpenter SR, Kitchell JF, Hodgson JR (1985) Cascading trophic interactions and lake productivity. BioScience 35:634–638.

Davis H, Weibe A (1930) Experiments in the culture of the black bass and other pondfish. U.S. Bureau of Fisheries Document 1085. 177–203.

Donaldson JR (1967) The phosphorus budget of Illiamna Lake, Alaska, as related to the cycle abundance of sockeye salmon. Ph.D. thesis, University of Washing-ton, Seattle.

Elser JJ, Goldman CR (1990) Experimental separation of the direct and indirect effects of herbivorous zooplankton and phytoplankton in a subalpine lake. Verh Int Verein Limnol 24:493–498.

Embody G (1921) The use of certain milk wastes in the propagation of natural fish food. Trans Am Fish Soc 51:76–79.

Gilbert CH, Rich WH (1927) Investigations concerning the red salmon runs to the Karluk River, Alaska. Fish Bull 43:1–69.

Gillespie DM, Benke AC (1979) Methods of calculating cohort production from field data—Some relationships. Limnol Oceanogr 24:171–176.

Harris GP (1980) Temporal and spatial scales in phytoplankton ecology: Mechanisms, methods, models, and management. Can J Fish Aquat Sci 37:877–900.

Hrbacek J, Dvorakova M, Korinek V, Prochazkova L (1961) Demonstration of the effect of the fish stock on the species composition of zooplankton and the intensity of metabolism of the whole plankton assemblage. Verh Int Verein Theor Angew Limnol 14:192–195.

Johnson WE (1965) On the mechanism of self-regulation of population abundance in *Oncorhynchus nerka*. Mitt Int Verein Limnol 13:66–87.

Juday C, Rich WH, Kemmerer GI, Mann A (1932) Limnological studies of Karluk Lake, Alaska, 1926–1930. Fish Bull 47:407–436.

Kitchell JF, O'Neill RV, Webb D, et al. (1979) Consumer regulation of nutrient cycling. BioScience 29:28–34.

Koenings JP, Burkett RD (1987) Population characteristics of sockeye salmon (*Oncorhynchus nerka*) smolts relative to temperature regimes, euphotic volume, fry density, and forage base within Alaska lakes. In Smith HD, Margolis L, Wood CC (Eds) Sockeye salmon (*Oncorhynchus nerka*) population biology and future management. Can Spec Publ Fish Aquat Sci, 216–234.

Koenings JP, Kyle GB (1991) Collapsed populations and delayed recovery of zooplankton in response to heavy juvenile sockeye salmon (*Oncorhynchus nerka*) foraging. Proc Int Symp Biol Interactions Enhanced Wild Salmonids. In Rev Spec Publ Can J Fish Aquat Sci.

Koenings JP, Burkett RD, Kyle GB, Edmundson JA, Edmundson JM (1986) Trophic level response to glacial meltwater intrusion in Alaska lakes. In Kane DL (Ed) Proc: Cold Regions Hydrol Am Water Resourc Assoc, Bethesda, MD, 179–194.

Koenings JP, Edmundson JE, Kyle GB, Edmundson JM (1987) Limnological field and laboratory manual: Methods for assessing aquatic production. Alaska Department of Fish and Game, Juneau, Alaska. FRED Div Rep Ser No 71.

Koenings JP, Geiger HJ, Hasbrouck JJ (1993) Smolt-to-adult survival patterns of sockeye salmon (*Oncorhynchus nerka*): Effects of smolt length and geographic latitude when entering sea. Can J Fish Aquat Sci 50:600–611.

Koenings JP, Kaill M, Clark J, Engel L, Haddix M, Novak P (1979) Policy and guidelines for lake fertilization. Alaska Department of Fish and Game, Juneau, Alaska. FRED Div Spec Rep.

Krohkin EM (1967) Influence of the intensity of passage of the sockeye salmon, *Oncorhynchus nerka* (Walbaum), on the phosphate content of spawning lakes. Izd. "Nauka," Leningrad. 15:26–30 [Fish Res Board Can Trans Ser 1273].

Kyle GB (1990) Aspects of the food habits and rearing behavior of underyearling coho salmon (*Oncorhynchus kisutch*) in Bear Lake, Kenai Peninsula, Alaska. Alaska Department of Fish and Game, Juneau, Alaska. FRED Div Rep Ser No 105.

Kyle GB (1994) Assessment of trophic-level responses and coho (*Oncorhynchus kisutch*) production following nutrient treatment (1981–1986) of Bear Lake, Alaska. Fish Res 20:243–261.

Kyle GB, Koenings JP, Barrett BM (1988) Density-dependent, trophic level responses to an introduced run of sockeye salmon (*Oncorhynchus nerka*) at Frazer Lake, Kodiak Island, Alaska. Can J Fish Aquat Sci 45:856–867.

LeBrasseur RJ, McAllister CD, Barraclough WE, et al. (1978) Enhancement of sockeye salmon (*Oncorhynchus nerka*) by lake fertilization in Great Central Lake: Summary report. J Fish Res Board Can 35:1580–1596.

Lloyd DS, Koenings JP, LaPerriere JD (1987) Effects of turbidity in fresh waters of Alaska. North Am J Fish Manage 7:18–33.

Mazumder A (1990) Ripple effects. Sciences Nov/Dec: 39–42.

Nelson PR (1958) Relationship between rate of photosynthesis and growth of juvenile red salmon. Science 128:205–206.

Nelson PR, Edmondson WT (1955) Limnological effects of fertilizing Bare Lake, Alaska. U.S. Fish Wildl Serv Fish Bull 56:413–436.

Ness J (1949) Development and status of pond fertilization in central Europe. Trans Am Fish Soc 76:335–358.

Olsson H, Blomqvist P, Olofsson H (1992) Phytoplankton and zooplankton community structure after nutrient additions to oligotrophic Lake Hecklan, Sweden. Hydrobiology 243/244:147–155.

Peltz L, Koenings JP (1989) Evidence for temperature limitation of juvenile sockeye salmon growth in Hugh Smith Lake, Alaska Alaska Department of Fish and Game, FRED Division Report Series 90, Juneau, AK.

Rogers BG (1979) Responses of juvenile sockeye salmon and their food supply to inorganic fertilization of Little Togiak, Alaska. M.S. thesis, University of Washington, Seattle.

Smith MW (1934) Physical and biological conditions in heavily fertilized water. J Biol Board Can 1:67–93.

Stockner JG (1987) Lake fertilization: The enrichment cycle and lake sockeye salmon (*Oncorhynchus nerka*) production. In Smith HD, Margolis L, Wood CC (Eds) Sockeye salmon (*Oncorhynchus nerka*) population biology and future management. Can Spec Publ Fish Aquat Sci, 198–215.

Vollenweider RA (1976) Advances in defining critical loading levels for phosphorus in lake eutrophication. Mem Ist Ital Idrabiol 33:53–83.

Weibe A (1930) Investigations on plankton production in fish ponds. U.S. Bur Fish Bull 16:137–176.

Weibe A, Radcliffe R, Ward F (1929) The effects of various fertilizers on plankton production. Trans Am Fish Soc 59:94–105.

Woods PF (1986) Deep-lying chlorophyll maxima at Big Lake: Implications for trophic state classification of Alaskan lakes. In Kane DL (Ed) Proceedings: cold regions hydrology symposium. American Water Resources Association, Bethesda, MD, pp 195–200

9.   Alaska Timber Harvest and Fish Habitat

Michael L. Murphy and Alexander M. Milner

Fishing and timber harvest are major industries in Alaska. If not carefully planned and conducted, timber harvest and associated road construction may adversely affect anadromous fish habitat, which has sometimes brought these industries into conflict.

This chapter reviews forestry–fisheries interactions in Alaska, from the early research on the effects of timber harvest to the consensus and compromise of today. The review is organized around two main parts of salmon freshwater life history: spawning and rearing. Spawning is primarily affected by physical habitat variables; rearing is affected by both physical and trophic variables, requiring a broader understanding of the stream ecosystem.

Timber Resources in Alaska

Most of the Alaska timber harvest comes from coastal forests of western hemlock and Sitka spruce. Historically, the timber industry centered on the Tongass National Forest in southeast Alaska (Fig. 9.1); but harvest has recently increased on private lands in both southeast Alaska and coastal forests of southcentral Alaska, primarily Afognak Island and western Cook Inlet (ADCED, 1990). In 1991, the total Alaska timber harvest was 921 million board feet (MBF): 63% from private land, 35% from the Tongass National Forest, and 2% from other public lands (Warren, 1992).

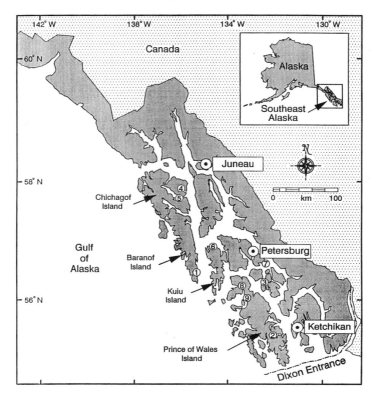

Figure 9.1. Map of southeast Alaska showing sites of research on timber harvest and fish habitat (1 = Sashin Creek; 2 = Hollis; 3 = Porcupine Creek; 4–9 = watersheds studied by Murphy et al. (1986): 4 = Kennel Creek; 5 = Corner Creek; 6 = Straight Creek, 7 = Big Creek; 8 = Gutchi Creek; 9 = Shaheen Creek). Most of southeast Alaska is in the Tongass National Forest.

In addition to its coastal forests, Alaska has extensive boreal forests in the interior, mostly on the intermontane plateau between the Alaska Range and the Brooks Range (Van Cleve et al., 1983). Boreal forests are dominated by white spruce, paper birch, poplars, and, on poorly drained soils, black spruce. These species are typically too small to be of commercial value, although the wood is good quality. The annual boreal timber harvest is small, about 13 MBF (Slaughter, 1990).

Timber management in coastal Alaska forests is based on even-aged silviculture on about a 100-year rotation (Harris and Farr, 1974). Trees are typically harvested by clear-cutting because this is economical and also favors regeneration of Sitka spruce. Logs are usually yarded to landings by high-lead systems and hauled by truck to facilities that transfer the logs to tidewater. Helicopter yarding is sometimes used where high-lead yarding is inappropriate and trees are of high value. Natural regeneration is usually relied on to restock cutover lands, which typically results in overly dense stands that benefit from thinning.

History of Forestry–Fisheries Research in Alaska

Forestry–fisheries research in Alaska has evolved through several phases (Gibbons et al., 1987). Before 1950, timber harvest was on a small scale and of negligible impact to fisheries. Since the 1950s, however, when timber harvest in southeast Alaska began to supply wood for the new pulp industry, effects of timber harvest on Alaska salmonid fisheries have become of increasing concern. From 1950 to the 1960s, research on effects of timber harvest focused on spawning habitat of pink and chum salmon, species that use freshwater streams solely for spawning. As timber harvest expanded from the 1960s through the 1970s, research shifted to the habitat requirements of species that spend several years in streams as juveniles before migrating to the sea (e.g., coho salmon and steelhead).

Since about 1980, forestry and fisheries groups have increased cooperation. The Alaska Working Group on Cooperative Forestry/Fisheries Research was formed in 1981. It consists of representatives of government resource agencies, the fishing and timber industries, and conservation groups, whose goal is to encourage studies to answer priority questions of streamside management (Simpson and Gibbons, 1984). Research and political compromise culminated in 1990 with the passage of the Tongass Timber Reform Act and revision of the Alaska Forest Resources and Practices Act, which prescribed new requirements for streamside buffer strips to protect salmonid habitat. Forestry–fisheries interactions now focus on implementing these new regulations and monitoring their effectiveness in protecting fisheries.

In the future, more commercial timber is likely to come from the coastal forests of southcentral Alaska and the interior boreal forests. Most research on the effects of timber harvest, however, has been conducted in southeast Alaska. Whether this research can be extrapolated to other regions of Alaska is questionable. Similar stream processes probably operate in these other regions, but forestry practices and the role of habitat in fish population dynamics could differ significantly. Consequently, further research is required to determine how timber harvest affects fish populations in the regions where harvest is expected to increase.

Effects of Timber Harvest on Spawning Habitat

Research on effects of timber harvest initially focused on spawning habitat. For successful spawning, adult salmon need access to gravel spawning beds with suitable permeability to water flow. Egg-to-fry survival is typically less than 25% in Alaska coastal streams, and increased mortality is principally caused by three factors: reduction in dissolved oxygen levels, freezing, and instability of spawning gravels (McNeil, 1964). Timber harvest can affect spawning success by altering numerous habitat variables, including sediment, temperature, streamflow, and channel morphology and stability, and by interfering with adult migrations (Gibbons and Salo, 1973).

Sedimentation

Increased sediment from timber harvest can significantly impact fish habitat (Everest et al., 1987; Iwamoto et al., 1978). Sediment is usually classified into two categories: suspended sediment in the water column (typically clay and silt <0.1 mm diameter) and bedload sediment in the streambed (typically >1 mm) that can move during storms by rolling and bouncing along the stream bottom. Particles between about 0.1 and 1 mm may be suspended or bedload depending on streamflow. The sediments that are most important to the spawning environment are the fine sediments, usually defined as particles less than 0.88 mm diameter[1], that move into or out of the streambed depending on flow.

Fine sediment deposited in the redd can reduce egg-to-fry survival and fry quality by suffocating eggs and larvae, and by forming a physical barrier to emergence (Everest et al., 1987). Permeability of stream gravel decreases sharply as fine sediment increases from 5 to 20% of the substrate material (Fig. 9.2); egg to fry survival declines similarly (Fig. 9.3; Everest et al., 1987) because reduced permeability decreases the availability of oxygen and the flushing of metabolic wastes from incubating eggs and alevins. Coarser sediment can also reduce spawning success by capping gravels and restricting the emergence of alevins. Egg-to-fry survival of coho salmon in Sashin Creek, Baranof Island (Fig. 9.1), was reduced by sediment <1.7 mm diameter (Crone and Bond, 1976). Mortality associated with low dissolved oxygen (<4 mg L^{-1}) in three Prince of Wales Island streams near Hollis (Fig. 9.1) ranged from 60 to 90% of deposited eggs (McNeil, 1964).

Logging roads are typically the greatest sediment source associated with timber harvest in Alaska. Sediment delivery to streams in two Chichagof Island watersheds during road construction was about 100 m^3 km^{-1}yr^{-1} of road compared to only 0.2 m^3 km^{-1} yr^{-1} from lightly used roads after construction (Swanston et al., 1990). Monitoring, however, indicated that best management practices (BMPs, regulations controlling the manner of harvest and road construction) limited sediment inputs from road construction to levels within the natural range; suspended sediment reached 277 mg L^{-1} at

---

[1] The 0.88 mm diameter size corresponded to a particular sieve size used in the research that demonstrated detrimental effects on egg-to-fry survival from particles passing through that sieve.

high streamflow but was generally lower (Paustian, 1987). In comparison, suspended sediments reached 320 mg L^{-1} in a Chichagof Island old-growth forest (Campbell and Sidle, 1985).

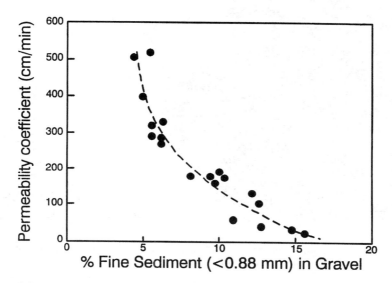

Figure 9.2. Relationship between gravel permeability and percentage of sediment less than 0.88 mm (after McNeil, 1964).

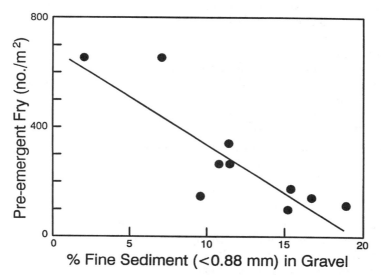

Figure 9.3. Relationship between pink salmon egg-to-fry survival and percentage of fine sediment. Correlation $r = 0.86$ (after Koski, 1972).

Figure 9.4. Log hauling on a gravel road during a dry spell on Kuiu Island, southeast Alaska (photo T.R. Merrell, Jr. 1982).

Regular road use can cause chronic sediment input to streams. Because of weak rock material, the gravel road surface in many areas of southeast Alaska breaks down with repeated heavy wheel loads of hauling trucks (Fig. 9.4), particularly under wet conditions, resulting in a continual source of fine sediment to streams (Powell, 1989). Sediment produced by log hauling is estimated at $50\,m^3\,km^{-1}\,yr^{-1}$ of road (Swanston et al., 1990). Lack of knowledge about the process of sediment delivery from roads to streams, however, makes the effects of road sediment on fish spawning success difficult to predict.

Road failure is a concomitant risk of road construction in mountainous terrain (Furniss et al., 1991). Severe gullying and landslides may result if roads intercept and concentrate runoff and discharge it into nondrainage areas. Stream crossings (i.e., culverts and bridges) pose the greatest risk to fish habitat of any road feature. When culverts are plugged by debris or overtopped by high streamflow, stream channels may realign and sedimentation may be severe. The risk of failure of a crossing structure depends on its size compared to flood events, and culverts should be sized to accommodate at least a 50-year flood (Furniss et al., 1991). Whatever the design life, any crossing structure is virtually certain to fail if it is not maintained or removed after the road is abandoned.

Compared to roads, tree felling and yarding away from streambanks usually produce negligible sediment due to several factors that limit surface erosion (Everest et al., 1987). In southeast Alaska, coarse-textured soils with thick organic surface layers, high soil permeability and infiltration, and rapid revegetation of disturbed soils tend to limit sediment production from tree felling and yarding.

Felling and yarding along streambanks, however, can produce sedimentation by mobilizing sediment stored in the stream channel and flood plain. Disturbing the stream channel during felling and yarding, introducing logging slash, and salvaging merchantable logs from the flood plain can cause significant scour and deposition (Fig. 9.5). Landslides also can increase because of tree removal (Schwan et al., 1985) because increased snow accumulation in clear-cuts increases downslope weight (Hartsog, 1990). Frequency of landslides in the Maybeso Creek watershed, Prince of Wales Island, increased 4.5-fold after timber harvest (Bishop and Stevens, 1964), and the increased risk of landsliding appears to be long term. Heavy rainfall in October 1993 triggered numerous landslides in second-growth areas of Prince of Wales Island at a frequency more than four times that in old-growth areas (Landwehr, 1994).

The effects of timber harvest on sedimentation are difficult to assess because of natural variation in sediment dynamics. Paustian (1987) highlighted the difficulty of detecting changes in suspended sediment after timber harvest because of extreme natural variation due to seasonal changes in discharge and storm events. Redd digging by salmon also re-

Figure 9.5. Destabilization and erosion of a stream channel after clear-cut logging (photo T.R. Merrell, Jr., 1982).

moves fine sediment from the streambed, so that sediment content in redds after digging is temporarily less than in the surrounding area (McNeil and Ahnell, 1964). Mean percentage of fine sediment in spawning gravels ranges from about 6 to 14% in undisturbed watersheds in southeast Alaska (Edgington, 1984; Kingsbury, 1973). Sheridan et al. (1984) found no significant difference in mean percentage of fine sediment in spawning riffles between six streams in logged areas (5 to 15% fines) and six in undisturbed areas (5 to 12% fines). Nevertheless, a 10% increase in fine sediments in spawning gravels may be undetectable because of natural variability, but could still reduce egg to fry survival (Edgington, 1984).

Early studies near Hollis found conflicting evidence concerning the effects of timber harvest on spawning gravel. The first studies indicated that sediment production was not significantly greater in logged watersheds (Alaska Forest Research Center, 1957, 1960). Later, however, Bishop and Stevens (1964) found that sedimentation rate was four times the natural rate. McNeil and Ahnell (1964) also reported two to four times more fine sediment in the Harris River during timber harvest, but could not conclusively identify timber harvest as the sediment source because of high variability. Sediment returned to normal 5 years after logging ceased (Sheridan and McNeil, 1968). In 108 Creek, fine sediment in spawning gravel increased 6 to 8% during timber harvest and remained elevated for 3 years (Kingsbury, 1973). Pella and Myren (1974) questioned the validity of the early Hollis studies because streambed sediment and egg-to-fry survival were not monitored long enough to evaluate long-term effects.

Although the effects of sediment from timber harvest are not well established for southeast Alaska, adverse effects have been documented by long-term studies in British Columbia (Holtby and Scrivener, 1989) and Washington (Cederholm et al., 1981). In Carnation Creek, British Columbia, for example, increased fine sediment in spawning gravels after timber harvest contributed to a 25% reduction in chum salmon escapement (Holtby and Scrivener, 1989). The Southeast Alaska Multiresource Model, developed by the U.S. Forest Service and based principally on Alaska research, indicated for a model 1,600-ha watershed that typical timber harvest and roading could increase fine sediment in spawning gravels by 160% and decrease pink salmon returns by 35% (from 275,000 to 180,000), with recovery taking 12 years (Garrett et al., 1990).

New road construction for timber harvest in Alaska could significantly impact spawning habitat. Under current projections, roads in the Tongass National Forest (now totaling 4800 km) would increase to over 14,000 km in 50 years (U.S. Forest Service, 1991). Impacts of sedimentation from these and other private logging roads are difficult to predict. Risks of impacts—from road failures, washout of culverts and bridges, and failure of culverts to pass fish—increase with the kilometers of roads constructed.

Temperature

Water temperature affects development and survival of salmonids at all life stages. Temperature regulates the development rate of eggs and larvae, and each life stage has an optimal temperature range with upper and lower lethal limits (Weber–Scannell, 1991). Removing the forest canopy by timber harvest along streams, especially where no buffer strips are present, can increase average stream temperature and its daily fluctuation. Stream size, gradient, orientation, and stream length open to sunlight determine the magnitude of temperature increases (U.S. Forest Service, 1986).

Research on effects of timber harvest on stream temperature in southeast Alaska began in 1949 near Hollis with intensive studies on four drainages: two to be harvested (Harris River and Maybeso Creek), and two unharvested controls (Indian Creek and Old Tom Creek). The first studies (Harris, 1960; James, 1956) reported no change in temperature after timber harvest. Further studies, however, showed that timber harvest increased the maximum summer temperature as much as 5°C, and the average increase in monthly maximum temperature was 2.2°C (Meehan et al., 1969).

Later, extensive studies found that temperature in small southeast Alaska streams in clear-cuts reached 23°C (Tyler and Gibbons, 1973), approaching the lethal levels for salmon of about 25 to 30°C (Bjornn and Reiser, 1991). The rate of warming per 100-m length of stream was 0.92°C in one clear-cut compared to 0.07°C in old-growth forest (Tyler and Gibbons, 1973). In a study of 13 southeast Alaska streams (Meehan, 1970), the largest increase in a clear-cut was 2.3°C in a 183-m reach (equal to 1.25°C $100 \, \text{m}^{-1}$), but temperature did not approach lethal levels. The increase in stream temperature after timber harvest is directly proportional to the area of stream exposed to solar radiation and inversely proportional to streamflow (Beschta et al., 1987).

Effects of increased stream temperature in summer are a lesser problem in most southeast Alaska streams than in more southerly states and other Alaska areas because of southeast Alaska's cooler and cloudier weather (Gibbons et al., 1987). Some southeast Alaska streams, however, are "temperature sensitive" because of their specific characteristics, and temperature can rise to undesirable levels if shade canopy is removed (Gibbons et al., 1987). Streams most likely to be temperature sensitive are relatively wide, shallow, low-gradient streams with lake or muskeg sources (U.S. Forest Service, 1986).

Temperature is particularly important for regulating the rate of egg development, and increased temperature during the incubation period can cause fry to emerge early. Thedinga et al. (1989) estimated that because of increased temperature, primarily in spring, coho salmon fry in two southeast Alaska watersheds could emerge 37 days earlier in clear-cut reaches

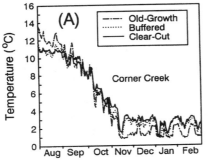 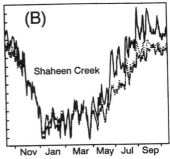

Figure 9.6. Mean daily water temperature in old-growth, buffered, and clear-cut reaches of streams in 1982–1983 in two watersheds: (**A**) Corner Creek, Chichagof Island, and (**B**) Shaheen Creek, Prince of Wales Island (after Thedinga et al., 1989).

and 22 days earlier in reaches with buffer strips than in old-growth reaches (Fig. 9.6). During the incubation period (November to May), clear-cut and buffered reaches accumulated 21 and 12%, respectively, more temperature units (degree-days) than old-growth reaches. Early emergence can be harmful because it exposes fry to floods in early spring (Hartman et al., 1984).

Timber harvest is also thought to decrease stream temperature in winter because insulating canopy is removed, but research results are equivocal. Meehan et al. (1969) reported little or no difference between winter temperature of southeast Alaska streams in logged and in forested areas. Thedinga et al. (1989) found higher winter temperature in an old-growth reach than in a clear-cut reach in one watershed, but the opposite in a second watershed (Fig. 9.6). Timber harvest may actually increase winter temperature by increasing soil insolation, as demonstrated for the Carnation Creek watershed in British Columbia (Hartman et al., 1984). Although the overall effects of timber harvest on winter temperature have not been clearly demonstrated, concern is warranted. Winter stream temperature is usually near 0°C and typically varies within a small range (Lamke et al., 1991). Small temperature shifts in winter can thus significantly influence incubating eggs and overwintering juveniles.

Streamflow

Timber harvest effects on streamflow can influence spawning success by altering the hydrologic regime. This regime responds to changed distribution of water and snow on the ground, rate of snowmelt, precipitation intercepted by and evaporated on foliage, amount of water transpired from soil by vegetation, and physical structure of the soil governing rate of water movement to stream channels (Chamberlin et al., 1991). The effects on streamflow that most concern fisheries managers are changes in base and peak flows.

Base flow usually increases during the first 10 to 20 years after timber harvest because removing trees decreases interception of precipitation and evapotranspiration (Chamberlin et al., 1991). Increases in base flow are greatest during the summer growing season. Later, in the rainy season (fall to winter), the soil is saturated and runoff is similar in both clear-cut and forested areas. After 10 to 20 years, base flows may actually decrease below preharvest levels because of greater evapotranspiration in rapidly growing second-growth forest (Harr, 1983; Hicks et al., 1991; Myren and Ellis, 1984). In southeast Alaska, base flow in Staney Creek (drainage area, 134 km^2), Prince of Wales Island, increased significantly after about 20% of the watershed had been harvested (Bartos, 1989).

Peak flow may also increase after timber harvest. Activities that disturb and compact the soil, such as road construction and yarding, reduce soil permeability causing more surface runoff and high peak flows. Ditches along roads collect surface runoff and intercept subsurface flow, also contributing to higher peak flows. More accumulated snow and faster melting rate in clear-cut areas causes higher peak flows during rain-on-snow events and spring snowmelt (Golding, 1987; Chamberlin et al., 1991). Higher peak flows are considered detrimental to fish because bed load overturn, which occurs during high flows, can kill many incubating eggs (McNeil, 1964; Murphy et al., 1990). Watershed geology may modify the effects of increased peak flow. Based on five gaged southeast Alaska watersheds, Bartos (1993) concluded that peak flows increased after timber harvest in three watersheds with metamorphic geology, but decreased in two watersheds with abundant limestone. The role of geology in determining effects of timber harvest on peak flows was unexplained.

Channel Morphology and Stability

Channel morphology in small streams is formed primarily by large woody debris (LWD), usually defined as tree boles, branches, and roots greater than 10 cm diameter and 1 m long (Bisson et al., 1987; Sullivan et al., 1987). Large woody debris plays an important role in creating spawning habitat in small streams. By forming pools, LWD increases the frequency of pool–riffle transition areas, where many salmonids prefer to spawn (Bjornn and Reiser, 1991). In gravel-poor streams, gravel deposits upstream from LWD can provide suitable spawning substrate (Everest and Meehan, 1981). In sediment-rich streams, the scouring caused by flow obstructions like LWD benefits spawning fish by sweeping fines out of spawning beds (Sedell and Swanson, 1984).

Timber harvest can affect spawning habitat significantly by altering LWD. Accumulations of LWD in Maybeso Creek decreased in size and number for 30 years after timber harvest (Bryant, 1980), and similar changes occurred in the Harris River (Bryant, 1985). Logging slash is less stable and less effective in forming fish habitat than LWD from natural sources; and when slash moves, it can destabilize natural LWD accumulations downstream (Bryant, 1980).

Channel stability is also important in determining egg-to-fry survival. Shifting of gravel and unstable LWD during floods is an important mortality factor in coastal streams (McNeil, 1964). Bedload overturn is a function of high water velocity during peak streamflows, and increased peak flows after timber harvest increase the risk of bedload overturn (Swanston et al., 1990).

Adult Migration

Timber harvest can interfere with adult migration by blocking migrations at stream crossings of logging roads, by causing logjams that block migration, by decreasing cover from predators, by decreasing the frequency of large pools used for resting, and by adversely affecting temperature and amount of dissolved oxygen.

Culverts, especially those installed above the grade of a stream, can be a barrier to upstream fish migration (Furniss et al., 1991). Culvert conditions that block fish passage include too fast water velocity, too shallow water depth, no resting pool below culvert, and too high a jump to the culvert. A single poorly installed culvert can eliminate the fish population of an entire stream system.

Before regulations were adopted to limit logging slash entering streams, accumulated slash from timber harvest could block upstream migrations of adult salmon (Fig. 9.7), and much effort was spent on removing these logjams (Bisson et al., 1987; Hall and Baker, 1982). The severity of these blockages has been reduced by the mandated removal of logging slash from streams after timber harvest, and today streamside buffer strips restrict logging slash from entering streams. Further, natural logjams that appear to be blockages often allow fish to pass at high flows during migration (Bryant, 1983).

Decreases in large pools and cover can also affect spawning migrations. Adult salmon often hold for several weeks in large pools when ascending a stream to spawn (Burger et al., 1985; Thedinga et al., 1993). The number of suitable large pools for adult holding can be limited along a stream. Timber harvest can decrease the number of large pools by removing, reducing, or destabilizing the large "key" pieces of LWD that create these pools (Sedell and Everest, 1990).

Prespawn die-offs of adult pink and chum salmon are not uncommon in southeast Alaska, particularly in drier summers (Murphy, 1985); and timber harvest has been implicated by increasing stream temperature. In 1989, for example, die-offs ranging from hundreds to 54,000 were documented in 19 streams (ADEC, 1990).

From the available record, prespawn die-offs occur in both logged and forested stream reaches. Most streams that have summer drought-related die-offs have several characteristics in common:

1. small drainage area ($<50\,km^2$);
2. headwater elevation from 300 to 450 m (hence, little snow pack);
3. absence of lakes, ponds, and beaver dams that would help to maintain streamflow during drought;
4. extensive low-elevation, shallow muskeg-pond tributaries that contribute to high water temperature during sunny periods; and
5. confined intertidal systems that restrict tidal exchange (Brown, 1989).

Die-offs can be triggered when returning adults follow the flood tide into a stream, and then are stranded in pools as the tide ebbs (Murphy, 1985).

Because of the public controversy about the role of timber harvest in prespawn die-offs, the Alaska Working Group on Cooperative Forestry/

Figure 9.7. Logging slash completely filling a stream channel (photo M.W. Oswood).

Fisheries Research initiated research to determine the cause of these summer die-offs. Murphy (1985) concluded that a die-off of pink and chum salmon in Porcupine Creek (an old-growth forest stream) was not directly due to high temperature (which did not exceed 19°C), but to suffocation from low dissolved oxygen ($<2\,mg\,L^{-1}$), a result of stranding and crowding in shallow water. After examining the concentration of dissolved oxygen in pools of seven streams in southeast Alaska in August 1990, Martin et al. (1991) also concluded that the most likely cause was oxygen depletion from fish respiration during low streamflow.

Effects of timber harvest on prespawn die-offs are still uncertain. High stream temperature from canopy removal may contribute to mortality by lowering the saturation concentration of oxygen in water and increasing fish respiration. Timber harvest also reduces evapotranspiration, increasing the summer base streamflow, that could make die-offs less severe (Murphy, 1985). However, a possible decrease in base streamflow 10 to 20 years after timber harvest could reverse this trend (Myren and Ellis, 1984; Hicks, Hall, et al., 1991).

Effects of Timber Harvest on Rearing Habitat

Research on spawning habitat addresses only one aspect of salmonid habitat requirements and does not examine the influence of timber harvest on species limited by the quality and quantity of rearing habitat: juvenile coho salmon, steelhead, cutthroat trout, and Dolly Varden. If spawning escapement is adequate, sufficient fry are usually produced to exceed the carrying capacity of available rearing habitat (e.g., Crone and Bond, 1976). Reduced spawning success in rearing-limited populations would not necessarily reduce smolt yield because of a compensatory density-dependent response at later stages (Bjornn and Reiser, 1991; Everest et al., 1987) as long as effects are not too severe and sufficient fry continue to be produced to fully seed available habitat.

Smolt yield is considered the key variable in assessing effects of timber harvest on rearing-limited species (Koski et al., 1985). The number of returning adults may be the ultimate measure of timber harvest effects, but it is confounded by widely varying marine survival factors (Pella and Myren, 1974). Timber harvest can still affect marine survival of smolts, however, particularly by altering a stream's trophic status and temperature regime. Increased temperature, for example, can cause smolts to migrate to sea early, as observed in a Carnation Creek study by Holtby and Scrivener (1989). They speculated that earlier smolt migration after timber harvest caused a net decrease in adult returns, even though it caused an increase in smolt yield. Earlier smolt migration is thought to reduce ocean survival because smolts enter saltwater before plankton food sources have bloomed, slowing their growth and increasing predation.

Effects of timber harvest on rearing habitat are more complicated to assess than effects on spawning habitat. Successful rearing depends on both physical and trophic factors and their interaction. As predators, salmonids are influenced by all trophic levels in the stream ecosystem, from primary production to decomposition, as well as by physical conditions of the habitat (Murphy and Meehan, 1991).

The predominant habitat features of streams in old-growth forest are an abundance of LWD, a mix of deciduous and coniferous leaf litter and woody debris, and canopy gaps that allow algal growth on the stream bottom (Sedell and Swanson, 1984; Fig. 9.8). Clear-cutting without buffer strips can reduce LWD by removing debris from stream channels after yarding, destabilizing LWD accumulations leading to their downstream export, salvaging downed logs from the flood plain, and reducing the supply of potential new LWD. Clear-cutting can severely reduce the diversity of stream habitats, with scant LWD in the stream channel and a long-term shortage of new supply (Fig. 9.9). Buffer strips (Fig. 9.10) were rare in southeast Alaska before the late 1980s because of the monetary value of streamside trees and the tendency of buffer strips to be blown down (Fig. 9.11). Concern about effects of timber harvest on rearing habitat prompted research to determine the effects of clear-cutting and the effectiveness of buffer strips in protecting stream habitat.

Figure 9.8. Typical section of a valley-bottom stream in old-growth forest in southeast Alaska (photo T.R. Merrell, Jr., 1982).

Figure 9.9. Stream flowing through a clear-cut area without a streamside buffer strip before the Tongass Timber Reform Act of 1990 (photo T.R. Merrell, Jr., 1982).

Figure 9.10. A streamside buffer strip used to protect fish habitat along a southeast Alaska stream before the Tongass Timber Reform Act of 1990 (photo T.R. Merrell, Jr., 1982).

Figure 9.11. Wind-thrown trees in a streamside buffer strip (photo T.R. Merrell, Jr., 1982).

To assess overall effects of timber harvest on rearing habitat of juvenile salmonids, one has to consider specific limiting factors ("bottlenecks") over the entire juvenile period. For the analysis of limiting factors, the rearing period is typically divided into summer and winter periods dominated by different habitat components (Reeves et al., 1989).

Summer Rearing

In summer, effects of timber harvest on juvenile salmon appear to be more related to food supply than to physical factors. Stream ecosystems have two energy sources: in-stream primary production (autochthonous sources) and out-of-stream sources of organic matter (allochthonous input) (Murphy and Meehan, 1991). Timber harvest can affect both these sources by altering riparian vegetation and other physical conditions in the stream, such as temperature, nutrients, stream morphology, and sediment. Effects vary from reach to reach, and through time as the riparian plant community passes through stages of recovery.

Removing the forest canopy increases insolation of the streambed, which can increase aquatic primary production (Gregory et al., 1987; Murphy and Meehan, 1991). Where light is limiting, the predominant energy source for the stream ecosystem shifts from allochthonous to autochthonous production for the first 10 to 20 years after timber harvest. For example, primary production was significantly higher in clear-cut reaches of tributaries of Staney Creek, Prince of Wales Island, than in old-growth Porcupine Creek, and was negatively correlated with percent canopy cover in Porcupine Creek (Walter, 1984). In three small streams on Prince of Wales Island, periphyton production and community respiration were higher in a recently clear-cut stream than in old-growth and second-growth streams (Duncan and Brusven, 1985a), and experimental removal of second-growth riparian alder increased primary production (Bjornn et al., 1992).

Although most Alaska studies have shown increased primary production after canopy removal, indicating light limitation, some streams show no response to increased light, probably because of nutrient limitations. Nitrate and phosphate levels in a Chichagof Island stream were only 0.07 and $0.009 \, mg \, L^{-1}$, respectively, levels not increased by clear-cutting and burning (Stednick et al., 1982). Rapid revegetation also helps to limit nutrient losses from the soil after timber harvest, and any increases in nutrient input to streams are usually small (Brown, 1972). In southeast Alaska watersheds underlain by sedimentary rocks such as limestone, streams are likely to be light limited; but where the parent material is igneous, streams may be nutrient limited (Murphy et al., 1986; Fig. 9.12).

The increased algal production after clear-cutting results in more abundant benthic invertebrates (Duncan and Brusven, 1985b). The density of invertebrates in old-growth, buffered, and clear-cut streams was directly related to biomass of algae (Fig. 9.13; Murphy et al., 1986). Removal of second-growth riparian alder increased both algal production and abundance of benthic invertebrates (Bjornn et al., 1992).

Although the predominant energy source changes after timber harvest, the dominant macroinvertebrates and functional feeding groups may remain unchanged (Duncan and Brusven, 1985b; Hawkins et al., 1982). Collector–gatherers dominate in both shaded and open stream reaches because much algal production is used as organic detritus after the algae sloughs from rocks (Murphy et al., 1981). Thus, canopy removal can increase the abundance of invertebrates by enhancing the food quality of detritus.

Density of coho salmon fry often increases during the first 10 to 15 years after timber harvest because of the increased production of invertebrates

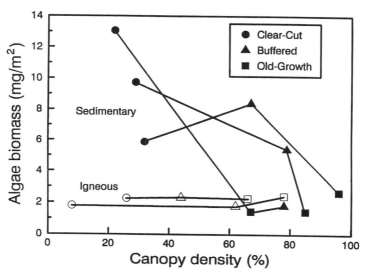

Figure 9.12. Relationship between algae biomass (ash-free dry matter) and canopy density in old-growth, buffered, and clear-cut reaches of streams in five areas of southeast Alaska. Closed symbols indicate areas with sedimentary geologic parent material; open symbols indicate areas with mostly igneous parent material (after Murphy et al., 1986).

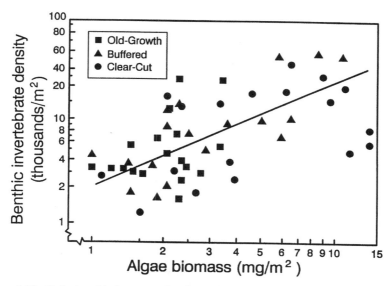

Figure 9.13. Relationship between density of benthic invertebrates and algae biomass (ash-free dry matter) in old-growth, buffered, and clear-cut reaches of streams in southeast Alaska. Both variables are in logarithmic scales. Correlation $r = 0.58$ (after Murphy et al., 1986).

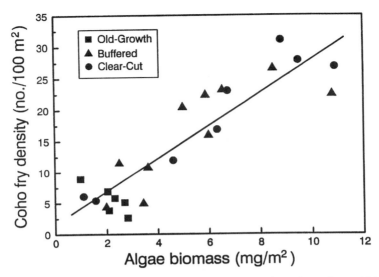

Figure 9.14. Relationship between summer density of coho salmon fry and biomass of algae (ash-free dry matter) in old-growth, buffered, and clear-cut reaches of streams in southeast Alaska. Correlation $r = 0.90$ (after Murphy et al., 1985).

(Murphy and Meehan, 1991). Where food is limiting, the summer density of coho salmon fry can be directly related to the abundance of invertebrates, amount of primary production, or algal biomass (Fig. 9.14). The higher density of coho salmon fry is probably the result of smaller feeding territories (Dill et al., 1981).

On the negative side, timber harvest along small streams may increase stream temperature enough to inhibit growth and cause mortality in juvenile salmonids. In Oregon, although stream temperature reached 30°C, coho salmon and cutthroat trout died only when slash burning quickly raised stream temperature from 13° to 28°C (Hall and Lantz, 1969). In southeast Alaska, however, stream temperature in clear-cuts rarely exceeds 26°C, even on sunny days, except in exposed, intermittent pools (Sheridan and Bloom, 1975). High temperature can also reduce fish growth if food is restricted (Bjornn and Reiser, 1991). However, diel temperature fluctuations simulating conditions in southeast Alaska streams in clear-cuts (maximum range, 6.5° to 20°C) did not influence mortality, growth, or energy reserves of juvenile coho salmon under laboratory conditions (Thomas et al., 1986).

Effects of timber harvest on a stream's physical habitat can also affect summer rearing. High densities of juvenile coho salmon and Dolly Varden are typically associated with LWD and pool habitat (Cardinal, 1980; Dolloff, 1986; Murphy et al., 1986). Loss of LWD can reduce cover and pool habitat, as well as reduce the stream channel's storage capacity for organic matter (Bilby and Likens, 1980). Increased fine sediment deposited on the streambed can reduce invertebrate production (Cordone and Kelley, 1961), and turbidity from suspended sediment can reduce primary production (Lloyd et al., 1987) and cause fish to emigrate (Bisson and Bilby, 1982).

Probably most important are the potential long-term effects of overshading from second-growth canopy that may outweigh the increased production in early successional stages (Murphy and Hall, 1981; Sedell and Swanson, 1984). Second-growth vegetation produces a denser shade and lacks the canopy gaps that are common in old-growth forest (Bjornn et al., 1992). Thus, increased stream production in the first 20 years after timber harvest may be followed by 100 years of depressed production (Sedell and Swanson, 1984).

Winter Rearing

Winter habitat, particularly pools with LWD, is critical to the winter survival of juvenile salmonids. Winter mortality, in contrast with summer mortality, appears to result mainly from density-independent factors (Hunt, 1969). Theoretically, carrying capacity of summer habitat sets the upper limit on smolt yield that is then reduced in relation to the severity of winter conditions and quality of winter habitat. Survival factors become more important as latitude increases, because winters are more severe and juvenile salmonids spend more years in freshwater. South of Alaska, most coho salmon spend only one winter in fresh water, but in southeast Alaska two-thirds spend two winters (Gray et al., 1981).

Winter motality in streams is usually substantial: 46 to 94% of summer populations (Hartman et al., 1987; Murphy et al., 1984). Mortality can usually be attributed to hazardous conditions during fall and winter freshets; stranding of fish by ice dams; and physiological stress from low temperature, oxygen depletion, or progressive starvation (Bryant, 1984; Harding, 1993; Murphy et al., 1984; and see Chapter 11). Winter mortality is inversely related to the amount of pool habitat and is usually greater in main stream reaches than in small tributary streams, sloughs, lakes, ponds, and other off-channel areas (Murphy et al., 1984).

Although timber harvest tends to increase fry abundance in summer by opening the canopy, this positive effect can be nullified by reduced winter habitat (Koski et al., 1985). In a study of six southeast Alaska watersheds (Fig. 9.1), summer density of coho salmon fry was significantly greater in both buffered and clear-cut streams than in old-growth streams (Murphy et al., 1986). The enhanced fry abundance in clear-cuts, however, was not maintained through the following winter. Presmolts in late winter were actually less abundant in clear-cuts than in old growth; however, buffered streams maintained higher presmolt density (Thedinga et al., 1989; Fig.

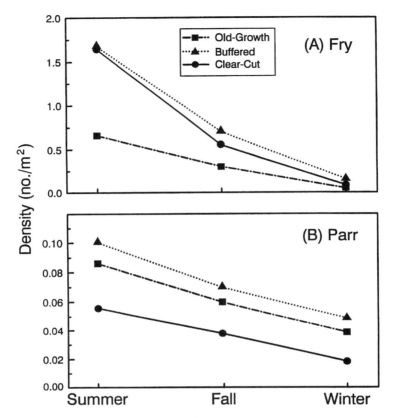

Figure 9.15. Decline in density of coho salmon (**A**) fry (age 0) and (**B**) parr (age 1 and 2) from August 1982 to March 1983 in old-growth, buffered, and clear-cut reaches of streams in southeast Alaska (after Murphy et al., 1985).

9.15). The disadvantage in clear-cuts was a reduction in pools and LWD; the advantage in buffered reaches was a combination of both enhanced food abundance because of more open canopy in summer and increased LWD cover in winter. Thus, winter habitat is frequently a bottleneck in freshwater production of salmon smolts, and timber harvest typically has its most detrimental effects on winter habitat.

LWD is clearly a key habitat feature for juvenile salmonids. It not only provides cover directly, but also forms 80 to 90% of pools in typical valley-bottom streams (Heifetz et al., 1986). LWD also helps maintain water levels in small streams during low-flow periods (Lisle, 1986), which occur in both summer and winter (Lamke et al., 1991). Removal of LWD causes loss of juvenile salmonids (Elliott, 1986).

Fish may travel long distances to suitable winter habitat, paticularly to small tributary streams, spring ponds, lakes, and stream reaches with abun-

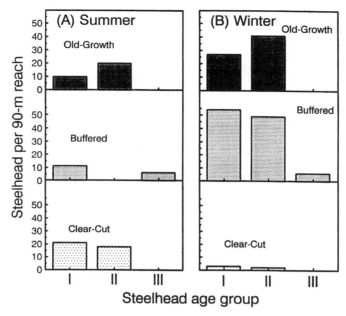

Figure 9.16. Comparison of (**A**) summer and (**B**) winter numbers of juvenile steelhead in old-growth, buffered, and clear-cut reaches of the Shaheen Creek drainage. Each bar shows the total number in three separate 30-m reaches (after Johnson et al., 1986).

dant LWD (Harding, 1993; Murphy et al., 1984). In Shaheen Creek, juvenile steelhead left clear-cut reaches where LWD had been depleted and moved into old-growth and buffered reaches where LWD cover was abundant (Johnson et al., 1986; Fig. 9.16). Logging roads can interfere with juvenile salmonid migrations to winter habitat if stream crossings (i.e., culverts) are not adequately designed and maintained for fish passage.

Blowdown of buffer strips, once considered damaging to habitat, can actually enhance winter cover for juvenile salmonids by providing additional LWD, particularly large trees with attached rootwards (Murphy et al., 1985). In a blowdown area of Shaheen Creek, the number of large pieces of LWD and rootwads more than quadrupled (Fig. 9.17). To our knowledge, no upper limit to the direct relationship between wintering juvenile salmonids and amount of LWD (Fig. 9.18) has yet been measured. In extreme cases of blowdown, however, habitat could be damaged by sediment introduced from upturned rootwads and increased channel instability.

Cumulative Effects of Timber Harvest

Cumulative effects are important considerations in timber harvest because individual harvest units must be considered in the context of all other

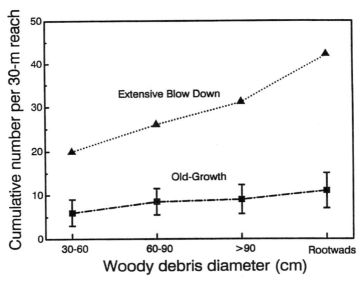

Figure 9.17. Cumulative mean number of pieces of woody debris by diameter size class per 30-m stream length in 18 old-growth reaches and in one buffered reach with extensive blowdown. Vertical bars denote 95% confidence intervals for old-growth (after Murphy et al., 1985).

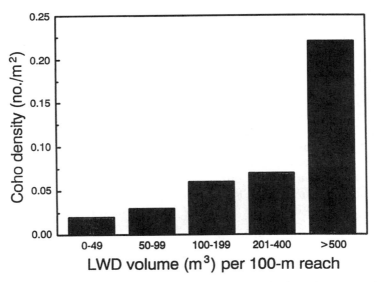

Figure 9.18. The relationship between winter density of coho salmon parr (age 1 and 2) and amount of large woody debris (LWD) (after Murphy et al., 1985).

activities in the watershed. Effects of localized small impacts can accumulate and interact, increasing the overall effect on the resource (Burns, 1991). Conventional impact assessments typically are bounded by the expected zone of influence of a single disturbance. Cumulative assessment takes a broader view, with wider boundaries in number of disturbances, geographic area, and time frame (Preston and Bedford, 1990).

Timber harvest effects can accumulate for several reasons. Changes in streamflow, water temperature, and sediment dynamics are not just local problems restricted to a particular stream reach, but problems that can have adverse cumulative effects throughout the entire downstream basin (Sedell and Swanson, 1984). For example, sediment from timber harvest, combined with altered sediment storage and transport due to changed LWD inputs and stability, may cause downstream streambed aggradation, channel widening, and degradation of fish habitat (Smith, 1989).

Wider temporal scales must be considered because habitat variables change systematically as the land and stream recover after timber harvest. Cumulative effects on streamflow, for example, depend on length of time since harvest, because the early decrease in evapotranspiration may be reversed after second-growth vegetation becomes established (Myren and Ellis, 1984). Similarly, the increased primary production after canopy removal can be reversed after second-growth vegetation shades the stream (Bjornn et al., 1992).

Cumulative effects on different habitat variables can also be synergistic or antagonistic. Increased streamflow, for example, may moderate effects of increased stream temperature from canopy removal (Chamberlin et al., 1991). Loss of essential winter cover because of LWD removal can nullify enhanced primary production in summer (Johnson et al., 1986; Murphy et al., 1986).

Identifying cumulative effects of timber harvest on fish habitat is therefore difficult, not only because of the technical complexity, but also because few studies have been sustained over the time required to elucidate the effects (Chamberlin et al., 1991). With the exception of Bryant's (1980, 1985) studies of LWD and channel morphology in Maybeso Creek, no major studies to date have addressed the cumulative effects of forest management on fish habitat in Alaska.

Regulatory Changes to Protect Fish Habitat

Research on the effects of timber harvest on fish habitat has highlighted the critical role of LWD in overwinter survival of juvenile salmonids, the inadequacy of clear-cutting in maintaining LWD, and the effectiveness of buffer strips in protecting fish habitat. The most important problem in streamside management has been how to provide for long-term maintenance of LWD. The best timber stands usually occur close to streams, and this wood is

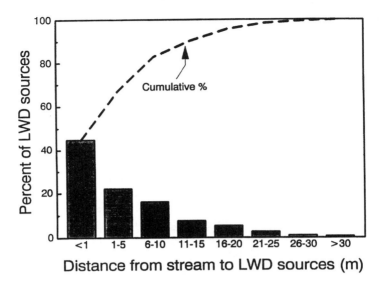

Distance from stream to LWD sources (m)

Figure 9.19. Distance from the stream bank to sources of large woody debris (LWD). Histogram bars show percentage of all identified sources ($n = 861$) at given distance from the stream for 32 stream reaches in old-growth forest, southeast Alaska (after Murphy and Koski, 1989).

valuable for both timber harvest and fish habitat. The important questions are how wide should buffer strips be and how many trees should be left along streams to maintain LWD input for the long term.

To establish requirements for buffer strips, studies were undertaken to determine the sources and depletion rate of LWD. Studies in old-growth forest streams showed that 99% of LWD comes from within 30m of the stream channel (Fig. 9.19) through stream undercutting, windthrow, mortality, landslides, and beaver activity (Murphy et al., 1987). By aging saplings growing on LWD, Murphy and Koski (1989) estimated the LWD's age and depletion rate. Longevity of LWD is proportional to bole diameter, increasing from about 100 years for small LWD (10 to 30cm diameter) to over 200 years for very large LWD (>90cm diameter). In most types of streams, LWD is depleted by decay, fragmentation, and export at 1 to 2% per year. A model of the long-term changes in LWD after clear-cutting without buffer strips predicted that LWD would be reduced by 70% at 90 years, and would take more than 250 years to recover (Fig. 9.20). This model made a strong case for the need for minimum 30-m buffer strips along streams to maintain LWD.

Based on this research, the National Marine Fisheries Service, Alaska Region, issued a policy statement in 1988 calling for minimum 30-m buffer strips along anadromous fish streams. This policy became the focal point for debate as the Congress and the Alaska Legislature considered reforming forest practices on Federal, State, and private lands.

Congress passed the Tongass Timber Reform Act (Public Law 101-626) in November 1990, reforming timber harvest practices on federal lands in the Tongass National Forest. The Act requires no-cut buffer strips at least 30 m wide on both sides of all class I streams (those with anadromous fish) and on those class II streams (those with resident fish only) that flow directly into a class I stream. No buffer strips are required along class III streams (those with no fish), but BMPs must be followed to prevent downstream sedimentation and excessive increases in temperature.

The Alaska Legislature amended the Alaska Forest Resources and Practices Act (AS41.17) in May 1990 to reform forest practices on State and private lands. The new standards differ between State and private lands. On State lands, timber harvest is prohibited within 30 m of anadromous fish streams; on private lands, standards differ according to stream type:

1. along a "type A" stream (anadromous fish stream not incised in bedrock), no timber may be harvested within 22 m of the stream bank;

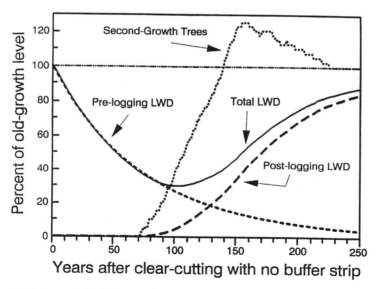

Figure 9.20. A model of changes in number of key pieces of large woody debris (LWD, >60 cm diameter) in a small valley-bottom stream channel after clear-cutting without a buffer strip. The model predicts the amount of LWD as a percentage of old-growth levels, based on estimates of the depletion rate of prelogging LWD, regrowth of streamside trees, and input of LWD from second-growth forest. Total LWD equals the sum of pre- and postlogging LWD (after Murphy and Koski, 1989).

2. along a "type B" stream (anadromous fish stream incised in bedrock), timber harvest within 30 m of the stream or to the slope break must comply with BMPs; and

3. along a "type C" stream (small tributary without anadromous fish), timber harvest within 15 m of the stream or to the slope break must comply with BMPs.

The Alaska Forest Resources and Practices Act grants variations from requirements in some cases. A landowner may propose a variation to the State Forester. The Department of Fish and Game has due deference in such requests concerning fish and wildlife habitat. There is also an automatic variation granted on small streams less than 1.5 m wide, which allows the landowner to harvest up to 25% of the trees within the area between 7.6 and 22 m of the stream.

These new laws enacted in 1990 substantially improved the protection of salmonid habitat in Alaska. Further research, however, is needed to assess effectiveness of BMPs in preventing downstream cumulative effects of timber harvest along small tributaries and in preventing sedimentation from construction and use of logging roads.

Summary

The interaction between timber and fisheries in Alaska has been evolving since the 1950s. Early research, focusing on spawning habitat, determined habitat requirements and mortality factors that timber harvest can affect, but impacts on salmon were not demonstrated. As awareness grew of possible impacts to salmon rearing habitat, timber and fisheries groups joined together to form the Alaska Working Group on Cooperative Forestry–Fisheries Research to foster research on priority issues. This research identified LWD as the most important variable in the timber–fisheries interaction in southeast Alaska, being both easily affected by timber harvest and critical for salmonid rearing habitat. Buffer strips were found to be effective in providing long-term sources of LWD after timber harvest. Consensus and compromise led to legislation in 1990 requiring buffer strips along anadromous fish streams on public and private lands. These laws provided the highest level of protection for anadromous fish habitat of any state.

The interaction between timber and fisheries continues today as land managers and regulators apply the new regulations in the field and monitor their effectiveness. Research continues to be needed to evaluate and refine BMPs and determine their applicability to other regions of Alaska where the functional role of LWD and other habitat variables in salmonid population dynamics may be different from that in southeast Alaska.

Acknowledgments. We thank William Meehan for his review of an earlier version of this manuscript. The U.S. Environmental Protection Agency provided a grant to The University of Alaska (A.M.M.) for a literature synthesis of timber harvest and water quality in Alaska that was the initial basis for this chapter.

References

ADCED [Alaska Department of Commerce and Economic Development] (1990) Alaska forest products manufacturer's directory—1990. ADCED, Division of Business Development, Juneau.

ADEC [Alaska Department of Environmental Conservation] (1990) Alaska nonpoint source pollution control strategy. ADEC, Division of Environmental Quality, Water Quality Management Section, Juneau.

Alaska Forest Research Center (1957) Pulp logging on salmon spawning streams. In Biennial report for 1956–1957. U.S. Forest Sevice, Alaska Forest Research Center, Juneau, 35–36.

Alaska Forest Research Center (1960) Watershed management research. In Biennial report 1958–1959. U.S. Forest Service, Alaska Forest Research Center, Juneau, 19–20.

Bartos L (1989) A new look at low flows after logging. In Alexander EB (Ed) Proc Watershed '89. U.S. Forest Service, Alaska Region, R10-MB-77, Juneau, Alaska, 95–98.

Bartos L (1993) Stream discharge related to basin geometry and geology, before and after logging. In Brock T (Ed) Proc Watershed '91. U.S. Forest Service, R10-MB-217, Juneau, Alaska, 29–32.

Beschta RL, Bilby RE, Brown GW, Holtby LB, Hofstra TD (1987) Stream temperature and aquatic habitat: Fisheries and forestly interactions. In Salo EO, Cundy TW (Eds) Streamside management: Forestry and fishery interactions. College of Forest Resources, University of Washington, Contribution 57, Seattle, 191–232.

Bilby RE, Likens GE (1980) Importance of organic debris dams in the structure and function of stream ecosystems. Ecology 61:1107–1113.

Bishop DM, Stevens ME (1964) Landslides on logged areas in southeast Alaska. Northern Forest Research Station, U.S. Forest Service, Research Paper NOR-1, Juneau, Alaska.

Bisson PA, Bilby RE (1982) Avoidance of suspended sediment by juvenile coho salmon. Nor Am J Fish Manage 4:371–374.

Bisson PA, Bilby RE, Bryant MD, et al. (1987) Large woody debris in forested streams in the Pacific Northwest: Past, present, and future. In Salo EO, Cundy TW (Eds) Streamside management: Forestry and fishery interactions. College of Forest Resources, University of Washington, Contribution 57, Seattle, 143–190.

Bjornn TC, Brusven MA, Hetrick NJ, Keith RM, Meehan, WR (1992) Effects of canopy alterations in second-growth forest riparian zones on bioenergetic processes and responses of juvenile salmonids to cover in small southeast Alaska streams. Idaho Cooperative Fish and Wildlife Research Unit, Technical Report 92–7, Moscow, ID.

Bjornn TC, Reiser DW (1991) Habitat requirements of salmonids in streams. In Meehan WR (Ed) Influences of forest and rangeland management on salmonid fishes and their habitats. Am Fish Soc Spec Publ 19:83–138.

Brown B (1989) Characteristics at streams with fish kills. Unpublished report. Alaska Department of Fish and Game, Juneau.

Brown GW (1972) Logging and water quality in the Pacific Northwest. In Csallany CS, McLaughlin TG, Striffeer WD (Eds) Watersheds in transition. Proc Sym Watersheds Transition. American Water Resources Association and Colorado State University. AWRA Proceedings Series 14, 330–334.

Bryant MD (1980) Evolution of large, organic debris after timber harvest: Maybeso Creek, 1949 to 1978. Pacific Northwest Forest and Range Experiment Station, U.S. Forest Service, General Technical Report PNW-101.

Bryant MD (1983) The role and management of woody debris in West Coast salmonid nursery streams. North Am J Fish Manage 3:322–330.

Bryant MD (1984) The role of beaver dams as coho salmon habitat in southeast Alaska streams. In Walton JM, Houston DB (Eds) Proc Olympic Wild Fish Conf [Available From Dr. J.M. Walton, Fisheries Technology Program, Peninsula College, 1502 E. Lauridsen Blvd., Port Angeles, WA 98362], 183–192.

Bryant MD (1985) Changes 30 years after logging in large woody debris, and its use by salmonids. In Johnson RR, Ziebell CD, Patton DR et al., (Eds) Riparian ecosystems and their management: Reconciling conflicting uses. First North American Riparian Conference. U.S. Forest Service, Rocky Mountain Forest and Range Experiment Station, General Technical Report RM-120, Fort Collins, 329–334.

Burger CV, Wilmot RL, Wangaard DB (1985) Comparison of spawning areas and times for two runs of chinook salmon (Oncorhynchus tshawytscha) in the Kenai River, Alaska. Can J Fish Aquat Sci 42:693–700.

Burns DC (1991) Cumulative effects of small modifications to habitat. AFS position statement. Fisheries 16:12–17.

Campbell AJ, Sidle RC (1985) Bedload transport in a pool-riffle sequence of a coastal Alaska stream. Water Resour Bull 21:579–590.

Cardinal PJ (1980) Habitat and juvenile salmonid populations in streams in logged and unlogged areas of southeast Alaska. M.S. thesis, Montana State University, Bozeman.

Cederholm CJ, Reid LM, Salo EO (1981) Cumulative effects of logging road sediment on salmonid populations in the Clearwater River, Jefferson County, Washington. In Salmon-spawning gravel: A renewable resource in the Pacific Northwest? State of Washington Water Research Center, Washington State University, Pullman, 38–74.

Chamberlin TW, Harr WD, Everest FH (1991) Timber harvesting, silviculture, and watershed processes. In Meehan WR (Ed) Influences of forest and rangeland management on salmonid fishes and their habitats. Am Fish Soc Spec Pub 19:181–205.

Cordone AJ, Kelley DW (1961) The influences of inorganic sediment on the aquatic life of streams. Calif Fish Game 47:189–228.

Crone RA, Bond CF (1976) Life history of coho salmon, Oncorhynchus kisutch, in Sashin Creek, southeast Alaska. Fishery Bulletin 74:897–923.

Dill LM, Ydenberg RC, Fraser AHG (1981) Food abundance and territory size in juvenile coho salmon (Oncorhynchus kisutch). Can J Zool 59:1801–1809.

Dolloff CA (1986) Effects of stream cleaning on juvenile coho salmon and Dolly Varden in southeast Alaska. Trans Am Fish Soc 115:743–755.

Duncan WFA, Brusven MA (1985a) Energy dynamics of three low-order southeast Alaskan streams: Autochthonous production. J Freshwater Ecol 3:155–166.

Duncan WFA, Brusven MA (1985b) Energy dynamics of three low-order southeast Alaska streams: Allochthonous processes. J Freshwater Ecol 3:233–248.

Edgington JR (1984) Some observations of find sediment in gravels of five undisturbed watersheds in southeast Alaska. In Meehan WR, Merrell TR Jr, Hanley

TA (Eds) Fish and wildlife relationships in old-growth forests. Proc Symp. American Institute of Fishery Research Biologists, 109–114 [Available from J.W. Reintjes, Rt. 4, Box 85, Morehead City, NC 28557].

Elliott ST (1986) Reduction of a Dolly Varden population and macrobenthos after removal of logging debris. Trans Am Fish Soc 115:392–400.

Everest FH, Beschta RL, Scrivener JC, Koski KV, Sedell JR, Cederholm CJ (1987) Fine sediment and salmonid production: A paradox. In Salo EO, Cundy TW (Eds) Streamside management: Forestry and fishery interactions. College of Forest Resources, University of Washington, Contribution 57, Seattle, 98–142.

Everest FH, Meehan WR (1981) Some effects of debris torrents on habitat of anadromous salmonids. National Council of the Paper Industry for Air and Stream Improvement, New York Tech Bull 353:23–30.

Furniss MJ, Roelofs TD, Yee CS (1991) Road construction and maintenance. In Meehan WR (Ed) Influences of forest and rangeland management on salmonid fishes and their habitats. Am Fish Soc Spec Publ 19:297–323.

Garrett LD, Fight RD, Weyerman DL, Mehrkens JR (1990) Using the model SAMM: Implications for management. In Fight RD, Garrett LD, Weyermann DL (Eds) SAMM: A prototype southeast Alaska multi-resource model. U.S. Department of Agriculture, General Technical Report PNW-GTR-255, Portland, OR, 78–89.

Gibbons DR, Meehan WR, Koski KV, Merrell TR Jr (1987) History of studies of fisheries and forestry interactions in southeastern Alaska. In Salo EO, Cundy TW (Eds) Streamside management: Forestry and fishery interactions. College of Forest Resources, University of Washington, Contribution 57, Seattle, 297–329.

Gibbons DR, Salo EO (1973) An annotated bibliography of the effects of logging on fish of the western United States and Canada. Pacific Northwest Forest and Range Experiment Station, U.S. Forest Service, General Technical Report PNW-10, Portland, OR.

Golding DL (1987) Changes in streamflow peaks following timber harvest of a coastal British Columbia watershed. Int Assoc Hydrolog Sci Publ 167:509–517.

Gray PL, Koerner JF, Marriott RA (1981) The age structure and length–weight relationship of southeastern Alaska coho salmon (*Oncorhynchus kisutch*), 1969–1970. Alaska Department of Fish and Game, Informational Leaflet 195, Juneau.

Gregory SV, Lamberti GA, Erman CD, Koski KV, Murphy ML, Sedell JR (1987) Influence of forest practices on aquatic production. In Salo EO, Cundy TW (Eds) Streamside management: Forestry and fishery interactions. College of Forest Resources, University of Washington, Contribution 57, Seattle, 233–256.

Hall JD, Baker CO (1982) Rehabilitating and enhancing stream habitat: 1. Review and evaluation. U.S. Forest Service, General Technical Report PNW-138, Portland, OR.

Hall JD, Lantz RL (1969) Effects of logging on the habitat of coho salmon and cutthroat trout in coastal streams. In Northcote TG (Ed) Symposium on salmon and trout in streams. H.R. MacMillan Lectures in Fisheries, University of British Columbia, Institute of Fisheries, Vancouver, 355–375.

Harding RD (1993) Abundance, size, habitat utilization, and intra-stream movement of juvenile coho salmon in a small Southeast Alaska stream. M.S. thesis, University of Alaska, Juneau.

Harr RD (1983) Potential for augmenting water yield through forest practices in western Washington and western Oregon. Water Resour Bull 19:383–393.

Harris AS (1960) The physical effect of logging on salmon streams in southeast Alaska [Abstract]. In Science in Alaska, Proc 11th Alaska Sci Conf, 143–144.

Harris AS, Farr WA (1974) The forest ecosystem of southeast Alaska: 7. Forest ecology and timber management. U.S. Forest Service, General Technical Report PNW-25, Portland, OR.

Hartman GF, Holtby LB, Scrivener JC (1984) Some effects of natural and logging-related winter stream temperature changes on the early life history of coho salmon (*Oncorhynchus kisutch*) in Carnation Creek, British Columbia. In Meehan WR, Merrell TR Jr, Hanley TA (Eds) Fish and wildlife relationships in old-growth forests. Proc Symp. American Institute of Fishery Research Biologists [Available from J.W. Reintjes, Rt. 4, Box 85, Morehead City, NC 28557], 141–149.

Hartman GF, Scrivener JC, Holtby LB, Powell L (1987) Some effects of different streamside treatments on physical conditions and fish population processes in Carnation Creek, a coastal rain forest stream in British Columbia. In Salo EO, Cundy TW (Eds) Streamside management: Forestry and fishery interactions. College of Forest Resources, University of Washington, Contribution 57, Seattle, 330–372.

Hartsog W (1990) Summary of slope stability and related programs on the Tongass National Forest. Unpublished report. U.S. Forest Service, Petersburg, AK.

Hawkins CP, Murphy ML, Anderson, NH (1982) Effects of canopy, substrate composition, and gradient on the structure of macroinvertebrate communities in Cascade Range streams of Oregon. Ecology 63:1840–1856.

Heifetz J, Murphy ML, Koski KV (1986) Effects of logging on winter habitat of juvenile salmonids in Alaskan streams. North Am J Fish Manage 6:52–58.

Hicks BJ, Hall JD, Bisson PA, Sedell JR (1991) Responses of salmonids to habitat changes. In Meehan WR (Ed) Influences of forest and rangeland management on salmonid fishes and their habitats. Am Fish Soc Spec Pub 19:483–518.

Hicks BJ, Harr RD, Beschta RL (1991) Long-term changes in streamflow following logging and implications in western Oregon. Water Resour Bull 27:217–226.

Holtby LB, Scrivener JC (1989) Observed and simulated effects of climatic variability, clear-cut logging, and fishing on the number of chum salmon (*Oncorhynchus keta*) and coho salmon (*O. kisutch*) returning to Carnation Creek, British Columbia. Can Spec Publ Fish Aquat Sci 105:62–81.

Hunt RL (1969) Effects of habitat alteration on production, standing crops and yield of brook trout in Lawrence Creek, Wisconsin. In Northcote TG (Ed) Symposium on salmon and trout in streams. H.R. MacMillan Lectures in Fisheries, University of British Columbia, Institute of Fisheries, Vancouver, 281–312.

Iwamoto RN, Salo EO, Madej MA, McComas RL (1978) Sediment and water quality: A review of the literature including a suggested approach for water quality criteria. U.S. Environmental Protection Agency, Region X, EPA 910/9-78-048, Seattle.

James GA (1956) The physical effect of logging on salmon streams of southeast Alaska. Alaska Forest Research Center, U.S. Forest Service, Station Paper 5, Juneau.

Johnson SW, Heifetz J, Koski KV (1986) Effects of logging on the abundance and seasonal distribution of juvenile steelhead in some southeastern Alaska Streams. North Am J Fish Manage 6:532–537.

Kingsbury AP (1973) Relationship between logging activities and salmon production. Alaska Department of Fish and Game, Completion Report 1970–1972, Project 5-24-R, Juneau.

Koski KV (1972) Effects of sediment on fish resources. Management Seminar. Washington State Department of Natural Resources, Seattle.

Koski KV, Heifetz J, Johnson S, Murphy M, Thedinga J (1985) Evaluation of buffer strips for protection of salmonid rearing habitat and implications for enhancement. In Hassler TJ (Ed) Proceedings: Pacific northwest stream habitat management workshop. American Fisheries Society, Humboldt State University, Arcata, CA, 138–155.

Lamke RD, Bigelow BB, Van Maanen JL, Kemnitz RT, Novcaski KM (1991) Water resources data—Alaska water year 1990. U.S. Geological Survey, Water Resources Division, Water-Data Report AK-90-1, Anchorage.

Landwehr DJ (1994) Inventory and analysis of landslides caused by the October 25, 26, 1993 storm event on the Thorne Bay Ranger District. Unpublished report. U.S. Forest Service, Ketchikan, Alaska.

Lisle TE (1986) Effects of woody debris on anadromous salmonid habitat, Prince of Wales Island, southeast Alaska. North Am J Fish Manage 6:538–550.

Lloyd DS, Koenings JP, LaPerriere JD (1987) Effects of turbidity in fresh waters of Alaska. North Am J Fish Manage 7:18–33.

Martin D, Edland S, Morrow R (1991) Pre-spawner mortality of adult salmon in southeast Alaska. Pentec Environmental, Inc., Final report to the Alaska Working Group on Cooperative Forestry/Fisheries Research, Juneau.

McNeil WJ (1964) Effect of the spawning bed environment on reproduction of pink and chum salmon. Fish Bull U.S. 65:495–523.

McNeil WJ, Ahnell WH (1964) Success of pink salmon spawning relative to size of spawning bed materials. U.S. Fish and Wildlife Service, Special Scientific Report on Fisheries 469. Washington, DC.

Meehan WR (1970) Some effects of shade cover on stream temperature in southeast Alaska. U.S. Forest Service, Pacific Northwest Forest and Range Experiment Station, Research Note PNW-113, Portland, OR.

Meehan WR, Farr WA, Bishop DM, Patric JH (1969) Some effects of clearcutting on salmon habitat of two southeast Alaska streams. U.S. Forest Service, General Technical Report PNW-82, Portland, OR.

Murphy ML (1985) Die-offs of pre-spawn adult pink salmon and chum salmon in southeast Alaska. North Am J Fish Manage 5:302–308.

Murphy ML, Hall JD (1981) Varied effects of clear-cut logging on predators and their habitat in small streams of the Cascade Mountains, Oregon. Can J Fish Aquat Sci 38:137–145.

Murphy ML, Hawkins CP, Anderson NH (1981) Effects of canopy modification and accumulated sediment on stream communities. Trans Am Fish Soc 110:469–478.

Murphy ML, Heifetz J, Johnson SW, Koski KV, Thedinga JF (1986) Effects of clear-cut logging with and without buffer strips on juvenile salmonids in Alaskan streams. Can J Fish Aquat Sci 43:1521–1533.

Murphy ML, Koski KV, Heifetz J, Johnson SW, Kirchhofer D, Thedinga JF (1985) Role of large organic debris as winter habitat for juvenile salmonids in Alaska streams. Proc Annu Conf West Assoc Fish Wild Agencies 64:251–262.

Murphy ML, Koski KV (1989) Input and depletion of woody debris in Alaska streams and implications for streamside management. North Am J Fish Manage 9:427–436.

Murphy ML, Lorenz JM, Heifetz J, Thedinga JF, Koski KV, Johnson SW (1987) The relationship between stream classification, fish, and habitat in Southeast Alaska. U.S. Forest Service, Wildlife and Fisheries Habitat Management Notes 12, Juneau.

Murphy ML, Koski KV, Elliott ST et al. (1990) The fisheries submodel. In Fight RD, Garrett LD, Weyermann DL (Eds) SAMM: A prototype Southeast Alaska Multiresource Model. U.S. Department of Agriculture, General Technical Report PNW-GTR-255, Portland, OR, 46–63.

Murphy ML, Meehan WR. (1991) Stream ecosystems. In Meehan WR (Ed) Influences of forest and rangeland management on salmonid fishes and their habitats. Am Fish Soc Spec Publ 19:17–46.

Murphy ML, Thedinga JF, Koski KV, Grette GB (1984) A stream ecosystem in an old-growth forest in southeast Alaska. Part 5: Seasonal changes in habitat utilization by juvenile salmonids. In Meehan WR, Merrell TR Jr, Hanley TA (Eds)

Fish and wildlife relationships in old-growth forests. Proc Symp. American Institute of Fishery Research Biologists [Available from J.W. Reintjes, Rt. 4, Box 85, Morehead City, NC 28557], 89–98.

Myren RT, Ellis RJ (1984) Evapotranspiration in forest succession and long-term effects upon fishery resources: A consideration for management of old-growth forests. In Meehan WR, Merrell JR Jr, Hanley TA (Eds) Fish and wildlife relationships in old-growth forests. Proc Symp. American Institute of Fishery Research Biologists [Available from J.W. Reintjes, Rt. 4, Box 85, Morehead City, NC 28557], 183–186.

Paustian SJ (1987) Monitoring nonpoint source discharge of sediment from timber harvesting activities in two southeast Alaska watersheds. In Huntsinger RG (Ed) Water quality in the great land, Alaska's challenge. Proc Univ Alaska, Fairbanks, IWR-109. American Water Resources Association, 153–168.

Pella JJ, Myren RT (1974) Caveats concerning evaluation of effects of logging on salmon production in southeastern Alaska from biological information. Northwest Sci 48:132–144.

Powell WO (1989) Low tire inflation pressure—A solution to breakdown of road surface rock. In Alexander EB (Ed) Proc Watershed '89. U.S. Forest Service, Alaska Region, R10-MB-77, Juneau, 121–123.

Preston EM, Bedford BL (1990) Evaluating cumulative effects on wetland functions: A conceptual overview and generic framework. Environ Manage 12: 565–583.

Reeves GH, Everest FH, Nickelson TE (1989) Identification of physical habitats limiting the production of coho salmon in western Oregon and Washington. U.S. Forest Service, General Technical Report PNW-GTR-245, Portland, OR.

Schwan M, Elliott S, Edgington J (1985) The impacts of clearcut logging on the fisheries resources of southeast Alaska. In Sigman MJ (Ed) Impacts of clearcut logging on the fish and wildlife resources of southeast Alaska. Alaska Department of Fish and Game, Technical Report 85–3, Juneau.

Sedell JR, Everest FH (1990) Historic changes in pool habitat for Columbia River basin salmon under study for TES listing. Unpublished report. U.S. Forest Service Pacific Northwest Research Station, Corvallis, OR.

Sedell JR, Swanson FJ (1984) Ecological characteristics of streams in old-growth forests of the Pacific Northwest. In Meehan WR, Merrell TR Jr, Hanley TA (Eds) Fish and wildlife relationships in old-growth forests. Proc Symp. American Institute of Fishery Research Biologists [Available from J.W. Reintjes, Rt. 4, Box 85, Morehead City, NC 28557].

Sheridan RL, Bloom AM (1975) Effects of canopy removal on temperature of some small streams in southeast Alaska. Unpublished report. U.S. Forest Service, Juneau.

Sheridan WL, McNeil WJ (1968) Some effects of logging on two salmon streams in Alaska. J Forestry 66:128–133.

Sheridan WL, Perensovich MP, Faris T, Koski K (1984) Sediment content of streambed gravels in some pink salmon spawning streams in Alaska. In Meehan WR, Merrell TR Jr, Hanley TA (Eds) Fish and wildlife relationships in old-growth forests. Proc Symp. American Institute of Fishery Research Biologists [Available from J.W. Reintjes, Rt. 4, Box 85, Morehead City, NC 28557].

Simpson R, Gibbons DR (1984) Alaska working group on cooperative forestry–fisheries research, 1984 annual report. Alaska Department of Fish and Game, Juneau.

Slaughter CW (1990) Sustaining watershed values while developing taiga forest resources. In Proceedings: New perspectives for watershed management: Balancing long-term sustainability with cumulative environmental change. University of Washington, Seattle, 10.

Smith RD (1989) Current research investigating channel unit distribution in streams of southeast Alaska. In Alexander EB (Ed) Proc Watershed '89. U.S. Forest Service, Alaska Region, R10-MB-77, Juneau, 91–92.

Stednick JD, Tripp LN, McDonald RJ (1982) Slash burning effects on soil and water chemistry in southeastern Alaska. J Soil Water Conserv 37:126–128.

Sullivan K, Lisle TE, Dolloff CA, Grant GE, Reid LM (1987) Stream channels: The link between forests and fishes. In Salo EO, Cundy TW (Eds) Streamside management: Forestry and fishery interactions. College of Forest Resources, University of Washington, Contribution 57, Seattle, 330–372.

Swanston DN, Webb TM, Bartos L, Meehan WR, Sheehy T, Puffer A (1990) The hydrology and soils submodel. In Fight RD, Garrett LD, Weyermann DL (Eds) SAMM: A prototype Southeast Alaska Multiresource Model. U.S. Department of Agriculture, General Technical Report PNW-GTR-255, Portland, OR, 28–45.

Thedinga JF, Johnson SW, Koski KV, Lorenz JM, Murphy ML (1993) Potential effects of flooding from Russell Fiord on salmonids and habitat in the Situk River, Alaska. U.S. Department of Commerce, Alaska Fisheries Science Center, National Marine Fisheries Service, Processed Report AFSC 93–01. [Available from Auke Bay Laboratory, 111305 Glacier Highway, Juneau, AK 99801–8626].

Thedinga JF, Murphy ML, Heifetz J, Koski KV, Johnson SW (1989) Effects of logging on size and age composition of juvenile coho salmon (*Oncorhynchus kisutch*) and density of presmolts in southeast Alaska streams. Can J Fish Aquat Sci 46:1383–1391.

Thomas RE, Gharrett JA, Carls MG, Rice SD, Moles A, Korn S (1986) Effects of fluctuating temperature on mortality, stress, and energy reserves of juvenile coho salmon. Trans Am Fish Soc 115:52–59.

Tyler RW, Gibbons DR (1973) Observations of the effects of logging on salmon-producing tributaries of the Staney Creek watershed and the Thorne River watershed and of logging in the Sitka district. Fisheries Research Institute, University of Washington, FRI-UW-7307, Seattle.

U.S. Forest Service (1986) Aquatic habitat management handbook. FSH 2609-.24. U.S. Forest Service, Juneau.

U.S. Forest Service (1991) Tongass land management plan revision. Supplement to the draft environmental impact statement—Summary. R10-MB-150. U.S. Forest Service, Juneau.

Van Cleve K, Dyrness, CT, Viereck, LA et al. (1983) Taiga ecosystems in interior Alaska. Bioscience 33:39–44.

Walter RA (1984) A stream ecosystem in an old-growth forest in southeast Alaska. Part 2: Structure and dynamics of the periphyton community. In Meehan WR, Merrell TR Jr, Hanley TA (Eds) Fish and wildlife relationships in old-growth forests. Proc Symp. American Institute of Fishery Research Biologists [Available from J.W. Reintjes, Rt. 4, Box 85, Morehead City, NC 28557].

Warren DD (1992) Production, prices, employment, and trade in Northwest forest industries, fourth quarter 1992. U.S. Forest Service, Pacific Northwest Research Station, Resource Bulletin PNW-RB-196, Portland, OR.

Weber–Scannell P (1991) Influence of temperatures on coldwater fishes: A literature review. Technical Report 91-1. Fairbanks, AK.

10. Gold Placer Mining and Stream Ecosystems of Interior Alaska

Jacqueline D. LaPerriere and James B. Reynolds

What Is Placer Mining?

Placer mining is primarily practiced in Alaska as a type of open pit mining that targets alluvial deposits called placers, which may be deeply buried under soil and gravel (overburden). Placer deposits of precious metals, including gold, occur when a lode or vein weathers and fragments and is moved by flowing water into an alluvial deposit. As the alluvium is re-worked and sorted by water flows, the heavy precious-metal ore tends to move downward until it is stopped by the bedrock. Sometimes the placer deposit is in an active stream channel, but often the alluvia are ancient and buried.

Methods of Placer Mining

Historically, overburden was removed with water pressurized through nozzles called hydraulic giants, a method called hydraulicing (Fig. 10.1). This practice is now somewhat rare because it is only practical where the effluent can be treated sufficiently to meet National Pollutant Discharge Elimination System (NPDES) standards. Large floating dredges that were once common are now rare due to their large energy costs and other factors such as unavailability of parts. In more recent times, small suction dredges are floated near a diver, who positions the hose to suck up the gravels; these

Figure 10.1. Hydraulic placer gold mining (courtesy of Rasmuson Library, University of Alaska Fairbanks).

are mainly used for recreational mining on instream placers. Current methods of removing overburden are mostly mechanical, using earth-moving machinery (Fig. 10.2).

The deep gravels containing the precious metal ore are sorted by gravity, using water processes. One process uses the sluice box, which is an inclined plane with sides and attached cross-bars (also called "riffles") that acts like an artificial stream, trapping the heavy gold ore at the bars by gravity while allowing the lighter gravel to wash through. The sorting can also be accomplished in a device like an inclined front-loading washing machine, called a trommel, that allows the larger gravel to wash through, while holes in the turning drum pass the smaller, heavier ore pieces and gravels of the same size. The gold ore is later separated from the gravel by a sluicing process (Bundtzen et al., 1994).

The process of separating gold from gravels, whether conducted with sluice boxes, or trommels and sluice tables, may add sediments to nearby streams. Settling ponds have long been used to remove sediments from the process water, particularly when water was in short supply and needed to be reused. Settling is the treatment method ordinarily used to produce an effluent with reduced sediments, but it may not be effective in removing very fine sediments, especially if the deposit contains clays. Coagulants can be added to enhance settling of fine materials.

Figure 10.2. Modern placer gold mining. Bulldozer loading paydirt onto sluice.

Placer Mining in Alaska

Enormous amounts of sediment are estimated to have flowed down nearby streams when floating dredges and hydraulicing were used. For example, mining engineers have estimated that more than 100 million cubic yards of sediment washed down the Chatanika River and Goldstream Creek into Minto Flats near Fairbanks during the first half of the 20th century (Wolf, 1982).

In interior Alaska, summer (the usual season for placer mining) is a time of low precipitation (Hartman and Johnson, 1978) and stream runoff; therefore, recycle of process waters is a historical practice. Reuse of process water is now encouraged under the NPDES effluent guidelines (CFR, 1989) that allow only excess water, such as that from storms, to be discharged.

In 1993, placer mining employed 1240 persons or 38% of all those employed by the mining industry in Alaska (Bundtzen et al., 1994). Gold was mined that year at 196 placer mines in the state, of which 112 were located in the eastern interior region, consisting of the Circle, Fairbanks, Forty-Mile, Kantishna, and Livengood–Tolovana mining districts. Other organized mining districts in the state include Haines, Iditarod, Juneau, Koyukuk, Valdez, and Yetna. Of these, the Circle district is currently one of the largest producers, estimated at having yielded 31,717 kg (1,019,725 troy oz) of gold in the first 100 years (1893 to 1993) (Bundtzen et al., 1994).

Other important mining areas include Chandelar, Innoko, Nome, and Wiseman.

Measurement of Sediments

Sediment in water can be divided into components using two systems of nomenclature, depending on the methods used (Fig. 10.3). When analyzed by filtration, drying, and weighing, the components are called total solids, total suspended solids, and total dissolved solids (APHA et al., 1989). The total solids and total suspended solids classifications can be further subdivided into fixed and volatile components if the dried and weighed components are then ignited and weighed. The fixed and volatile components are approximations of inorganic and organic fractions when the ash is considered equal to the inorganic solids and the volatile portion is assumed to yield the weight of the organic solids that burned off.

Water-borne sediment can also be separated into settleable and nonsettleable solids (Fig. 10.3). Settleable solids are measured by allowing a liter of water and sediments to stand quietly for an hour (tapping or stirring after 45 minutes to consolidate the solids at the bottom) in an Imhoff cone and measuring the volume of solids that settle to the bottom in milliliters per liter. Nonsettleable solids are measured gravimetrically on samples siphoned off the top after settling in an appropriate cylinder for an hour, and are therefore expressed in units of milligrams per liter. Settleable solids can then be calculated in units of milligrams per liter by subtracting nonsettleable solids, measured as above, from total suspended solids measured on the same sample (APHA et al., 1989). The Imhoff cone measurement of settleable solids is traditionally used by engineers to help design various settling processes, but has a specified value of $0.2\,mL\,L^{-1}$ in effluents allowed for placer mining (CFR, 1989).

Turbidity is an approximation of the amount of suspended sediments in water. It is a measure of the amount of light absorbed and scattered by particles, and because it varies not only with the concentration of the particles, but with such variables as the particles' refractive index and size distribution and the wavelength and pathlength of the light, it is only used to *estimate* the concentration of suspended sediments. To control the variation among instruments, the standard method (APHA et al., 1989) now specifies many characteristics of the instrument for measuring turbidity and requires calibration with a polymer of uniform particle size called formazin. Therefore, turbidity measurements have units of formazin turbidity units (FTU) or, more commonly, nephelometric turbidity units (NTU), because the specified standard instrument is a nephelometer that measures light scattered at 90° from the incident beam.

Turbidity measurements provide a quick estimation of suspended solids that can be conducted on site; measurement of suspended solids takes more

GRAB SAMPLE

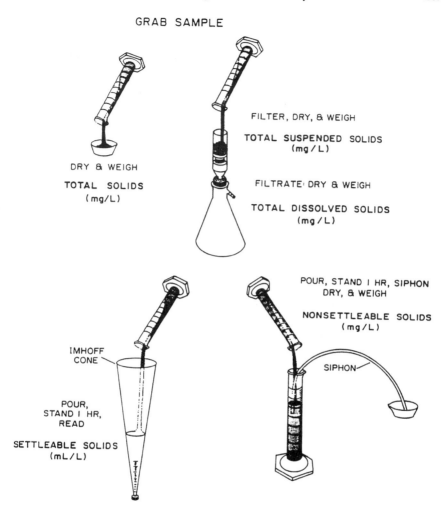

FILTER, DRY, & WEIGH

TOTAL SUSPENDED SOLIDS
(mg / L)

DRY & WEIGH

TOTAL SOLIDS
(mg/L)

FILTRATE: DRY & WEIGH

TOTAL DISSOLVED SOLIDS
(mg / L)

POUR, STAND I HR, SIPHON
DRY, & WEIGH

NONSETTLEABLE SOLIDS
(mg/L)

IMHOFF
CONE

SIPHON

POUR,
STAND I HR,
READ

SETTLEABLE SOLIDS
(mL/L)

Figure 10.3. Analysis of sediments in water. The sample can be processed to give total solids, total suspended solids, and total dissolved solids; or settleable and nonsettleable solids. All these solids measurements are defined by the methods used to analyze them. (Illustration by E. Sturm from sketch by L. Simmons.)

than 1 hour and must be done in a laboratory equipped with ovens and balances. Strong correlations can be expected between measurements of turbidity and suspended solids as long as the particles causing the turbidity are from similar sources (Lloyd et al., 1987). However, the relationship between turbidity and suspended solids should be evaluated when turbidity will be used to estimate suspended solids among mine sites with different soil characteristics.

Physical Effects of Sediments

Two interior Alaska river systems, the Chatanika River and Birch Creek, each of which had a mined and unmined tributary, were studied intensively in 1982 and 1983 (Bjerklie and LaPerriere, 1985; LaPerriere et al., 1985; Reynolds et al., 1989; Van Nieuwenhuyse and LaPerriere, 1986; Wagener and LaPerriere, 1985) (Fig. 10.4). The streams were studied near the confluence of the mined and unmined tributaries and were hydrologically similar (Table 10.1). Tributaries with mines (Upper Birch Creek and Faith Creek) had higher measured sediment concentrations than those without mines (Twelvemile Creek and McManus Creek) (Table 10.2). The difference in the sediment concentrations reflects that Upper Birch Creek had more active mines than Faith Creek in 1983. At that time, some mines in interior Alaska apparently had no settling ponds, or inadequate ones, for the removal of sediments, which may have also contributed to the relatively high sediment conditions in Birch Creek. The mean value for turbidity of 75 NTU of Faith Creek is close to the maximum mean value (89 NTU) for placer mined streams studied in 1989 on the west coast of the South Island of New Zealand (Davies–Colley et al., 1992).

When sediment settles out of the water onto the streambed, the bed may become sealed, causing a separation of the visible surface flow and the

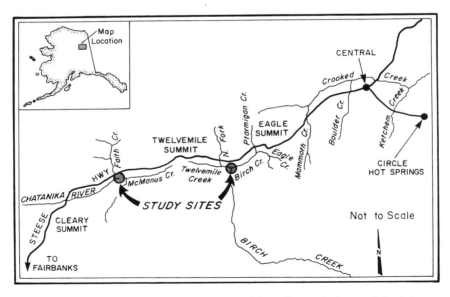

Figure 10.4. Sketch map of the region studied for effects of placer gold mining on streams in the 1980s. The major sites marked are treated in Bjerklie and LaPerriere (1985), LaPerriere et al. (1985), Reynolds et al. (1989), and Van Nieuwenhuyse and LaPerriere (1986). Wagener and LaPerriere (1985) concerns nine sites on this map.

Table 10.1. Selected Basin and Hydrologic Characteristics of Paired Watersheds With and Without Gold-Placer Mining, June–August 1983

Stream	Mean Flow $(m^3 s^{-1})$	Flow per Unit Area $(m^3 s^{-1} km^{-2})$	Basin Area (km^2)	High Flow $(m^3 s^{-1})$	Date	Low Flow $(m^3 s^{-1})$	Date
Pair 1							
Faith Creek (M)	2.7	0.02	156	9.4	7/17	1.1	8/11
McManus Creek (C)	2.4	0.01	201	13.2	7/17	1.1	8/13
Pair 2							
Upper Birch Creek (M)	4.2	0.02	231	13.9	6/27	1.3	8/11
Twelvemile Creek (C)	1.7	0.01	122	9.6	6/27	0.6	8/12

M, mined; C, control.

underflow within the substrate (Fig. 10.5). Depression of the piezometric water level was too low to be measured in Twelvemile Creek, which had essentially no measurable settleable solids, but was severe in the otherwise matched heavily mined Upper Birch Creek, which was carrying about $1000 \, mg \, L^{-1}$ of suspended solids (Bjerklie and LaPerriere, 1985). In the mined Upper Birch Creek, where a 0.3 m depression of the piezometric level was measured, the bottom felt like concrete underfoot. In the unmined Twelvemile Creek, streambed gravel moved freely underfoot.

Chemical Effects of Sediment

Sediment added to streams by placer mining activity may have associated heavy metals that are mostly adsorbed to the surface of the sediment particles. Because much of the gold in interior Alaska is in the form of

Table 10.2. Summer Average Waterborne Sediment Measurements of Paired Watersheds With and Without Gold-Placer Mining, June–August 1983

	Suspended Solids $(mg \, L^{-1})$	Turbidity (NTU)	Detectable (Trace Amounts) Settleable Solids (% Readings)
Pair 1			
Faith Creek (M)	83.1	75	2
McManus Creek (C)	8.9	2.7	0
Pair 2			
Upper Birch Creek (M)	655	749	83
Twelvemile Creek (C)	11.6	1.4	3

M, mined; C, control.

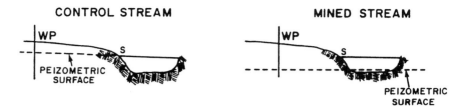

Figure 10.5. Illustration of depression of the elevation of the underflow of a mined stream caused by sealing of the bed substrate by sedimentation (WP, well point; S, surface water elevation).

arsenopyrites, action of iron bacteria can free arsenic and other associated metals from the placer when it is exposed to weathering by mining activity (Luong et al., 1985). Arsenic in the groundwater of areas near Fairbanks is associated with dredging for gold prior to World War II, which left tailing deposits (materials left after ore is removed) at the surface (Wilson and Hawkins, 1978). In the Circle mining district, gold mining was associated with increases in arsenic, copper, lead, zinc (Tables 10.3 and 10.4) (LaPerriere et al., 1985), gold, and silver (Scannell, 1988) in the receiving streams. Additional metals such as mercury, nickel, antimony, and cadmium have been found (as total metals) in streams near placer mining activity in the Kantishna Hills mining district, a recent addition to Denali National Park and Preserve (West and Deschu, 1984).

Most heavy metals are transported while adsorbed to sediment particles, and relatively less is moved in the ionic form. However, if flow increases,

Table 10.3. Total Recoverable Heavy Metals ($\mu g\,L^{-1}$)

Date 1982	Upper Birch Creek (Mined Watershed)				Twelvemile Creek (Unmined Watershed)			
	As	Pb	Zn	Cu	As	Pb	Zn	Cu
7/30	66	40	200	160	<5	<5	<10	<5
7/31	53	32	180	140	<5	<5	<10	<5
8/01	41	23	150	150	<5	<5	<10	<5
8/02	34	15	110	87	<5	<5	<10	<5
8/10	41	23	160	130	<5	<5	<10	14
8/11	38	24	120	100	<5	<5	<10	<5
8/12	28	8	90	69	<5	<5	<10	<5
8/17	57	26	160	130	<5	<5	<10	<5
8/18	72	23	210	170	<5	<5	<10	<5
8/19	41	26	120	95	<5	<5	<10	<5
8/20	29	15	100	68	<5	<5	<10	<5
8/25	32	18	100	73	<5	<5	<10	<5
8/26	23	13	80	50	<5	<5	<10	<5
8/27	15	8	60	37	<5	<5	<10	<5

From LaPerriere et al. (1985).

diluting particulate concentrations, metals can dissolve into the water as ions. Release of bound metals from adsorption to clays is possible when the water becomes acidic (Förstner and Wittman, 1981). Mobilization of trace metals from hydrous Fe/Mn oxides is high under reducing conditions, particularly true when there are sufficiently high dissolved organic material concentrations (Förstner and Wittman, 1981). This is likely in winter in interior Alaska when streams may go anoxic under prolonged ice cover.

Sealing of the streambed eventually causes depletion of dissolved oxygen in the water flowing through the bottom sediments (the underflow) (Bjerklie and LaPerriere, 1985). This effect is more severe as the stream receives more sediment and thus there may be a cumulative effect of additional mines along a stream. Unmined Twelvemile Creek (1983 summer average of $4.2\,mg\,L^{-1}$ suspended solids) had oxygen in the underflow reduced to about 58% of saturation in autumn; the underflow of mined Upper Birch Creek (1983 summer average of $655\,mg\,L^{-1}$ suspended solids) then had only 3% of full oxygen saturation. The separation of the underflow from the surface flow in mined tributaries also causes surface water to be lower in dissolved materials than if mixing with the underflow occurred; underflow is correspondingly higher in dissolved solids than it would be if diluted by surface flow (Table 10.5) (Bjerklie and LaPerriere, 1985). The surface waters of streams with active placer mines also have increased total iron content compared to their unmined controls (Table 10.5), probably due to the disturbance of pyrites by the mining activity. Iron concentrations are not increased in the underflow of streams with sealed beds.

Table 10.4. Ranges of Heavy Metals ($\mu g\,L^{-1}$) for Paired Watersheds With and Without Gold-Placer Mining, July and August 1983, Circle Quadrangle, Alaska

	As	Cu	Hg	Pb	Zn
Pair 1					
Faith Creek (M)					
Dissolved	2–7	<2–8	<0.2–0.3	<0.1–49	<2
Total	3–59	<1–34	<0.2–0.5	<2.2–40	<2–29
McManus Creek (C)					
Dissolved	<0.2–0.4	<2	<0.2	28[a]	<2
Total	1–3	<1–18	<0.2–0.7	<0.1–3	<2
Pair 2					
Upper Birch Creek (M)					
Dissolved	<0.2–7	<2–1	<0.2	<0.1	3–5
Total	21–67	9–69	<0.2–0.9	11–36	12–68
Twelvemile Creek (C)					
Dissolved	<0.2–0.2	<2	<0.2–0.3	—[b]	<2
Total	<0.2–3	<1–2	<0.3–0.7	<0.1	<2

During that period, stream discharge ($m^3 s^{-1}$) varied between 1.1 and 9.4 on Faith Creek, 1.1 and 13.2 on McManus Creek, 1.3 and 5.8 on upper Birch Creek, and 0.6 and 2.1 on Twelvemile Creek. From LaPerriere et al. (1985). M, mined; C, control.
[a] Contamination suspected.
[b] (—) Analysis not run.

Table 10.5. Summer Average Values of Selected Water Quality Constituents of Paired Watersheds With and Without Gold-Placer Mining, 1983

	Conductivity (μS cm^{-1})	Alkalinity (mg L^{-1} as CaCO$_3$)	Hardness (mg L^{-1} as CaCO$_3$)	Total Iron (mg L^{-1})	pH
Pair 1					
Faith Creek (M)					
Surface flow	67	22.0	32.0	0.71	6.7
Underflow	147	50.4	69.9	0.54	6.6
McManus Creek (C)					
Surface flow	128	39.2	63.2	0.06	6.8
Underflow	140	46.7	69.2	0.19	6.7
Pair 2					
Upper Birch Creek (M)					
Surface flow	113	38.0	60.0	1.20	7.1
Underflow	210	83.0	119.0	0.32	6.7
Twelvemile Creek (C)					
Surface flow	121	39.5	60.7	0.06	6.8
Underflow	142	51.1	69.4	0.09	6.5

M, mined; C, control. From Bjerklie and LaPerriere (1985).

Biological Effects

The most serious effect of increased sediments in normally clear water streams can occur at the base of the food chain when the turbidity attenuates light necessary for plant photosynthesis. In interior Alaska streams, the most important plants are attached algae, or periphyton. Input of particulate matter from terrestrial vegetation (allochthonous inputs) is extremely low in these streams (Cowan and Oswood, 1983) and does not contribute as much carbon to the stream food webs when compared to most temperate streams.

Van Nieuwenhuyse and LaPerriere (1986) describe the relationship between turbidity exerted by placer mining sediments in streams of interior Alaska and the resultant decrease in light penetration into the water:

$$N_t = 2.83 + 0.056 \left(T\right) \quad \left(r^2 = 0.99, \, n = 8\right)$$

where N_t = extinction coefficient for light and T = turbidity in nephelometric turbidity units (NTU). Further, the decrease in light penetration is related to a decrease of light at depth:

$$PAR_{\bar{z}} = PAR\o e^{-N_t \bar{z}}$$

where $PAR_{\bar{z}}$ = photosynthetically active radiation, in kcal m^{-2} d^{-1}, reaching the mean depth, $\bar{z}$; and $PAR_{\o}$ is the incident light at the top of the water column.

Finally, the benthic algal productivity is reduced as the light penetration is reduced:

$$GPP = 0.0021 \times PAR_{\bar{z}} \quad (r^2 = 0.67, n = 48)$$

where GPP is gross primary productivity, in $kcal\,m^{-2}\,d^{-1}$.

Therefore, 5 NTUs of turbidity can decrease the primary productivity of a 0.5 m deep stream by about 13% from clear conditions (0 NTU), and 25 NTU could decrease primary productivity 50% (Lloyd et al., 1987). In Birch Creek, when turbidity averaged 749 NTU in 1983, primary productivity was unmeasurable.

Placer mining may affect the standing crop of benthic algae both by the decrease in productivity due to light limitation and by mechanical effects due to smothering and scouring (Van Nieuwenhuyse and LaPerriere, 1986; Ryan, 1991). No standing crop of benthic algae was found on Plexiglas™ substrates after several weeks of exposure in Birch Creek in 1983 (Van Nieuwenhuyse and LaPerriere, 1986). Also, the disturbance associated with placer mining activity reduces or eliminates riparian vegetation that is slow to recover (Weber and Post, 1985). Riparian vegetation is, of course, important to the biota that live in and along the stream.

Interior Alaska streams have a low diversity of aquatic invertebrates compared to temperate streams (Oswood, 1989). In subarctic streams, densities of benthic invertebrates may be high, but the associated biomass can be relatively low because of the small size of these subarctic stream organisms (Cowan and Oswood, 1984). Nevertheless, these invertebrates form an important link between algae and fish in these streams.

Placer mining sediments cause decreased abundance and biomass of aquatic invertebrates in proportion to any measure: turbidity, suspended sediments, or settleable solids (Table 10.6). Turbidity is the best predictor of the reduction (Wagener and LaPerriere, 1985). In New Zealand placer mined streams, extreme decreases of invertebrate density were associated with slight increases in turbidity of only about 6 NTU (Ryan, 1991). Studies

Table 10.6. Sample Means ($n = 20$) of Invertebrate Density and Biomass and Associated Sediment Measurements at Times of Collection ($n = 4$; Summer 1983), for Paired Watersheds With and Without Gold-Placer Mining

	Turbidity (NTU)	Total Suspended Solids (mg L^{-1})	Settleable Solids (mL L^{-1})	Invertebrate Density (N 0.1 m^{-2})	Invertebrate Biomass (mg 0.1 m^{-2})
Pair 1					
Faith Creek (M)	56	62	<0.1	18	2.5
McManus (C)	3.4	24	<0.1	68	4.5
Pair 2					
Upper Birch Creek (M)	947	443	<0.6	8	0.7
Twelvemile Creek (C)	1.3	21	<0.1	56	3.7

M, mined; C, control. From Wagener and LaPerriere (1985).

in Alaska have not yet addressed the effects on invertebrate abundance of such small increases in turbidity due to placer mining.

Placer mining sediments may reduce aquatic invertebrate density in streams through smothering or scouring, through decrease of food provided by benthic algae, and through avoidance of direct sediment effects via catastrophic drift (Rosenberg and Wiens, 1975). Some aquatic invertebrates may serve as indicator species of placer mining effects in interior Alaska streams. For example, orthoclad chironomids and chloroperlid stoneflies suffer reductions in density, and water mites remain missing from the community for many years following the cessation of mining in the watershed (Wagener and LaPerriere, 1985).

Fish suffer both direct and indirect effects when exposed to sediment from placer mining (Reynolds et al., 1989). Bioassays involving all life stages of Arctic grayling (*Thymallus arcticus*) resulted in mortality (acute effects) only among sac fry (Table 10.7). However, older age-0 fish (young-of-year) caged in mined streams suffered gill abrasion and starvation (chronic effects) during a 16 day exposure (Reynolds et al., 1989). Many studies have shown that fish can tolerate sediment exposure for short periods (McLeay et al., 1983); but when duration is considered as well as concentration, a duration times exposure limit appears to apply to most fish (Newcombe and MacDonald, 1991).

Laboratory bioassays have also identified concentrations of heavy metals associated with placer mining that cause acute and chronic toxicities to Arctic grayling, coho salmon (*Oncorhynchus kisutch*), and rainbow trout (*O. mykiss*) (Buhl and Hamilton, 1990). In the experiments of Buhl and Hamilton (1990), arsenic, zinc, lead, and copper were lethal to fish in 4 day tests when exposed at concentrations measured in Birch Creek in 1982 (LaPerriere et al., 1985). Bioassays including both sediment and heavy metals have not yet been conducted.

The main indirect effect of stream sediment on Arctic grayling is avoidance (Simmons, 1984). Arctic grayling are sight feeders on drift; their ability to feed depends on light level (Schmidt and O'Brien, 1982). Increased turbidity decreases their reaction distance to food items and therefore their

Table 10.7. Mean Survival (%) of Caged Sac Fry Arctic Grayling July 3–7, 1985

Stream	Experiment Duration (h)			
	24	48	72	96
Upper Birch Creek (M)	85	74	59	53
Twelvemile Creek (C)	94	86	85	87

M, mined; C, control. Graylings caged as replicates of 100 individuals; three replicates were sampled each day in Upper Birch Creek (mined; 526 NTU) and in Twelvemile Creek (control; 0.7 NTU). From Reynolds et al. (1989).

feeding volume; for example, Scannell (1988) estimated that detection of food supply of small Arctic grayling is reduced by 90% from maximum when turbidity increases from 0 to 10 NTU.

Because of the spacing in pools caused by the territorial interactions of Arctic grayling (Hughes and Dill, 1990), fish that avoid turbid streams or reaches probably cannot crowd into unaffected streams without losses in feeding efficiency, because behavioral interactions associated with defending a feeding position limit the density of Arctic grayling in a clearwater stream. Loss in a watershed's carrying capacity to support Arctic grayling is probably directly related to the proportion of stream length in which turbidity eliminates sight feeding (Reynolds et al., 1989).

Summary

Placer mining has historically been an economically important part of the mining industry in Alaska. The discovery of gold brought many settlers to the state, and access to the gold fields provided impetus for expansion of the transportation systems.

Because water is used in most processes that separate placer gold from lighter materials, there is the possibility of sediment pollution of nearby water bodies. When we conducted our research on the effects of placer-gold mining on streams (1982 to 1987), we studied several streams receiving heavy sediment loads. Emphasis was placed on Birch Creek, where the heavy sedimentation at the main research site was primarily due to the cumulative effects of numerous mines above, and partially due to their inadequate or even nonexistent settling ponds.

Today, Birch Creek apparently receives far less sediment from mining, and the water quality is visibly improved. This improvement is undoubtedly partially due to decreased industrial activity because of the lower price of gold. The annual value per refined ounce peaked at $455 in 1987, falling to $359 in 1993 (Bundtzen et al., 1994). However, another factor in water quality improvement was undoubtedly the completion of the effluent guidelines (CFR, 1989) issued by announcement in 53 FR 18788, May 24, 1988, which essentially requires total recycle of the process water, with only excess stormwater allowed to be discharged. Discharged water must be treated if necessary to reduce settleable solids to $0.2\,mL\,L^{-1}$, which the U.S. EPA considers to be the detection limit of the method.

Best management practices are also stipulated in NPDES permits developed from the effluent guidelines, requiring the following: diversion of surface waters away from the plant site; construction of water retention structures "such that they are reasonably expected to reject the passage of water"; assurance that removed pollutant materials will be retained in storage areas (i.e. settling ponds); limitation of new water to the minimum amount required as make-up water for processing operations; and maintenance of water control devices to prevent "unexpected and catastrophic failure."

In the early 1980s, when effluent guidelines for the writing of NPDES permits were not yet finalized, some streams in the Circle mining district were drastically impacted by excess sediment. Sediment released by placer mining deposited in the beds of the receiving streams, and when the concentrations were high the underflow beneath the sediment–water interface was isolated from the surface water flowing in the open channel. The isolation of the underflow caused depletion of dissolved oxygen, with potentially harmful effects on streambed organisms.

We developed a mathematical-statistical model to explain the relationship between turbidity, light penetration, and algal production in interior Alaskan streams. Increased sediment, in what were normally clear water streams, caused turbidity, thereby decreasing light penetration and consequently primary production by benthic algae. Reduction of algal standing crop was thought to also be caused by the mechanical effects of sediment scouring and deposition.

Increased sediment lowered the abundance and biomass of benthic aquatic invertebrates in affected streams in proportion to sediment concentrations in the water column or filling the streambed. Turbidity was the best descriptor of these reductions. Chloroperlid stoneflies, orthoclad chironomids (nonbiting midge flies), and water mites were sensitive to placer mining sedimentation, and could be used in the future as indicators of sediment effects.

Placer mining released significant amounts of arsenic, lead, zinc, and copper into receiving streams. Even when a mined stream (Faith Creek) carried no settleable solids, the lead bound to the fine sediments exceeded the water quality criteria for the protection of aquatic life.

Arctic grayling reacted to increased sedimentation of normally clear water streams by avoidance of mined stream reaches. This is probably due to the fact that they sight feed on drifting invertebrates. Bioassay experiments showed that grayling forced to remain in sediment-laden waters were killed as sac fry, and older stages suffered chronic effects such as starvation and abrasion of skin and gill clumping. Laboratory experiments quantified the avoidance response with test fish, showing a marked aversion to turbid water above 20 NTU. Other laboratory experiments and subsequent modeling demonstrated that at 10 NTU only 10% of an Arctic grayling's food supply is probably visible to these sight-feeding fish. Grayling attempting to feed in sedimented waters would also have their food supply reduced by the sediment-caused reduction in the standing crop of benthic invertebrates.

Aquatic life in Alaskan clear water streams is currently assumed to be protected by a criterion of 25 NTU above normal conditions. However, because of the marked effects on stream life of sediment in the range of 10 to 20 NTU, enforcement of the State of Alaska water quality criterion specific to drinking water supply streams of 5 NTU above natural conditions would provide more realistic protection for clear water streams receiving placer mining effluent.

References

Alaska Administrative Code, Title 18, Chapter 70. January, 1995. Alaska Water Quality Standards Document. Alaska Department of Environmental Conservation. Water Quality Management Section Juneau, AK.

American Public Health Association [APHA], American Water Works Association, and Water Environment Federation (1989) Standard methods for the examination of water and wastewater, 17th ed. Washington, DC.

Bjerklie DM, LaPerriere JD (1985) Gold-mining effects on stream hydrology and water quality, Circle Quadrangle, Alaska. Water Resour Bull 21:235–243.

Buhl KJ, Hamilton SJ (1990) Comparative toxicology of inorganic contaminants released by placer mining to early life stages of salmonids. Ecotoxicol Environ Safety 20:325–342.

Bundtzen TK, Swainbank RC, Clough AH, Henning MW, Hansen EW (1994) Alaska's mineral industry 1993. Alaska Department of Natural Resources, Division of Geological and Geophysical Surveys and Division of Mining, and Alaska Department of Commerce and Economic Development, Division of Economic Development, Fairbanks.

Code of Federal Regulations [CFR] (1989) Title 40. Protection of environment. Part 440. Ore mining and dressing point source category. Subpart M—Gold placer mine subcategory, 440.140–440.148.

Cowan CA, Oswood MW (1983) Input and storage of benthic detritus in an Alaskan subarctic stream. Holarct Ecol 6:340–348.

Cowan CA, Oswood MW (1984) Spatial and seasonal associations of benthic macroinvertebrates and detritus in an Alaskan subarctic stream. Polar Biol 3:211–215.

Davies–Colley RJ, Hickey CW, Quinn JW, Ryan PA (1992) Effects of clay discharge on streams. 1. Optical properties and epilithon. Hydrobiologia 248:215–234.

Förstner U, Wittman GTW (1981) Metal pollution in the aquatic environment, 2nd rev ed. Springer–Verlag, New York.

Hartman CW, Johnson PR (1978) Environmental atlas of Alaska, 2nd ed. University of Alaska.

Hughes NF, Dill LM (1990) Position choice by drift-feeding salmonids: Model and test for Arctic grayling (*Thymallus arcticus*) in subarctic mountain streams, interior Alaska. Can J Fish Aquat Sci 47:2039–2048.

LaPerriere JD, Wagener SM, Bjerklie DM (1985) Gold-mining effects on heavy metals in streams, Circle Quadrangle, Alaska. Water Resour Bull 21:245–251.

Lloyd DS, Koenings JP, LaPerriere JD (1987) Effects of turbidity in fresh waters of Alaska. North Am J Fish Manage 7:18–33.

Luong HV, Braddock JF, Brown EJ (1985) Microbial leaching of arsenic from low-sulfide gold mine material. Geomicrobiol J 4:73–86.

McLeay DJ, Knox AJ, Malick JB, Birtwell IK, Hartman G, Ennis GL (1983) Effects on Arctic grayling (*Thymallus arcticus*) of short-term exposure to Yukon placer mining sediments: Laboratory and field studies. Canadian Technical Report of Fisheries and Aquatic Sciences No. 1171. Fisheries and Oceans, Ottawa, Canada.

Newcombe CP, MacDonald DD (1991) Effects of sediments on aquatic ecosystems. North Am J Fish Manage 11:72–82.

Oswood MW (1989) Community structure of benthic invertebrates in interior Alaska (USA) streams and rivers. Hydrobiologia 172:97–110.

Reynolds JB, Simmons RC, Burkholder AR (1989) Effects of placer mining discharge on the health and food of Arctic grayling. Water Resour Bull 25:625–635.

Rosenberg DM, Wiens AP (1975) Experimental sediment addition studies on the Harris River, N.W.T. Canada: The effect on macroinvertebrate drift. Verh Int Verein Theor Angew Limnol 19:1568–1574.

Ryan PA (1991) Environmental effects of sediment on New Zealand streams: A review. NZ J Mar Freshwater Res 25:207–221.

Scannell PM (1988) Effects of elevated sediment levels from placer mining on survival and behavior of immature Arctic grayling. Unpublished M.S. thesis, University of Alaska Fairbanks, Rasmuson Library.

Schmidt D, O'Brien WJ (1982) Planktivorous feeding ecology of Arctic grayling (*Thymallus arcticus*). Can J Fish Aquat Sci 39:475–482.

Simmons RC (1984) Effects of placer mining sedimentation on Arctic grayling of interior Alaska. Unpublished M.S. thesis, University of Alaska Fairbanks, Rasmuson Library.

Van Nieuwenhuyse EE, LaPerriere JD (1986) Effects of placer gold mining on primary production in subarctic streams of Alaska. Water Resour Bull 22:91–99.

Wagener SM, LaPerriere JD (1985) Effects of placer mining on invertebrate communities of interior Alaskan streams. Freshwater Invertebr Biol 4:208–214.

Weber P, Post R (1985) Aquatic habitat assessment in mined and unmined portions of the Birch Creek drainage. Alaska Department of Fish and Game, Technical Report No. 85-2. Juneau.

West RL, Deschu NA (1984) Kantishna Hills heavy metals investigations Denali National Park, 1983. National Park Service and U.S. Fish and Wildlife Service. Interagency Agreement No. 14-16-007-82-5524.

Wilson FE, Hawkins DB (1978) Arsenic in streams, stream sediments, and ground water, Fairbanks area, Alaska. Environ Geol 2:195–202.

Wolf EN (1982) The effects of placer mining on the environment in central Alaska. Report No. 48, Mineral Industry Research Laboratory, University of Alaska, Fairbanks.

11. Ecology of Overwintering Fishes in Alaskan Freshwaters

James B. Reynolds

Sixty years have passed since Hubbs and Trautman (1935) made their plea for winter studies of freshwater fishes. They perceived a serious neglect of such investigations and concluded that the state of affairs had resulted because of lack of winter-trained personnel, limited funding for winter research, and—probably most importantly—a preference among biologists for summer field work. The reasons for concern are the same now as they were then: winter may be a critical period controlling or limiting freshwater fish production. Many of the questions posed by Hubbs and Trautman are framed in an ecological context: Do fish experience food shortages during winter? Is ice formation an important factor in fish mortality? Are habitat requirements similar between winter and summer?

Hubbs and Trautman would be pleased to know that winter research in fisheries and freshwater ecology has significantly increased since the 1970s, particularly in Alaska and Canada, where the severity of winter has long been regarded as a significant factor in freshwater fish ecology. In Alaska, this increase occurred because of increased State funds (oil development), changes in priorities (environmental concerns) and improved technology (e.g., biotelemetry). Alaskan winter fisheries studies have centered in three areas of the state: northern or arctic Alaska, including the North Slope and Brooks Range, where winter extractions of gravel and water from streams are needed for site development by the oil and gas industry; central Alaska, including the interior and Alaska Range, due to human population growth and its related impacts (e.g., proposed hydroelectric dams on the Susitna River); and southeast Alaska, the coastal rain forest, spurred by potential impacts of mining and timber harvest on salmonid production in coastal streams.

This chapter describes Alaskan freshwaters as winter habitat for fish, and summarizes the results of Alaskan studies of freshwater fish populations during winter. Scientific and common names of fishes referenced in this chapter are listed in Table 11.1. Much of the work by government agencies is excellent, but the resulting reports remain part of the "gray" literature; these sources have been used only when no published source was available to support a particular point. Fortunately, a number of key studies have been published in the peer-reviewed literature; these serve as the primary source of information for this chapter. In addition, relevant studies in Canada, the continental United States, and elsewhere are cited, not as an exhaustive review, but as needed to support and complement the purpose of this review (i.e., to synthesize what is known about overwintering fishes in Alaskan freshwaters).

Table 11.1. Scientific and Common Names (American Fisheries Society, 1991) of Overwintering Fishes

Scientific Name	Common Name	Life History
Coregonus autumnalis	Arctic cisco	AP
C. nasus	Broad whitefish	AP, R
C. pidschian	Humpback whitefish	AP, R
C. sardinella	Least cisco	AP, R
Dallia pectoralis	Alaska blackfish	R
Esox lucius	Northern pike	R
Lota lota	Burbot	R
Oncorhynchus gorbuscha	Pink salmon	AD
O. keta	Chum salmon	AD
O. kisutch	Coho salmon	AD
O. mykiss	Rainbow trout[a]	AP, R
O. nerka	Sockeye salmon[b]	AD, R
O. tshawytscha	Chinook salmon	AD
Prosopium cylindraceum	Round whitefish	R
Salmo salar	Atlantic salmon	AD
Salvelinus alpinus	Arctic char	AD, R
S. fontinalis	Brook trout	AP, R
S. malma	Dolly Varden	AP, R
S. namaycush	Lake trout	R
Thymallus arcticus	Arctic grayling	R

AD, anadromous; AP, amphidromous; R, freshwater resident.
[a] Sea-run (anadromous) rainbow trout are called steelhead.
[b] Lake-resident (land-locked) sockeye salmon are called kokanee.

Winter Habitats

Freshwater habitats vary in their importance to overwintering fish, depending on their water source (see Chapter 1 by Milner et al. for a description of freshwater types). Most streams and rivers—particularly glacial, clear water, and spring-groundwater—offer overwintering habitat for fish. However, brown water streams draining tundra, bogs, and other low-lying wetlands, have low discharge in winter, usually freeze solid, and offer little habitat to fish during winter. Lakes and ponds of thermokarst, fluvial, and volcanic origin are generally inaccessible to fish (no connecting tributaries) or inhabitable (too shallow) during winter. Glacial-origin lakes and beaver ponds are lentic habitats most used by Alaskan fishes in winter. The characteristics of these important overwintering habitats—glacial, clear water, and spring-groundwater streams, glacial lakes, and beaver ponds—are reviewed in terms of their suitability as habitat for overwintering fishes.

Glacial Rivers

Rivers fed *only* by glaciers (e.g., braided rivers of Denali National Park) cease to flow in winter and support few fish, if any. Large glacial rivers (e.g., Copper, Susitna, Tanana) with substantial base flows of groundwater are essential to fishes year-round. The discharge of these rivers peaks in July to August and is quite turbid (>30 NTU; Milner and Petts, 1994) during the open water season. During autumn and early winter, September to October, turbidity decreases as discharge drops to base flow. By November, and for the remainder of winter until breakup (April to May), discharge is clear and low but steady, because the base flow is entirely from groundwater. The decrease in turbidity occurs during freeze-up when glacial flow ceases upstream. In the Tanana River during 1974, for example, turbidity was near 130 NTU in early October as freeze-up began, then steadily decreased by an order of magnitude to about 10 NTU in mid-October when freeze-up was mostly complete (Osterkamp, 1975).

Glacial rivers with significant groundwater sources are important winter habitats in Alaskan lotic systems, particularly in the Yukon and southcentral hydrologic regions where these rivers are large enough to provide substantial winter flows. Rock-cobble substrate and large woody debris (sunken, dead trees) probably provide cover for smaller fish, and deep holes below river bluffs provide areas for large fish. Sites at or just below the mouths of large clear water tributaries also support overwintering fish. These observations are mostly anecdotal because systematic information on fish habitat use during winter in these rivers is quite limited (A. Ott, Alaska Department of Fish and Game, personal communication).

Because large glacial rivers are often sealed by thick ice cover during winter, dissolved oxygen may decrease from freeze-up to breakup, and from upstream to downstream sites. Dissolved oxygen reaches minimal levels in March or April, just before breakup. In March 1971, dissolved oxygen was $10.5\,mg\,L^{-1}$ (73% saturation) in the Yukon River at Eagle, Alaska (1664 km upstream of the mouth) and decreased linearly with distance to $1.9\,mg\,L^{-1}$ (13% saturation) at Alakanuk near the mouth (Shallock and Lotspeich, 1974). Other northern rivers may exhibit spatial, as opposed to seasonal, depressions in dissolved oxygen due to reduced aeration, photosynthesis, and nutrient dilution; these features give rise to concerns for the ecological effects of waste discharge in arctic and subarctic rivers (Prowse and Gridley, 1993).

Clear Water Streams

Clear water streams are common throughout all hydrologic regions of Alaska. In contrast to glacial streams, summer water clarity is high, and flow is derived principally from precipitation. However, clearwater and glacial systems are similar in that seasonally low flows occur mostly in winter from groundwater. Clear water streams, at least those of the Yukon hydrologic region, also exhibit decreasing oxygen content temporally (as winter advances) and spatially (from headwaters to mouth). Studies in the late 1960s and early 1970s showed that minimal dissolved oxygen levels were reached earlier and sustained longer (December to February) in the Chena, Salcha, and Chatanika rivers, all clear water rivers near Fairbanks (Shallock and Lotspeich, 1974) than in glacial rivers. The lowest recorded concentration in the lower Chena River was $1.5\,mg\,L^{-1}$ (10% saturation), apparently much lower than minimal values measured in the Salcha and Chatanika (6.5 to $7.0\,mg\,L^{-1}$); this difference may have been due to domestic effluents discharged into the Chena at Fairbanks during that period (Oswood et al., 1992). Dissolved oxygen $<1\,mg\,L^{-1}$ has been recorded in the lower Chatanika (Burkholder and Bernard, 1994) and probably occurs in many other clear water rivers, depending on stream discharge and winter severity (as indicated by duration of ice cover and minima of air temperature) in a given year. Glacial rivers are larger and freeze up later than smaller clear water streams, probably accounting for the earlier, longer period of oxygen depression in clear water streams (Shallock and Lotspeich, 1974).

Due to the smaller volume of winter habitat available in clear water streams, fish overwintering is probably less extensive (i.e., more restricted) compared to that in glacial rivers. In the lower reaches, deep pools near cutbanks and large woody debris are important because substrate tends to be dominated by fine sediments. Winter habitat in upper reaches is limited to sites of groundwater discharge or springs (see below); otherwise, these areas freeze solid. Streambeds that normally remain unfrozen during winter are likely to freeze in winters, even mild ones, following dry summers because groundwater discharge is too low (Irons et al., 1989). In two tributaries of the Chena River, most aquatic insects avoided the freezing margin of the streambed by migrating, instead of depending on physiological adaptations for survival (Irons et al., 1992); benthic fishes would likely respond in the same way, but overwintering fish eggs could not exercise a migratory option and would suffer high mortality.

Arctic mountain streams (Craig and McCart, 1975), prevalent in northeastern Alaska, originate in the Brooks Range and include most of the overwintering habitat (lakes, springs, deep pools, and delta channels) on the North Slope. These streams are least productive (invertebrate densities) in summer and may be partly fed by glaciers. In winter, they frequently freeze to a depth of 2 m and temperature of flowing water seldom exceeds 1°C; small changes in ice thickness may drastically reduce overwintering habitat. The Canning, Colville, and Sagavanirktok rivers are examples of mountain streams. Tundra streams flow through the coastal plain, primarily on the western North Slope, often meandering through lakes; these streams are warmer and more productive in summer, and thus contain a greater diversity of feeding fishes compared to mountain streams. Fish vacate tundra streams in winter because these streams freeze solid.

Groundwater-Fed Systems

Groundwater-fed streams and perennial springs are among the most important winter habitats for fishes in Alaskan freshwaters, particularly in arctic areas. These systems are dominated by groundwater sources throughout the year. Clearwater Creek (also known as the Delta Clearwater River), a tributary of the Tanana River near Delta Junction (eastern Yukon hydrologic region), is the best known example of a groundwater-fed stream in Alaska (LaPerriere, 1994).

Clearwater Creek is fed by a massive aquifer recharged by the glacial Gerstle, Tanana, and Delta rivers draining east-central Alaska. Flow remains at approximately $20 \, m^3 s^{-1}$ (±10%) throughout the year. Despite winter air temperatures often reaching –40°C, ice cover is seldom complete. Water temperature rarely exceeds 8°C in the summer and probably averages 1° to 2°C during winter, due to groundwater at 4°C welling up over the entire stream. The stable flow and thermal regimes, coupled with high alkalinity, make Clearwater Creek among the most productive of interior Alaskan streams (LaPerriere, 1994). However, Arctic grayling of all ages use this stream only for summer feeding, not overwintering (Ridder, 1991). One explanation for this paradox: invertebrate drift decreases to low densities in winter (LaPerriere, 1994), but the stream remains open to diurnal lighting albeit low light intensity. If Arctic grayling remained, trying to visually drift feed as they do in summer, they would not meet their metabolic costs (Hughes and Dill, 1990). Frequent production of frazil ice has also been proposed as a reason for absence of Arctic grayling in Clearwater Creek (Armstrong, 1986); the presence of other species, such as slimy sculpin, may be explained by their residence within the substrate where the effects of frazil ice would be lessened.

Perennial springs are critical overwintering habitat in arctic regions because much of the remaining drainage freezes solid. These sites are usually quite localized, providing small, stable discharges ($\leq 1\,m^3\,s^{-1}$) of groundwater well above freezing (4° to 6°C). Perennial springs, described in detail by Craig and McCart (1975), are the most productive of arctic freshwaters in terms of invertebrate densities (Craig, 1989). Arctic char and Dolly Varden rely on these areas for autumn spawning, then remain for the duration of winter; these fish may be permanent spring residents (McCart and Craig, 1973) or annual immigrants (Craig, 1978). Thermal springs, releasing warm (14° to 16°C) water with high concentrations of dissolved solids (<2000 ppm), are an unusual winter habitat for Arctic char in the Northwest Territories (McCart and Bain, 1974).

Glacial Lakes

Most of the larger lakes of Alaska, particularly those in the southwest and southcentral hydrologic regions, were formed by glaciation and are important to both resident and anadromous fish species for overwintering. These lakes are typically deep, steep-sided, oligotrophic systems. Some, like Kenai Lake, have glacial tributaries, and others, like Iliamna Lake, have clear water tributaries. Oxygen depletion, a cause of fish "winterkill" in temperate lakes, is generally not a problem in these lakes because their mean depths are greater than the 5 m minimum threshold identified by Mathias and Barica (1980) as the depth required for low oxygen depletion rates in lakes of North America. Many glacial lakes have outlets connected to marine waters and support abundant populations of juvenile sockeye salmon. Sockeye salmon generally spawn during summer in lake tributaries; their progeny spend the first winter as embryos in stream gravels, then emerge the following spring as fry and migrate to lakes where they spend one to two winters before emigrating to sea as smolts. Production of sockeye fry is largely light limited, depending on the extent of glacial inflow (Lloyd et al., 1987); but other factors, such as nutrient levels, are also important (See Chapter 8). Although oxygen and temperature are optimal for overwintering, juvenile sockeye salmon are subject to predation by overwintering resident fishes, particularly rainbow trout and Dolly Varden; protection from predation is afforded by rock substrates along the lake margin. Other species of Pacific salmon do not require lake residence during their early life history; their young spend winters in natal streams before going to sea.

Beaver Ponds

Beavers (*Castor canadensis*) typically inhabit riparian zones in forested catchments, particularly in the Yukon, southcentral, and southeast hydrologic regions, where they often construct dams to impound water. Fisheries managers view these impoundments either with favor or disfavor, depending on their objectives (reviewed by Swanston, 1991). Beaver ponds trap sediment, increase temperature and biological productivity, and provide fish habitat, but may also block fish passage, reduce dissolved oxygen, and eliminate spawning areas. There can be no doubt, however, that beaver ponds can provide important overwintering habitat, particularly in headwaters that freeze solid, or nearly so, during winter.

Bryant (1984) has demonstrated that beaver ponds in southeast Alaska are especially important for the overwintering of juvenile salmonids, particularly coho salmon. Increased discharge during fall, a characteristic of southeast coastal streams, tends to flush sediments from these ponds, assuring adequate water quality during winter. Beaver ponds significantly improve overwinter survival of coho salmon because these areas are deeper and have more cover than adjacent stream habitat.

In the Tiekel River, a small mountain tributary of the Copper River, beaver ponds probably account for most of the production of resident Dolly Varden in that system (Gregory, 1988). This braided, second-order stream is quite cold and has low productivity; but an abundance of beaver dams in the upper reaches has altered the stream, providing higher temperatures and productivity in summer and deeper, ice-covered pools in winter (Martin, 1988). Abundance and survival of Dolly Varden in a beaver pond seem directly related to the age of the dam (Gregory, 1988). Were it not for these ponds, Dolly Varden would have only a few spring-fed reaches to escape the extensive icing that occurs in winter; subsequent abundance and survival would be quite low.

Overwintering Fishes

Migration occurs when an animal population can no longer meet its survival needs in its present environment (Taylor and Taylor, 1977). Ontogenetic niche shifts, during development from egg to adult, result in migrations because the environment no longer satisfies the needs of the developing fish (Groot, 1982). However, seasonal migrations occur because the environment changes, not the fish. Winter's onset is a period of environmental change that almost always results in migration, either within a given habitat (e.g., shift of residence from the water column to the underlying substrate), among habitats of the same system (e.g., movement from upper reaches to lower reaches of a river), or among systems (e.g., movement from marine to freshwater). Winter conditions affect all age and size groups in a fish population, although the migrational response may vary among these groups. Conditions of general significance to an entire population include decreased temperature, solar radiation, dissolved oxygen, food supply, and space; these reductions are largely the result of ice formation and associated snow cover.

Although migration is a general response by fish populations to survive the rigors of winter, physiological and behavioral adaptations by individuals to low temperature and reduced food are also critical to winter survival. Ulsch (1989) reviewed the physiology and behavior of overwintering fishes in freshwater. Physiological adaptations are required to cope with anoxia, hypoxia, and energy deficits. Behavioral adaptations include microhabitat selection to escape predation and reduce energy demands, and air breathing, in some species, to survive anoxia. Adaptations often have both physiological and behavioral characteristics that are interrelated.

Most Alaskan studies of overwintering fishes have been based on tracking of migrating individuals through radiotelemetry or detection of overwintering fish by capture. Biological data (e.g., growth and survival) have usually been obtained by comparing population status in the late summer or fall and the following spring; such data on fish during winter are sparse. Studies involving direct observation of overwintering fish in Alaskan freshwaters are rare.

Following is a summary of Alaskan studies of overwintering fishes, organized by migrational life history, based on the terminology of Myers (1949). *Anadromous* fishes (i.e., Pacific salmon in Alaska) spend most of their lives at sea, returning to freshwater only once to spawn, then die; only the eggs and juveniles of these species overwinter in freshwater. *Amphidromous* fishes (i.e., char, most whitefishes, steelhead), after spending the summer feeding in marine waters, move to freshwater in late summer or fall to spawn and overwinter; they are capable of spawning more than once. *Resident* fishes remain in freshwater during their entire life cycle, some venturing briefly into brackish water to reach overwintering areas in adjacent coastal tributaries; examples are Alaska blackfish, Arctic grayling, burbot, northern pike, round whitefish, and slimy sculpin. Classification of the migrational life history for species covered in this summary are provided in Table 11.1. A conceptual summary of movements by anadromous, amphidromous, and resident fishes is presented in Fig. 11.1.

Anadromous Fishes

Salmonids employ a wide range of strategies for overwintering in interior and southcentral freshwaters. Pink and chum salmon fry migrate to sea after emerging from gravels in the spring; only the egg stage overwinters in these species. Adult pink salmon and chum salmon have been observed in Beaufort coastal areas, but no populations are established in North Slope streams (Morrow, 1980). Spawners straying from the western Alaska coast may reach North Slope streams and spawn there in small numbers. Most eggs are likely frozen during winter, making the establishment of a natural spawning stock a difficult proposition. Pink salmon spawn in coastal streams where winter temperatures are more moderate (Morrow, 1980); freezing air temperatures and high stream discharge during winter are

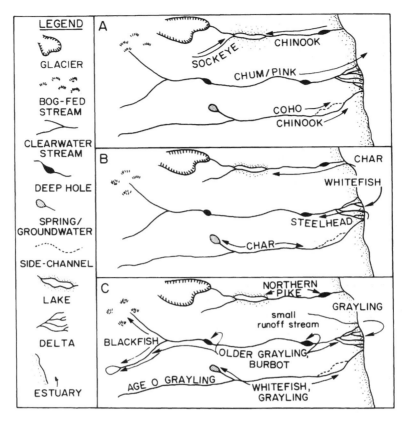

Figure 11.1. Summary of fish movements to overwintering areas (A, anadromous fishes; B, amphidromous fishes; and C, resident fishes).

common causes of high mortality among embryos in the streambed (McNeil, 1966). Chum salmon, particularly in rivers of the interior, seek groundwater upwellings as areas for spawning; the large upwelling area in the Tanana River at Big Delta is well known locally for its aggregation of spawning chum salmon in late fall to early winter (A. Ott, Alaska Department of Fish and Game, personal communication). Although egg survival of chum salmon exceeded 80% in the Chena River, the weight of fish at hatching increased from 0.1 to 2.0mg over a range of 2 to 4mgL^{-1} in dissolved oxygen (Kogl, 1965). As a result, winter oxygen levels could be expected to have a marked effect on the ability of young fish to survive downstream migration, with the larger individuals surviving better.

Chinook and coho salmon remain in rivers for one or more winters after hatching and emergence during their first winter; they then migrate to sea as smolts (Morrow, 1980). During winter, juvenile chinook tend to remain in main channels, occupying spaces within large substrate (Everest and

Chapman, 1972). Juvenile coho move to side channels and backwaters (Bustard and Narver, 1975a). However, in the Kenai River during the fall, a large proportion of the age-0 cohort of chinook salmon migrated over 50 km upstream to overwinter in or near the outlet of Skilak Lake (Bendock, 1989). Sockeye salmon overwinter in lakes for one or more winters before migrating to sea.

Winter is a critical time for juvenile salmonids in coastal streams of southeast Alaska and British Columbia; production of juveniles reaching the smolt stage is probably determined by winter conditions (Bustard, 1986). However, winter conditions, and the responses of juvenile salmonids to these conditions, is markedly different among rivers, depending on whether the catchment basin is coastal or interior in location (Fig. 11.2). In rivers draining interior catchments, the seasonal hydrograph is similar to streams of northern Alaska; discharge peaks during summer and low flow occurs in winter. The seasonal discharge of rivers draining coastal catchments peaks during the winter rainy season with low flows in summer. Juvenile salmon in interior rivers seek the protection of main channel substrate (chinook salmon) or side channel cover (coho salmon). Low winter flow can cause significant mortality among fish in side channels due to freezing, stranding, low oxygen, and predation (Bustard, 1986). In coastal source rivers, high winter flows force juveniles into tributaries to avoid high energy costs and injuries from abrasion.

The effects of forestry practices on anadromous salmonids and their coastal stream habitat has been the subject of much scrutiny and investigation in southeast Alaska (see Chapter 9) and British Columbia. Much of the early work centered on cutthroat trout, steelhead, and coho salmon in Carnation Creek, British Columbia (Bustard and Narver, 1975a,b). These studies showed that stream disturbances from logging significantly reduced

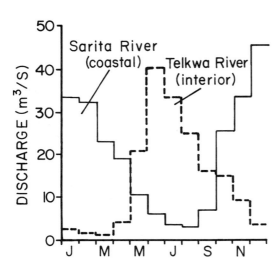

Figure 11.2. Seasonal discharge of an interior river (Telkwa) and coastal river (Sarita) in British Columbia, Canada (modified from Bustard, 1986).

winter survival of juvenile salmonids through increases in siltation, water velocity, and water temperature, and decreased cover due to losses of large woody debris. Further work by Peterson (1982a,b) in Washington demonstrated that juvenile coho salmon in large rivers move from main channels to riverine ponds and side channels for winter residence; these areas are more productive and protective than the main river. These winter habitat preferences have been further demonstrated elsewhere in British Columbia (Swales et al., 1986, 1988) and in several streams of southeastern Alaska (Heifetz et al., 1986).

Amphidromous Fishes

Amphidromous species in arctic Alaska are Dolly Varden, and a complex of coregonids or whitefishes: Arctic cisco, least cisco, broad whitefish, and humpback whitefish. It is generally recognized that amphidromous char in arctic Alaska and western arctic Canada are a northern form of Dolly Varden; those in eastern arctic Canada are Arctic char (Morrow, 1980). However, these names continue to be used interchangeably by some authors (Craig, 1989). For simplicity in this chapter, both species are collectively called char. These species spawn in the fall, usually September and October, but may enter freshwater in late summer, several weeks before spawning. Craig (1989) has summarized the conditions faced by arctic amphidromous fishes: very cold water, with annual averages of 1°C in coastal waters and 2.5°C in larger rivers; extreme icing that can reduce available stream habitat by 95% in winter; low densities of forage fish in arctic rivers; a brief, 3 month summer period to feed and store the energy needed for both fall spawning (adults) and winter survival (all fish). Arctic amphidromous fishes have adapted behaviorally to these conditions by occupying a nearshore zone of relatively warm, brackish (5° to 10°C, 10 to 25‰) water that is distinctly different from offshore, marine water (−1° to 3°C, 27 to 32‰) during summer (Fig. 11.3). This nearshore zone is also more productive of invertebrates, mostly amphipods, compared to North Slope streams (Fig. 11.4). However, during winter all coastal waters are supercooled (i.e., <0°C), and most freshwaters are frozen, leaving sites of flowing freshwater few and far between (Fig. 11.5). Arctic amphidromous fishes have incorporated fall spawning in their life histories to coincide with their return to freshwater for winter; energetic costs may prevent them from spawning every year (Craig, 1989).

Char often spawn in spring-fed areas (McCart, 1980), then remain there to overwinter with juveniles of the same population. In small North Slope drainages, where only a few springs may be available for overwintering, virtually all members of a char population (eggs, juveniles, and adults) could conceivably occupy a single spring during the 8 to 9 months of winter (Craig, 1978). They also overwinter in deep river pools, lakes, and large deltas, but do not spawn in these habitats (various authors cited by Craig, 1989). Char feed at minimal rates during winter, as shown by McCart

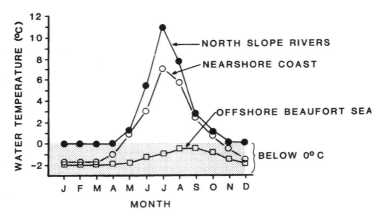

Figure 11.3. Annual cycle of water temperature in stream, nearshore, and offshore environments of the Beaufort Sea coast (reprinted from Craig, 1989, with permission).

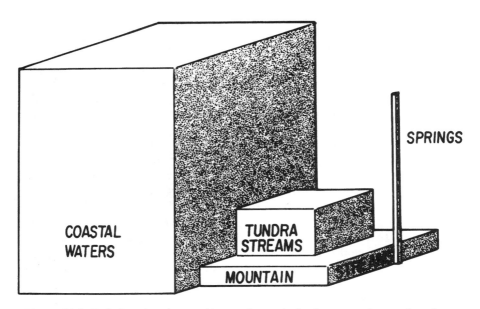

Figure 11.4. Relative abundance of invertebrates in freshwater and coastal environments of the North Slope. Volume of boxes is proportional to total invertebrate abundance, where upper surface area (length times width) represents area of available habitat in summer, and height represents average invertebrate density (reprinted from Craig, 1989, with permission).

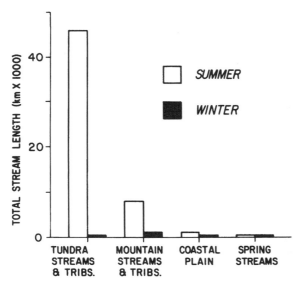

Figure 11.5. Linear distance of stream habitat available to North Slope fishes in summer and winter (reprinted from Craig, 1989, with permission).

(1980), who found empty stomachs at an incidence of 27 to 100%. Loss of energy reserves (lipids and proteins) among char is severe, the extent depending on maturity status the previous fall. In Nauyuk Lake, Northwest Territories, Canada, char that did not spawn lost nearly one-third of their total reserves during winter; those that did spawn lost nearly half their reserves and did not spawn again for at least 2 years (Dutil, 1986).

In rivers of northwestern Alaska, overwintering strategies of char are related to feeding and spawning patterns the following year (DeCicco, 1989). Fall spawners feed at sea and migrate up rivers to springs where they spawn and overwinter; they spend one winter in freshwater and are able to return to the sea each summer. Summer spawners either migrate from the lower river where they overwintered to upper reaches of the same river to spawn in June, or they migrate from another river in which they overwintered, arriving in the upper reaches to spawn in July. In either case these fish return to the lower reaches of the spawning river to spend a second winter, going to sea the following summer to feed after spending nearly 2 years in freshwater.

Amphidromous char in southeastern Alaska exhibit complex migrational patterns that include lakes for overwintering (Armstrong and Morrow, 1980). After a summer of feeding in marine waters, adult fish enter streams during fall to spawn. After spawning, they leave those streams if no lake can be reached in that system, and enter streams connecting to lakes for winter residence. Char overwintering in Buskin Lake, Kodiak Island, represent three stocks; two spawn in nearby rivers and one spawns in the lake itself; all remain in the lake for winter's duration (Whalen, 1993).

Amphidromous coregonids (Arctic cisco, least cisco, and broad white-fish) are weak swimmers (Bernatchez and Dodson, 1985); on the North Slope, they tend to overwinter in the lower reaches and deltas of larger rivers. Schmidt et al. (1989) conducted an intensive study in the Colville and Sagavanirktok deltas during November 1985 to May 1986; results differed significantly between the two areas. Winter habitat in the Sagavanirktok delta was small, less than 2 km of pools in one channel, and supported only a few thousand large (>250 mm long) Arctic cisco. As winter progressed, ice formation isolated the pools, restricted fish, and depressed dissolved oxygen (range 0.3 to 2.7 mg L^{-1}), resulting in a fish kill under the ice during April. Fish apparently did not feed, and the surviving fish were in poor condition. In contrast, the Colville delta offered about 220 km of winter habitat to an estimated 1 to 2 million large Arctic cisco. The deeper channels in the Colville delta remained open to fish movement, despite continued ice formation, allowing them to feed (mostly upon amphipods) and survive with adequate levels of dissolved oxygen.

Steelhead enter the Copper River during September and October and remain there or in the lower Gulkana River, a tributary, during winter. In May, they rapidly ascend the Gulkana to spawn in the headwaters of one of its tributaries, the Middle Fork (Burger et al., 1983); the spawning area is well defined by groundwater discharge dominated by upwelling. However, steelhead (and rainbow trout) of all ages do not use the groundwater areas for overwintering; they apparently prefer deeper reaches in the lower Gulkana River (personal observation).

Resident Fishes

Resident fishes in arctic Alaska include Arctic grayling, Alaska blackfish, burbot, lake trout, and round whitefish. Resident populations of Arctic char also occur in springs (McCart and Bain, 1974). Of these, the best-studied species during winter is the Arctic grayling (Armstrong, 1986).

Arctic grayling are successful in arctic freshwaters because they spawn in streams immediately after breakup, then move to freshwater habitats throughout the North Slope to feed during summer. Because they feed primarily on aquatic insects throughout their life, most freshwaters are exploitable to them if the water clarity permits sight feeding. Fry emerge from gravels only a few weeks after egg deposition and are able to feed continuously during most of the summer, rapidly attaining a size adequate for overwinter survival.

Overwintering is probably the most critical period for Arctic grayling. As feeding waters cool in August and September, all fish must leave these areas to reach a limited number of overwintering sites. If migration to overwintering areas is learned, not genetically programmed, stranding is probably frequent among first-year fish and mortality high (Craig, 1989). Adults migrate at rates up to 5 to 6 km d^{-1} in early September (Fig. 11.6) to reach

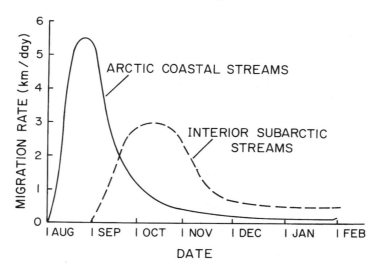

Figure 11.6. Known migration rate and timing of Arctic grayling in Arctic coastal streams (West et al., 1992), compared with estimated rate and timing in subarctic streams (Armstrong, 1986).

deep pools, river deltas, springs, and lakes; they may move 100 km or more, changing river systems through brackish coastal water to do so (West et al., 1992). Upon reaching an overwintering area, an entire population or mixed populations may be dependent upon a single spring for 7 to 8 months; such appeared to be the case for the population of a small tributary of the Kavik River overwintering in a small spring 85 km distant (Craig and Poulin, 1975).

In the broad expanse of interior Alaska (Yukon hydrologic region), resident species are more numerous than anadromous ones. For example, only 2 of the 13 fish species in the Chena River, chinook salmon and chum salmon, are anadromous (Oswood et al., 1992). Populations of some other species in the Chena (e.g., inconnu) are residents, but coastal populations exist elsewhere that use brackish water for summer feeding and coastal deltas and lagoons for overwintering (Morrow, 1980). Winter studies of interior fishes have focused on Arctic grayling, burbot, lake trout, northern pike, and rainbow trout. In addition, some information is available on Alaska blackfish.

Rainbow trout do not occur naturally in the interior, but are found in the Copper River drainage and other waters of southcentral and southwestern Alaska. Rainbow trout (and steelhead) use both rivers and lakes for over-wintering, but their winter habitats are not documented in Alaska. Over-wintering conditions are considered to be the primary set of factors limiting northern distribution of rainbow trout (Morrow, 1980). M. Merritt (Alaska Department of Fish and Game, Fairbanks, unpublished manuscript) used the analytic hierarchy process, a quantitative decision-making technique, to show that overwintering capacity of interior streams accounted for most of the limitations (others were spawning and growth potentials) in evaluating interior streams for stocking rainbow trout.

Arctic grayling in subarctic systems migrate from summer feeding areas to overwintering areas in the same general pattern as do those in arctic systems (Armstrong, 1986). However, subarctic adults migrate later in the fall (Fig. 11.6) and seem to have less dependence on springs than their arctic counterparts; they rely more on deep reaches in lower clear water rivers and adjacent reaches of glacial rivers. Juveniles in headwater reaches appear to avoid the energetic costs and predatory risks of migrating long distances downstream by ascending a groundwater-fed tributary and spending the winter confined in pipelike channels in *aufeis* (N. Hughes, Alaska Cooperative Fish and Wildlife Research Unit, personal communication). Fish of all ages forego their summer territorial behavior (Hughes, 1992), crowding together in small spaces, swimming in slow currents ($<0.2 \, m \, s^{-1}$) and feeding benthically, rather than seeking shelter in substrate. Arctic grayling may be vulnerable to predation by river otters (*Lutra canadensis*); they remain under ice near open leads rather than taking positions in ice-free sections. Long-term tracking of Arctic grayling with implanted radio transmitters indicates that these fish tend to return to the same overwintering areas in successive winters (B. Lubinski, Alaska Cooperative Fish and Wildlife Research Unit, personal communication).

Northern pike and Alaska blackfish spend winter in lakes and rivers, usually in slack water (Morrow, 1980). Alaska blackfish overwinter in ponds and small lakes; they are also capable of surviving as air breathers while "nesting" in the damp vegetation of dewatered, ice-covered side channels near large rivers (personal observation). Their notoriety for returning to life after being frozen is widespread; in fact, they die even if only partially frozen but can survive body cooling just above freezing and near-zero levels of dissolved oxygen (Morrow, 1980). Northern pike also tolerate low levels of dissolved oxygen ($<1 \, mg \, L^{-1}$) but can detect and avoid these levels (Casselman, 1978). Higher oxygen concentrations are probably necessary to support the active feeding and associated metabolism of northern pike during winter (Diana and Mackay, 1979). In Minto Flats, a subarctic wetland near Fairbanks, adult northern pike avoid the extensive freezing of shallow lakes and tributaries by migrating to the main channels of the Chatanika, Tolovana, and Tanana rivers during September (Burkholder and Bernard, 1994). Instead of losing energy reserves over winter, as do most fresh water fishes, northern pike in the Chatanika River gained energy reserves because they actively fed on least cisco and humpback whitefish returning downstream from fall spawning (Murphy, 1989). Females were particularly active in feeding as a means to prepare ovarian tissue for spring spawning. Winter predation by northern pike has also been documented in Canada (Casselman, 1978; Diana and Mackay, 1979). In 1985, a winter sport fishery developed on a concentration of northern pike in a short reach of the Chatanika River. An annual harvest of 2300 to 5000 fish, the majority females, was considered excessive, and the fishery was closed in 1987 by the Alaska Department of Fish and Game (Holmes and Burkholder, 1988).

Burbot spawn in midwinter, usually January and February while congregated in rivers (Chen, 1969; Morrow, 1980). These winter-active fish occupy lakes and rivers during winter, feeding actively on invertebrates while young and gradually changing to a diet of fish as they grow (Chen, 1969). Burbot radiotagged in the Tanana River near Fairbanks apparently formed numerous, localized spawning aggregations over a reach exceeding 100 km and remained there for much of winter (Evenson, 1993). In the upper Tanana drainage (Nabesna and Chisana rivers), burbot moved extensively during November to March, apparently to reach spawning areas upstream (Breeser et al., 1988).

Lake trout usually spend their entire lives in lakes where overwintering presents less of a challenge. In Alaska, they feed on a wide variety of invertebrates (especially snails) and fish, probably because lake productivity is so low and omnivory is essential to survival (Morrow, 1980). Overwintering success of juvenile lake trout is probably more dependent on food and subsequent growth, than on physical limits of habitat. Rates of sexual maturity and recruitment into the adult population depend on attainment of minimal sizes to compete successfully with older lake trout (Johnson, 1976).

Fish Overwintering Strategies

Although environmental conditions during winter in Alaskan freshwaters are extreme, many fish species are well-adapted for surviving winter. The most general adaptation is behavioral: movement from inadequate to adequate habitat as required by the environmental limits imposed by winter's onset (Thorpe, 1994). These adaptations are the result of a relatively predictable (i.e., stable) environment and flexible life history strategies to overcome environmental variations that inevitably occur.

The seasonally predictable conditions in Alaskan fresh waters, especially in arctic areas, are availability and location of food, low water temperature, and habitat reduction in winter (Craig, 1989). Food is abundant only during a 3 to 5 month summer, depending on latitude, and feeding must be intense for a fish to build its body reserves to survive a winter with little or no food. If food is not adequate to support both somatic and gondal growth, energy will be diverted to metabolic maintenance; spawning is deferred to another year. Food is most available only in certain areas, requiring annual feeding migrations. Thus, arctic anadromous fishes enter coastal areas to feed, and resident fishes move from groundwater upwelling areas to warmer, shallow streams and lakes. Declining temperatures in autumn require fish to move to overwintering areas; juvenile fish (age 0) avoid predation by moving less distance than adults. In Canadian studies, juvenile Atlantic salmon moved into stream substrate rather than emigrate from summer feeding areas (Rimmer et al., 1983); this behavior may occur in southeastern Alaska but seems much less likely in subarctic and arctic Alaska. A more likely, and promising, avenue for juvenile stream fishes in central and arctic Alaska is movement upstream or downstream to groundwater areas to avoid anchor ice, then moving into substrate or remaining in the water column, depending on size (personal observation). Small, resident brook trout in a subarctic Quebec river were able to use substrate for winter residence, thus conserving energy; their anadromous counterparts were too large to do so and had to migrate to larger, deeper areas (Cunjak and Power, 1986). Extreme icing and resultant reduction in fish habitat requires that fish rely on virtually the same overwintering areas year after year. Many fish populations appear to depend on "traditional" overwintering areas; although this dependence is apparently learned, not genetic, enough fish do learn in each generation to ensure species success. While in winter residence, fish undergo behavioral and physiological changes to minimize energy loss. Territoriality, evident in summer feeding and spawning areas, is repressed and aggregation is much more evident. Metabolic costs are high in early winter during initial acclimitization; thereafter, metabolism is lower and more stable, conditioning fish for the slow habitat deterioration over the remainder of winter (Cunjak and Power, 1987; Cunjak, 1988).

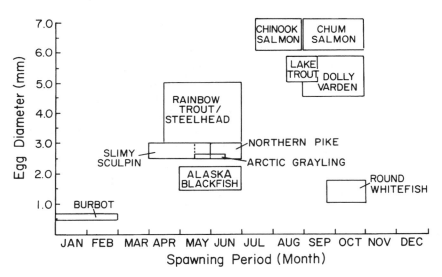

Figure 11.7. Egg diameter and period of spawning for selected species of fish in Alaskan freshwaters (data from Morrow, 1980). The vertical dashed line is the left side of the "box" for northern pike.

Despite the predictable aspects of the Alaskan freshwater environment, variations in winter severity occur that require flexibility to assure species survival; this flexibility is evident in life history strategies, particularly with respect to reproduction. Two general approaches are used: spring spawning and summer to fall spawning (Fig. 11.7). Spring spawners have devoted energy to gamete production during the previous summer or winter, but have invested less energy in doing so; their eggs are smaller and hatch in early summer. Spring spawners are free to spend most of the summer feeding to restore and perhaps add to body reserves lost during winter. However, their young must grow enough during summer to survive their first winter. Summer to fall spawners devote more energy to egg production, generally producing larger eggs that will overwinter and hatch in the substrate, living on yolk before emergence; despite feeding advantages in spring and early summer, which they pass on to their young as energy for winter survival, adults must face winter in reduced body condition. Their response is to spawn once and die (Pacific salmon), spawn once every 2 to 3 years (char), or produce large quantities of small eggs (round whitefish).

Winter is likely the limiting, or most critical, period for fishes in Alaskan freshwaters. They primarily rely on behavioral adaptations to survive and ensure reproductive success in these extreme conditions. Their basic life history plan is that of a K-strategist (Craig, 1989): long-lived adults with high survival rates who spawn several times at low reproductive rates, producing young with low survival and recruitment rates. As a result, Alaskan freshwater fishes are like climax forests (Johnson, 1976): their populations are stabilized at maximum biomass and low turnover, consisting of multiple age groups. The risks of winter are probably spread, therefore, among the juveniles in most populations and cohorts of a species.

References

American Fisheries Society (1991) Common and scientific names of Fishes, 5th ed. American Fisheries Society, Bethesda, MD, Special Publication 20.

Armstrong RH (1986) A review of Arctic grayling studies in Alaska, 1952–1982. Biol Pap Univ Alaska 23:3–17.

Armstrong RH, Morrow JE (1980) The Dolly Varden charr. In Balon EK (Ed) Charrs, salmonid fishes of the genus *Salvelinus*. Dr W. Junk bv Publishers, The Netherlands, 99–140.

Bendock T (1989) Lakeward movements of juvenile chinook salmon and recommendations for habitat management in the Kenai River, 1986–88. Alaska Department of Fish and Game, Fishery Manuscript Series Number 7, Anchorage, AK.

Bernatchez L, Dodson J (1985) Influence of temperature and current speed on the swimming capacity of lake whitefish (*Coregonus clupeaformis*) and cisco (*C. artedi*). Can J Fish Aquat Sci 42:1522–1529.

Breeser SW, Stearns FD, Smith MW, West RL, Reynolds JB (1988) Observations of movements and habitat preferences of burbot in an Alaskan glacial river system. Trans Am Fish Soc 117:506–509.

Bryant MD (1984) The role of beaver dams as coho salmon habitat in southeast Alaska streams. In Walton JM, Houston DB (Eds) Proceedings, Olympic Wild Fish Conference, Peninsula College, Port Angeles, WA, 183–192.

Burger C, Scott M, Small M (1983) Overwintering and spawning areas of steelhead trout (*Salmo gairdneri*) in tributaries of the upper Copper River, Alaska. Final Report, Bureau of Land Management, Anchorage.

Burkholder A, Bernard DR (1994) Movements and distribution of radio-tagged northern pike in Minto Flats. Alaska Department of Fish and Game, Fishery Manuscript 94-1, Anchorage.

Bustard D (1986) Some differences between coastal and interior stream ecosystems and the implications to juvenile fish production. In Patterson JH (Ed) Proceedings of the workshop in habitat improvements. Canadian Fisheries and Aquatic Sciences Technical Report 1483. Nanaimo, British Columbia, Canada.

Bustard DR, Narver DW (1975a) Aspects of the winter ecology of juvenile coho salmon (*Oncorhynchus kisutch*) and steelhead trout (*Salmo gairdneri*). J Fish Res Board Can 32:667–680.

Bustard DR, Narver DW (1975b) Preferences of juvenile coho salmon (*Oncorhynchus kisutch*) and cutthroat trout (*Salmo clarki*) relative to simulated alteration of winter habitat. J Fish Res Board Can 32:681–687.

Casselman JM (1978) Effects of environmental factors on growth, survival, activity, and exploitation of northern pike. Am Fish Soc Spec Publ 11:114–128.

Chen LC (1969) The biology and taxonomy of the burbot, *Lota lota leptura*, in interior Alaska. Biol Pap Univ Alaska 11.

Craig PC (1978) Movements of stream-resident and anadromous Arctic char (*Salvelinus alpinus*) in a perennial spring on the Canning River, Alaska. J Fish Res Board Can 35:48–52.

Craig PC (1989) An introduction to anadromous fishes in the Alaskan Arctic. Biol Pap Univ Alaska 24:27–54.

Craig PC, McCart P (1975) Classification of stream types in Beaufort Sea drainages between Prudhoe Bay, Alaska and the Mackenzie Delta, N.W.T. Arct Alpine Res 7:183–198.

Craig PC, Poulin VA (1975) Movements and growth of Arctic grayling (*Thymallus arcticus*) and juvenile Arctic char (*Salvelinus alpinus*) in a small arctic stream, Alaska. J Fish Res Board Can 32:689–698.

Cunjak RA (1988) Physiological consequences of overwintering in streams: The cost of acclimatization? Can J Fish Aquat Sci 45:443–452.

Cunjak RA, Power G (1986) Seasonal changes in the physiology of brook trout, *Salvelinus fontinalis* (Mitchill), in a sub-Arctic river system. J Fish Biol 29: 279–288.

Cunjak RA, Power G (1987) The feeding and energetics of stream-resident trout in winter. J Fish Biol 31:493–511.

DeCicco AL (1989) Movements and spawning of adult Dolly Varden charr (*S. malma*) in Chukchi Sea drainages of northwestern Alaska: Evidence for summer and fall spawning populations. Physiol Ecol Jpn Spec Vol 1:229–238.

Diana JS, Mackay WC (1979) Timing and magnitude of energy deposition and loss in the body, liver and gonads of northern pike (*Esox lucius*). J Fish Res Board Can 36:481–487.

Dutil J-D (1986) Energetic constraints and spawning interval in the anadromous Arctic charr (*Salvelinus alpinus*). Copeia 1986:945–955.

Evenson MJ (1993) Seasonal movements of radio-implanted burbot in the Tanana River drainage. Alaska Department of Fish and Game, Fishery Data Series 93–47, Anchorage, AK.

Everest FH, Chapman DW (1972) Habitat selection and spatial interaction of juvenile chinook salmon and steelhead trout in two Idaho streams. J Fish Res Board Can 29:91–100.

Gregory LS (1988) Population characteristics of Dolly Varden in the Tiekel River, Alaska. M.S. thesis, University of Alaska, Fairbanks.

Groot C (1982) Modifications on a theme—A perspective on migratory behavior of Pacific salmon. In Brannon EL, Salo EO (Eds) Salmon and trout migratory behavior symposium, University of Washington, Seattle, 1–21.

Heifetz J, Murphy ML, Koski KV (1986) Effects of logging on winter habitat of juvenile salmonids in Alaskan streams. North Am J Fish Manage 6:52–58.

Holmes RA, Burkholder A (1988) Movements and stock composition of northern pike in Minto Flats. Alaska Department of Fish and Game, Fishery Data Series 53, Anchorage, AK.

Hubbs CL, Trautman MB (1935) The need for investigating fish conditions in winter. Trans Am Fish Soc 65:51–56.

Hughes NF (1992) Ranking of feeding positions by drift-feeding Arctic grayling (*Thymallus arcticus*) in dominance hierarchies. Can J Fish Aquat Sci 49:1994–1998.

Hughes NF, Dill LM (1990) Position choice by drift-feeding salmonids: Model and test for Arctic grayling (*Thymallus arcticus*) in subarctic mountain streams, interior Alaska. Can J Fish Aquat Sci 47:2039–2048.

Irons JG III, Miller LK, Oswood MW (1992) Ecological adaptations of aquatic macroinvertebrates to overwintering in interior Alaska (U.S.A.) subarctic streams. Can J Zool 71:98–108.

Irons JG III, Ray SR, Miller LK, Oswood MW (1989) Spatial and seasonal patterns of streambed water temperatures in an Alaskan subarctic stream. Headwaters Hydrol Am Water Res Assoc June: 381–390.

Johnson L (1976) Ecology of arctic populations of lake trout, *Salvelinus namaycush*, lake whitefish, *Coregonus clupeaformis*, Arctic char, *S. alpinus* and associated species in unexploited lakes of the Canadian Northwest Territories. J Fish Res Board Can 33:2459–2488.

Kogl DR (1965) Springs and ground-water as factors affecting survival of chum salmon spawn in a sub-arctic stream. M.S. thesis, University of Alaska, Fairbanks.

LaPerriere JD (1994) Benthic ecology of a spring-fed river of interior Alaska. Freshwater Biol 32:349–357.

Lloyd DS, Koenings JP, LaPerriere JD (1987) Effects of turbidity in fresh waters of Alaska. North Am J Fish Manage 7:18–33.

Martin DC (1988) Aquatic habitat of the Tiekel River, southcentral Alaska, and its utilization by resident Dolly Varden (*Salvelinus malma*). M.S. thesis, University of Alaska, Fairbanks.

Mathias JA, Barica J (1980) Factors controlling oxygen depletion in ice-covered lakes. Can J Fish Aquat Sci 37:185–194.

McCart P (1980) A review of the systematics and ecology of Arctic char, *Salvelinus alpinus*, in the western Arctic. Canadian Technical Report, Fisheries and Aquatic Sciences 935. Nanaimo, British Columbia, Canada.

McCart P, Bain H (1974) An isolated population of Arctic char (*Salvelinus alpinus*) inhabiting a warm mineral spring above a waterfall at Cache Creek, Northwest Territories. J Fish Res Board Can 31:1408–1414.

McCart P, Craig P (1973) Life history of two isolated populations of Arctic char (*Salvelinus alpinus*) in spring-fed tributaries of the Canning River, Alaska. J Fish Res Board Can 30:1215–1220.

McNeil WJ (1966) Effect of the spawning bed environment on reproduction of pink and chum salmon. Fish Bull 65:495–523.

Milner AM, Petts GE (1994) Glacial rivers: Physical habitat and ecology, Freshwat Biol 32:295–307.

Morrow JE (1980) The freshwater fishes of Alaska. Alaska Northwest Publishing Company, Anchorage.

Murphy RL (1989) Seasonal allocation of energy in four tissues of northern pike from Minto Flats, Alaska. M.S. thesis, University of Alaska, Fairbanks.

Myers G (1949) Usage of anadromous, catadromous and allied terms for migratory fishes. Copeia 1949:89–96.

Osterkamp TE (1975) Observations of Tanana River ice. In Frankenstein GE (Ed) Proc Third Int Symp Ice Problems, Hanover, NH, 201–208.

Oswood MW, Reynolds JB, LaPerriere JD, et al. (1992) Water quality and ecology of the Chena River, Alaska. In Becker CD, Neitzel DA (Eds) Water quality in North American river systems, Battelle Press, Columbus, OH. 7–27.

Peterson NP (1982a) Population characteristics of juvenile coho salmon (*Oncorhynchus kisutch*) overwintering in riverine ponds. Can J Fish Aquat Sci 39:1303–1307.

Peterson NP (1982b) Immigration of juvenile coho salmon (*Oncorhynchus kisutch*) into riverine ponds. Can J Fish Aquat Sci 39:1308–1310.

Prowse TD, Gridley NC (Eds) (1993) Environmental aspects of river ice. National Hydrology Research Institute, Science Report 5, Saskatoon, Saskatchewan.

Ridder WP (1991) Summary of recaptures of Arctic grayling tagged in the middle Tanana River drainage, 1977 through 1990. Alaska Department of Fish and Game, Fishery Data Series 91-34, Anchorage.

Rimmer DM, Paim U, Saunders RL (1983) Autumnal habitat shift of juvenile Atlantic salmon (*Salmo salar*) in a small river. Can J Fish Aquat Sci 40:671–680.

Schmidt DR, Griffiths WB, Martin LR (1989) Overwintering biology of anadromous fish in the Sagavanirktok River delta, Alaska. Biol Pap Univ Alaska 24: 55–74.

Shallock EW, Lotspeich FB (1974) Low winter dissolved oxygen in some Alaskan rivers. EPA-660/3-74-008, U.S. Environmental Protection Agency, Corvallis, OR.

Swales S, Caron F, Irvine JR, Levings CD (1988) Overwintering habitats of coho salmon (*Oncorhynchus kisutch*) and other juvenile salmonids in the Keogh River system, British Columbia. Can J Fish Aquat Sci 66:254–261.

Swales S, Lauzier RB, Levings CD (1986) Winter habitat preferences of juvenile salmonids in two interior rivers in British Columbia. Can J Zool 64:1506–1514.

Swanston DN (1991) Natural processes. In Meehan WR (Ed) Influences of forest and rangeland management on salmonid fishes and their habitat. American Fisheries Society, Bethesda, MD, Special Publication 19, 139–179.

Taylor LR, Taylor RAJ (1977) Aggregation, migration and population mechanics. Nature (Lond) 265:415–421.

Thorpe JE (1994) Salmon flexibility: Responses to environmental extremes. Trans Am Fish Soc 123:606–612.

Ultsch GR (1989) Ecology and physiology of hibernation and overwintering among freshwater fishes, turtles, and snakes. Biol Rev 64:436–516.

West RL, Smith MW, Barber WE, Reynolds JB, Hop H (1992) Autumn migration and overwintering of Arctic grayling in coastal streams of the Arctic National Wildlife Refuge, Alaska. Trans Am Fish Soc 121:709–715.

Whalen ME (1993) Dynamics of a super-population of Dolly Varden in the Chiniak Bay system, Kodiak Island, Alaska. M.S. thesis, University of Alaska, Fairbanks.

12. Glacial Recession and Freshwater Ecosystems in Coastal Alaska

Alexander M. Milner

At present, approximately 75,000 km^2 or 5% of Alaska is covered by glaciers (Post and Meier, 1980) that occur principally as extensions of large ice fields centered in mountain ranges that reach 4000 to 6000 m (Calkin, 1988). The estimated average stream and river flow derived from glaciers is 220 km^3 yr^{-1}, which is equivalent to 35% of the total runoff in Alaska, 14% of the runoff from the conterminous states, and 10% of the total runoff in the United States (Mayo and Trabant, 1986). Glaciers influence the flow and water quality of all major Alaska rivers except the Colville (Benson et al., 1986).

During the past two centuries, significant retreat of a number of glaciers from their Neoglacial maxima has occurred in Alaska, particularly those at tidewater (Field, 1979; Miller, 1964). Miller (1964) indicated that of 174 glaciers observed in southcentral and southeast Alaska, 106 were significantly receding and 33 were gradually receding. Glacial recession of tidewater glaciers is indirectly linked to global climate change, a topic examined in more detail at the end of this chapter. Rapid recession of tidewater glaciers may create remnant ice sheets in valleys upon separation from the actively retreating glacier fronts. Meltwater from these remnant ice sheets may form new streams (see Fig. 12.1) that are subsequently colonized by biotic communities. These streams are very different from streams issuing from active glacier fronts terminating on land. Here streambeds are typically braided and show little sign of definite channel development and stabilization.

Figure 12.1. "The Resurrected Forest"—a new glacial stream cuts through morainic gravels and reveals remnants of the forest in Glacier Bay before the ice advanced (photo from Harriman Expedition, 1899, University of Alaska, Fairbanks).

In contrast, streams derived from remnant ice sheets frequently have a proglacial lake at their margin that may buffer flow variations and result in a more stable stream channel. As the ice ablates, these streams may eventually become clear water fed by snowmelt and rainfall in the valley.

Gore and Milner (1990) proposed five levels of disturbance of lotic systems, each leading to different recovery regimes of biotic communities. Level 1 covers catastrophic disturbances, wherein communities are completely destroyed along the entire stream length leaving no upstream or downstream sources of colonizers. Recovery is dependent upon primary successional processes involving colonization and subsequent change in an area where no trace of a previous community exists (Fisher, 1990). Studies of the colonization of newly created stream channels following glacial recession in coastal Alaska provide a unique opportunity to examine community development in lotic freshwater habitats under an equivalent scenario to level 1 disturbances and thereby provide benchmark recovery rates for many catastrophic disturbances. Primary succession has rarely been studied in running waters, particularly where upstream sources of potential drift colonizers are absent (Cushing and Gaines, 1989).

Although examples of site-specific temporal succession in streams have been suggested by a number of authors (Downes and Lake, 1991; Hemphill and Cooper, 1983; Lake and Doeg, 1985; Peckarsky, 1986), the spatial and temporal scales of these studies have been small (10^{-2} to $10\,m$) and short (weeks). These studies all describe "patch colonization" involving short-term selection from a pool of available colonizers. In "true site-specific temporal succession" a change in the composition and/or dominance of the pool of potential colonizing species occurs over time that is unrelated to seasonal influences (Milner, 1994). In studies of site-specific temporal succession, colonization and succession should preferably be examined on a spatial scale of stream segments or entirely new stream systems (10^{-1} to 10km) (Malmqvist et al., 1991) and for several years (Sheldon, 1984).

This chapter overviews studies principally up to 1991, of community development in new streams following recession of tidewater glaciers in two principal areas: Glacier Bay National Park in southeast Alaska and Kenai Fjords National Park in southcentral Alaska. As mentioned previously, it should be emphasized that glacial recession is occurring at other sites along the coastline of Alaska and new stream habitat has been created. These studies offer insights into postglacial colonization patterns, processes that must have similarly occurred following retreat of the glacial ice sheets after the Wisconsin Ice Age when half of North America and northern Europe were covered in ice.

Glacier Bay National Park

Glacier Bay National Park, covering 11,030 km², lies in southeastern Alaska encompassing a fjord over 100km long and 20km wide with two major arms, the northwest arm and Muir Inlet. A Neoglacial ice advance reached its maximum near the mouth of Glacier Bay close to 1700 AD after which a major recession commenced sometime between 1735 and 1785 (Goldthwait, 1966; Miller, 1964). This glacial retreat has continued to the present, interrupted only by several periods of decreased rate of retreat (Miller, 1964) (Fig. 12.2).

From 1977 to 1979, biological communities were examined along a chronosequence of five streams from newly formed to approximately 150 years old to infer temporal changes in biotic community structure (Milner, 1987; Milner and Bailey, 1989). Overall, biotic community development in these postglacial streams appeared to be controlled principally by abiotic factors, especially flow and sedimentation processes. In the absence of large woody debris following glacial recession, the presence of a lake that reduces flow variations, and enables coarser sediment (>0.1 mm) to settle out, is an important factor in the development and productivity of biotic communities (Sidle and Milner, 1989).

However, site-specific community changes have now been observed to occur in one stream, Wolf Point Creek, since 1979. This stream, along with a number of other recently formed streams, lies in upper Muir Inlet (Fig. 12.3). Wolf Point Creek, approximately 2km long, is a first-order stream originating from a meltwater lake (unofficial name Lawrence Lake) near the southeast margin of the Muir Glacier remnant ice sheet. Lawrence Lake began to form in the late 1960s (Fig. 12.4) and the lake gradually increased in surface area as the remnant ice ablated attaining 1.35km² by 1991 (Fig. 12.5). A 1991 bathymetric survey indicated that two principal subbasins (maximum depths 30 to 35 and 35 to 40m, respectively) exist. The lake

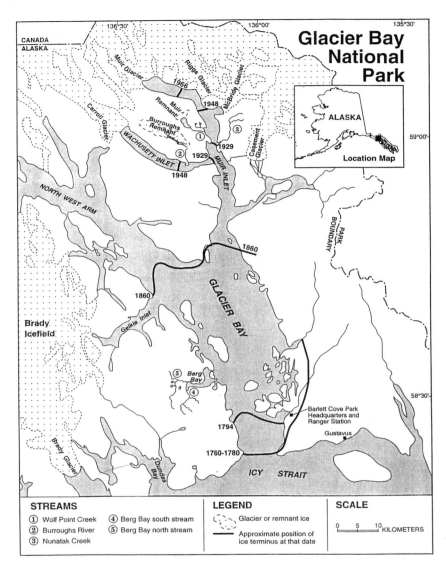

Figure 12.2. Map of Glacier Bay National Park showing ice recession.

outlet flows initially over a bedrock falls (almost certainly a barrier to migration of anadromous salmon) and then over glacial moraine, till, and outwash deposits. Although the mouth of this stream was uncovered in the late 1940s, the study reach (approximately 0.9 km from the mouth) was deglaciated in approximately 1960. However it is likely that the channel was ill-defined and braided at that time prior to the formation of Lawrence Lake. One sampling station has been considered representative of the stream community due to the homogenous nature of the substrate and flow

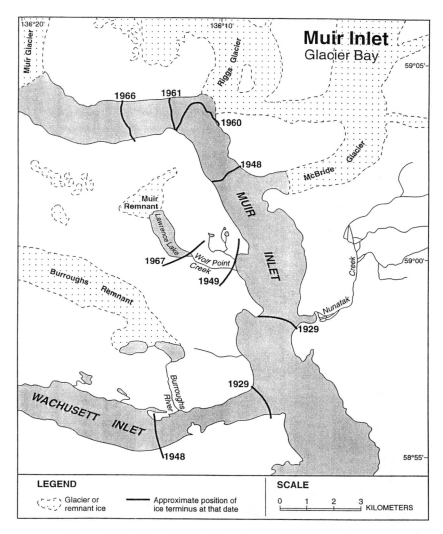

Figure 12.3. Map of Muir Inlet showing Wolf Point Creek.

regime along the stream length and that invertebrate sampling indicated no significant longitudinal differences in community structure (Milner, 1983).

Studies of the stream channel development at the sampling reach from 1982 to 1985 indicate only slight channel changes, mostly related to minor erosion and deposition of silt released from the remnant glacial ice (Sidle and Milner, 1989). Larger sediment particles settle in the meltwater lake, and thus the major sediment loading to this stream is composed of fine silts and clays (<0.002 mm) that, at most stages of flow, is transported in suspension out of the system. The lake also acts to buffer discharge fluctuations

Figure 12.4. Muir Remnant in 1968 (photo by D.B. Lawrence).

Figure 12.5. Retreat of Muir Remnant and formation of Lawrence Lake 1993.

thereby resulting in relatively stable discharges. Hence the detrimental effects of sediment deposition and discharge variations observed in other streams without lakes (Milner, 1987) are ameliorated in Wolf Point Creek. The stabler discharges have allowed in recent years the development of a riparian zone composed principally of alder (*Alnus crispa* Pursh) and willows (*Salix* spp.) 1.5 to 2 m in height (Figs. 12.6, 12.7). Turbidities over the study period to 1990 typically remained in the 100 to 150 NTU range, but August water temperatures increased from 2° to 3°C in 1978 to 9°C in 1990 as the source lake increased in size, thereby significantly enlarging the surface area available for receiving solar radiation.

Macroinvertebrates

The macroinvertebrate community displayed distinct changes with stream development from 1978 to 1990 as outlined in Milner (1994) (see Fig. 12.8). Possible seasonal differences in community structure were avoided by only comparing late summer samples (mid-August to mid-September). Samples were collected at other time periods but not consistently from year to year. In 1978, when August water temperatures were 2° to 3°C, six species of larval chironomid were the only invertebrates collected. Chironomid densities were positively associated with filamentous algal growth (Milner, 1987). Five of the six chironomid species belonged to the genus *Diamesa*, the most abundant from the *Diamesa davisi* group. Two species from this group have been found in Wolf Point Creek: *Diamesa lupus*, a new species presently known only from Glacier Bay and the Jasper–Banff area of Alberta, Canada, and *Diamesa alpina*, a first record for North America (Willassen, 1985). The larvae are presently undescribed and hence cannot be distinguished. *Diamesa sp.* B (probably *sommermanni*) was the second most common species with *Orthocladius sp.* A the other principal species. Low densities of *Diamesa sp.* C and *Diamesa sp.* D were also collected. Although unquantified, relatively intensive macroinvertebrate collections were made concurrent with stream channel investigations during the period 1981 to 1983 to ascertain if taxa other than chironomids were present. However none were found.

Quantitative samples in 1986 revealed an increase in species diversity and the incursion into the late summer community of the mayfly *Baetis* sp. and, in May, the stonefly *Capnia* sp. These taxa presumably colonized between 1984 and 1986. Chironomid species increased to seven with *Orthocladius* sp. A (not the sp A of Pinder, 1978) becoming codominant with the *Diamesa davisi* group. *Diamesa* spp. B and D were absent. In 1986 a few *Diamesa* sp. E and *Pagastia* sp. A (similar to *Pagastia partica* of Oliver and Roussel, 1982) were collected for the first time. Densities and biovolume of the macroinvertebrate community were still dominated by chironomids (Fig. 12.9).

Significant changes in the chironomid community were evident from 1988 to 1990. By 1990, August water temperature had reached 9°C. *Diamesa* spp.

Figure 12.6. Sampling station in Wolf Point Creek in 1983.

Figure 12.7. Sampling station in Wolf Point Creek in 1992. Note the dramatic development of riparian vegetation.

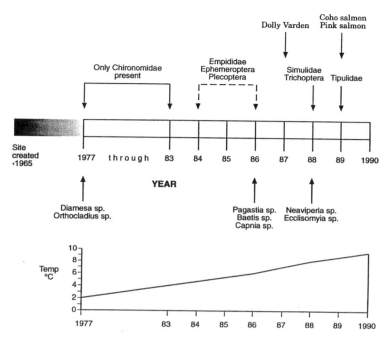

Figure 12.8. Sequence of invertebrate colonization and water temperature in Wolf Point Creek from 1977 to 1990. Arrow indicates first records of named taxa. (– – –) probable period of colonization (after Milner, 1994).

C and E were not collected after 1986 and densities of the *Diamesa davisi* group and *Orthocladius* sp. A were significantly lower (Milner, 1994). Densities of *Pagastia* sp. A increased and by 1990 it was the dominant chironomid exceeding 70 individuals 0.1 m⁻². This species had previously been found to be the dominant member of the chironomid community in Berg Bay north stream (see Fig. 12.2; a system approximately 150 years old and one of the most stable in Glacier Bay) (Milner, 1987). *Orthocladius mallochi* appeared in 1988, was the dominant chironomid in 1989, and occurred in similar densities in 1990.

In 1988, a predatory stonefly, *Neaviperla forcipata* (Chloroperlidae), first appeared in the late summer community. *N. forcipata* showed significant increases in abundance in 1989 and made up 17% by biovolume of the major families in 1990 (Fig. 12.4). *Capnia* was still found in small numbers in early summer samples. *Prosimulium* sp. (Simuliidae) was first collected in 1988 and in 1989 early instar larvae made up over 50% by density of the principal families present, although only 17% by biovolume. A trichopteran, *Ecclisomyia* sp., was found only in 1988 and *Limnophila* sp. (Tipulidae) in 1990. Total Tipulidae densities were <1% (Milner, 1994).

Total densities were relatively similar from 1986 to 1990 although mark-

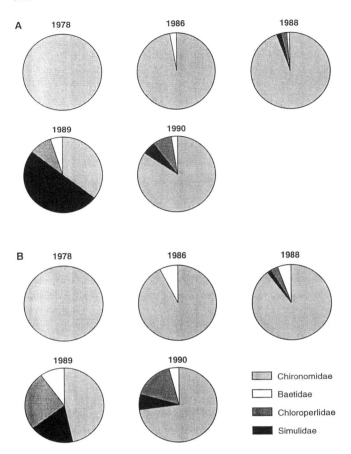

Figure 12.9. Pie diagrams showing (**a**) relative abundances, and (**b**) relative biovolumes of the four major insect families in Wolf Point Creek in 5 years between 1978 and 1990 (from Milner, 1994).

edly lower than densities found in 1978 (Fig. 12.10). Invertebrate biovolumes increased in 1990 due to the abundance of *Pagastia sp.* A, a large chironomid species (up to 11 mm in length), but did not attain 1978 levels (Fig. 12.10) (Milner, 1994).

Invertebrate species diversity, as estimated by the Shannon Index and the reciprocal of Simpson's Index, increased with community development over time reaching a maximum in 1988. Total invertebrate density was greatest in 1978, 10 years earlier. Species diversity then decreased in 1989 and 1990 due to the increased dominance of *Pagastia sp.* A and the overall reduced diversity of the chironomid community.

Salmonids

Extensive minnow trapping for juvenile salmonids in Wolf Point Creek from 1978 first revealed the presence of young of the year Dolly Varden char (*Salvelinus malma* Walbaum) in 1988, indicating this species probably colonized the stream the previous year. In Nunatak Creek, a nearby clear

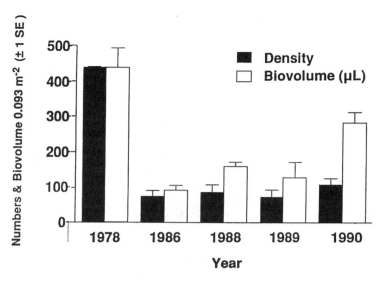

Figure 12.10. Total densities and biovolumes of invertebrates in Wolf Point Creek in 5 years between 1978 and 1990 (from Milner, 1994).

water stream, chironomids were shown to contribute nearly 90% of the diet of juvenile Dolly Varden that, using the selectivity index of Gabriel (1978), exhibited a strong positive selection for them compared to their abundance in the benthos (Milner and Bailey, 1989). In August 1990, over 76% of the diet of juvenile Dolly Varden from Wolf Point Creek were chironomids followed at 14% by simuliids for which there was a positive selection (Milner, 1994). Although a positive selection for smaller chironomid larvae was evident, there was a negative selection for *Pagastia* sp. A. The positive selection for simuliid larvae may have arisen because they were easier for fish to see on the upper surfaces of rocks, whereas *Pagastia* sp. A were typically associated with filamentous algae.

Coho salmon (*Onchoryhnchus kisutch* Walbaum) are also an early colonizer of new streams in Glacier Bay along with sockeye salmon (*Onchoryhnchus nerka*) where significant lakes occur in the developing stream system (Milner and Bailey, 1989). Small numbers of juvenile coho salmon were collected in Wolf Point Creek in 1990, indicating that coho salmon first spawned in this stream in the autumn of 1989. However in these new stream habitats, juvenile Dolly Varden typically dominate over juvenile coho salmon as their juveniles are primarily bottom feeders and are not at a disadvantage in habitats dominated by riffles and glides with an absence of pools. Juvenile coho salmon typically favor pools and less turbid water to facilitate sight feeding on drifting organisms.

In September of 1989, approximately 20 pink salmon carcasses (*Onchoryhnchus gorbuscha*) were found in Wolf Point Creek. In 1991 over 1000 pink salmon were estimated to have spawned in this stream and in

Table 12.1. Nutrient and Chlorophyll *a* Data for Lawrence Lake, 1991

Date	Depth (m)	Color (Pt)	TP (μg L^{-1})	TN (TKN + NO$_3$ + NO$_2$) (μg L^{-1})	TKN (μg L^{-1})	NH$_3$ (μg L^{-1})	NO$_3$ NO$_2$ (μg L^{-1})	Chlorophyll *a* (μg L^{-1})	N : P Ratio
6/27	1	19	80.4	78.2	65.0	15.0	13.2	1.32	2.1
7/30	1	22	44.1	109.0	109.0	2.6	<3.4	0.12	5.5
	10	21	63.1	47.5	47.5	<1.1	<3.4	0.16	1.65

1993, when stream turbidity became markedly reduced (<30 NTU), approximately 3000 spawners were counted. The clearing of the stream was due to improved clarity of Lawrence Lake as inputs of suspended sediment from remnant ice became reduced. Stream temperature maxima increased to 14°C in 1993.

Limnological studies of Lawrence Lake in 1991 indicated a Secchi disk value of 0.2 m but in 1993 this had increased to 1.8 m with the improvement of lake clarity. Nutrient data for June and July 1991 indicates high total phosphorus (TP) levels (presumably derived from glacial sediments) and hence very low atomic N–P ratios (<5.5) (Table 12.1). The June value of 80.4 μg L^{-1} is close to the maximum of 84 μg L^{-1} reported for 14 turbid lakes in other areas of Alaska (G. Kyle, unpublished data). Zooplankton were virtually absent with levels of <10 organisms m^{-3} (Table 12.2). However coincident with the improvement of clarity in September 1993, zooplankton densities exceeded 10,000 organisms m^{-3}. Although 20 to 30 sockeye salmon were observed milling at the base of the falls in 1993, there is no feasible passage to this lake and hence it will prove interesting to monitor future development of the zooplankton community within a new lake basin in the absence of predation by juvenile sockeye salmon.

Limnological analyses were also undertaken of the small lakes in Berg Bay north stream (Fig. 12.2) during 1990 and 1991. This stream supports a small run of sockeye salmon with approximately 500 to 1000 spawners and is one of the oldest post Neoglacial systems (150 years+) in Glacier Bay (Milner and Bailey, 1989). The upper lake is the larger of the two with an area of 0.15 km^2 and maximum depth of 12 m. In 1991, chlorophyll *a* values (indicative of primary production) were very low (<0.30 μg L^{-1}) indicating

Table 12.2. Zooplankton Density and Biomass in Lawrence Lake, 1991.

Taxa	Numbers (m^3) 6/27	Seasonal Mean	Mean Biomass (mg m^{-3})
Cyclops	7	7	0.03
Bosmina sp.	1	1	0.001
TOTALS	8	8	0.031

nutrient limitation to phytoplankton growth and hence limiting available forage for zooplankton (Table 12.3). Total nitrogen (TN) values at 1 m increased throughout the summer reaching 171 μg L^{-1} towards the end of August. These TN values are lower than the 200 μg L^{-1} typically required in clear water lakes at spring overturn for phytoplankton growth not to be nutrient limited (Mason, 1991). TP values at 1 m were typically less than the limiting 10 μg L^{-1} value, increasing from 3.8 μg L^{-1} in early July to 9.6 μg L^{-1} by August 20. The 1 m seasonal mean TP value of 7.0 μg L^{-1} is slightly lower than the average seasonal mean of 8.1 μg L^{-1} recorded for 18 other stained lakes in Alaska (range 1.2 to 23.2 μg L^{-1}). The TN seasonal mean value of 88.2 μg L^{-1} was lower than seasonal means for the 18 stained Alaskan lakes that averaged 221.3 μg L^{-1} (range 32.2 to 1172 μg L^{-1}) (G. Kyle, unpublished data). TP values were similar in 1990 but 1 m TN values were higher ranging from 159.3 μg L^{-1} in August to 176.8 μg L^{-1} in May (Table 12.4).

Forage was clearly limiting sockeye salmon production in these ultraoligotrophic lakes. Zooplankton in 1991 were virtually nonexistent with densities <13 organisms m^{-3}. The mean biomass for the summer was 0.009 mg m^{-3} (Table 12.5). Only small species of zooplankton were found but none of the larger favored prey (e.g. *Daphnia* spp.) of juvenile sockeye salmon were present. No organisms collected in May 1990 but in August, 178 organisms m^{-3} were found with *Bosmina* spp. accounting for over 60% by number. The August biomass was 0.29 mg m^{-3}.

Similar low levels of chlorophyll *a* were found in the lower lake (area = 0.05 km^2, maximum depth 4 m) in 1991 with values <0.40 μg L^{-1} (Table 12.3). TN and TP values were higher than the upper lake but still limiting to phytoplankton production. The seasonal mean TP value at 1 m was 10.5 μg L^{-1} and the TN value was 130.2 μg L^{-1}, respectively slightly higher and markedly lower than the seasonal mean for 18 other stained Alaskan lakes. In 1990, TP values ranged from 5.6 to 8.9 μg L^{-1} and TN values at 1 m were 64.9 μg L^{-1} in May and 77.5 μg L^{-1} in August (Table 12.4). For both lakes, typical atomic N–P ratios exceeded 16.1 indicating that phosphorus was the more limiting of the two nutrients to phytoplankton growth.

Again, forage in this smaller lake was extremely low. Highest zooplankton densities in 1991 were a depauperate 63 organisms m^{-3} composed of smaller species, principally copepods and chydorids (Table 12.5), which are not the preferred food of juvenile sockeye salmon. The mean biomass was a meager 0.063 mg m^{-3}. In 1990, no organisms were found in this lake in May but in August, 515 organisms m^{-3} were collected with *Bosmina* spp. and ostracods numerically accounting for over 80%. The total biomass was higher than the mean for 1991 at 0.328 mg m^{-3}.

These data suggest that in these two sockeye salmon lakes, zooplankton productivity is low due to a possible combination of nutrient limitations and predation by sockeye salmon fry (i.e. both bottom-up and top-down effects). It would also appear to indicate that the sockeye fry, from an estimated escapement for Berg Bay North Stream of approximately 500 to

Table 12.3. Nutrient and Chlorophyll *a* Data for Berg Bay North Stream Lakes, 1991

Date	Depth (m)	Color (Pt)	TP ($\mu g\,L^{-1}$)	TN (TKN + NO$_3$ + NO$_2$) ($\mu g\,L^{-1}$)	TKN ($\mu g\,L^{-1}$)	NH$_3$ ($\mu g\,L^{-1}$)	NO$_3$ NO$_2$ ($\mu g\,L^{-1}$)	Chlorophyll *a* ($\mu g\,L^{-1}$)	N : P Ratio
				Upper Lake					
7/1	1	26	3.8	91.5	52.0	4.7	39.5	0.04	53.3
	6	27	3.9	93.9	58.0	3.1	35.9	0.12	53.3
7/21	1	29	5.1	107	80.0	<1.1	27.1	0.28	46.5
	10	18	13.5	173	105	58.0	68.0	1.06	28.4
8/12	1	29	9.3	167	102	11.9	65.1	0.06	39.8
	10	36	8.2	104	104	2.0	<3.4	0.35	28.1
8/20	1	40	9.6	171	119	<1.1	52.6	0.29	39.4
	10	30	21.8	207	207	22.2	<3.4	0.27	21.0
				Lower Lake					
7/1	1	22	14.9	110	89.1	3.1	21.3	0.12	16.4
	2	14	13.2	143	118.0	2.6	25.7	0.09	24.1
7/21	1	36	5.4	95	85.3	<1.1	9.6	0.20	38.9
	3	37	20.6	151	140	1.5	11.1	0.29	16.1
8/9	1	37	11.4	187	131	7.8	56.4	0.02	36.3
	3	35	10.8	157	157	19.7	<3.4	0.24	32.3
8/20	1	38	10.4	129	129	<1.1	<3.4	0.20	27.4
	3	40	12.2	138	135	<1.1	3.8	0.21	25.0

Table 12.4. Nutrient Data for Berg Bay North Stream Lakes, 1990

Date	Depth (m)	Color (Pt)	TP (µg L⁻¹)	TN (TKN + NO₃ + NO₂) (µg L⁻¹)	TKN (µg L⁻¹)	NH₃ (µg L⁻¹)	NO₃, NO₂ (µg L⁻¹)	N : P Ratio
				Upper Lake				
5/8	1	19	5.8	176.8	54.3	13.6	122.5	95.3
	6	18	3.1	163.8	41.8	14.7	122.0	88.4
8/23	1	20	5.7	159.3	98.7	11.8	60.6	61.7
	6	23	13.9	193.7	129.3	25.4	64.4	30.7
				Lower Lake				
5/8	1	21	2.6	150.9	48.0	7.0	102.9	35.6
	3	18	2.5	292.1	182.1	8.7	110.0	258.1
8/23	1	26	5.4	159.1	109.9	12.6	49.2	65.0
	3	33	24.7	107.2	61.8	9.8	45.4	9.5

Table 12.5. Zooplankton Density and Biomass in Berg Bay North Stream Lakes in 1991

Taxa	Numbers (m^{-3}) 7/1	7/21	8/12	8/20	Seasonal Mean (n)	Mean Biomass ($mg\,m^{-3}$)
			Upper Lake			
Cyclops sp.	1	1	1		1	0.001
Bosmina sp.		2	4	10	4	0.005
Chydorinae	1	1	2	3	1	0.003
Totals	2	4	7	13	6	0.009
			Lower Lake			
	7/1	7/21	8/9	8/20		
Diaptomus	3				1	0.003
Cyclops sp.	14	22			9	0.04
Bosmina sp.			1	2	1	
Chydorinae	1	41	1		11	0.02
Totals	18	63	2	2	33	0.063

1000 adult sockeye salmon, must be utilizing other food sources (e.g. insects or benthic microcrustacea) in the lakes to sustain a run of this size. Other studies will concentrate on examining the stomach contents of juvenile sockeye salmon and determining the length of freshwater residence.

Kenai Fjords National Park

Kenai Fjords National Park lies approximately 800 km northwest of Glacier Bay on the Kenai Peninsula in southcentral Alaska (Fig. 12.11) and covers an area of approximately 2430 km². McCarty Glacier in McCarty Fjord has receded from a Neoglacial maxima in 1840 in a similar manner to glaciers in Muir Inlet (Fig. 12.3; dates from Post, 1980) creating three streams on the eastern side of the fjord, all possessing lakes as their principal source. The youngest of these lakes are the Delusion Lakes (unofficial name) that began to form in 1974. The upper lake is still fed by a stream from a small pocket of remnant ice (Fig. 12.12). From the lower of these lakes a stream flows southwest across unconsolidated glacial deposits for 2 km into McCarty Fjord. This is predominantly a first-order stream (but with width and discharge characteristic of higher order streams) but becomes second order in the lower 1.2 km. The two lakes are connected by a short stream segment <300 m in length. The older lakes, Desire and Delight, were formed between 1935 and 1940, and 1920 and 1925, respectively.

Macroinvertebrates

Preliminary studies commenced on Delusion Creek in 1989 and invertebrate samples were taken between April to September. Juvenile fish collec-

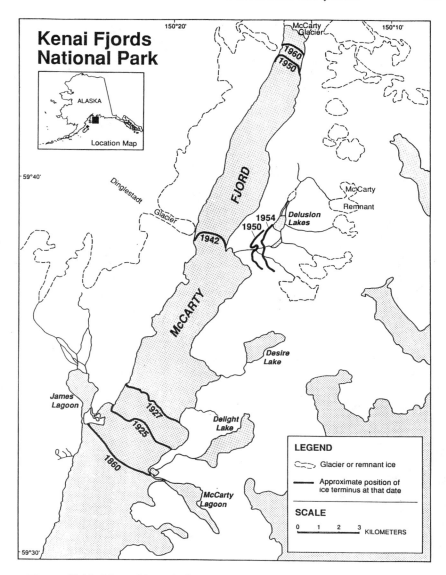

Figure 12.11. Map of Kenai Fjords National Park (dates from Post, 1990).

tions (minnow trapping) were also made in study reaches. May and June invertebrate samples in 1989 were dominated by chironomids and the stonefly *Neaviperla forcipata* although overall densities were low (<800 m²). Over 90% of the chironomids were individuals of the *Diamesa davisi* group. Small numbers of *Baetis* and Simuliidae were also found but no Trichoptera were collected. August stream temperature was 9°C. However the lakes in

Figure 12.12. Aerial photograph of the Delusion Creek drainage in 1994 (Scale 1: 24,000 approx).

this system do not function in the same manner as Lawrence Lake with respect to Wolf Point Creek in Glacier Bay. Although the Delusion lakes act as sediment traps at most times of the year (stream turbidity at base flow <30 NTU), the heavy rainfall characteristic of this area and the steep terrain restrict their ability to act as buffers to reduce discharge variations. Hence no close border of riparian vegetation exists and the channel is unstable and frequently shifting. Stream discharges have been observed to rise from a base flow of under 5 to $25\,m^3s^{-1}$ after less than 24 hours of heavy rainfall (G. York and A. Milner, unpublished data).

Salmonids

Juvenile Dolly Varden were found to be the dominant rearing salmonid in Delusion Creek along with small numbers of juvenile coho salmon in 1989 and 1991. In 1992, a small number (<100) of pink salmon were observed spawning in the system and an aerial survey by the Alaska Department of Fish and Game (ADFG) estimated that approximately 900 adult sockeye salmon spawned in gravels along shallow margins of the upper lake (ADFG, 1992). Sockeye salmon were first observed from these aerial surveys in 1987.

Fisheries data from 1979 through 1989 (ADFG, 1992) indicate that in less than 100 years, sufficient numbers of sockeye salmon have colonized Desire and Delight lake systems to support a commercial fishery. In high years, these two systems have supported commercial catches exceeding 50,000 red salmon, although since 1986 catches have declined significantly (Fig. 12.13) and in 1992 low numbers of sockeye spawners prevented a commercial opening (Fig. 12.13). Studies of otoliths by ADFG from adult sockeye

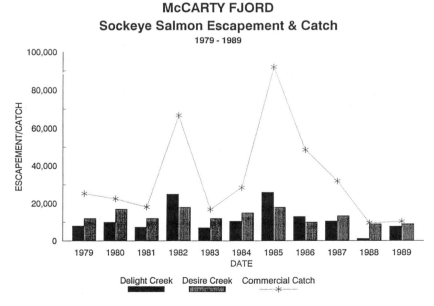

Figure 12.13. Sockeye salmon escapement and commercial catch for Delight and Desire creeks in McCarty Fjord, Kenai Fjords from 1979 to 1989.

Table 12.6. Nutrient and Chlorophyll *a* Data for Delusion Lakes in 1991

Date	Depth (m)	Color (Pt)	TP ($\mu g\,L^{-1}$)	TN (TKN + NO$_3$ + NO$_2$) ($\mu g\,L^{-1}$)	TKN ($\mu g\,L^{-1}$)	NH$_3$ ($\mu g\,L^{-1}$)	NO$_3$ NO$_2$ ($\mu g\,L^{-1}$)	Chlorophyll *a* ($\mu g\,L^{-1}$)	N : P Ratio
				Lower Lake					
6/14	1	6	7.0	143	89.9	<1.1	53.4	<0.01	45.4
				Upper Lake					
6/15	1	26	77.2	11.1	11.1	20.2	<3.4	0.08	0.32

spawners in the three systems during 1992 were used to determine overall age and compare the numbers of years spent in freshwater to the number of years spent in saltwater. Thus a spawning sockeye salmon aged 4 with 1 year of growth in freshwater and 3 years of growth in saltwater would be categorized as 1:3. The majority of juvenile sockeye salmon in Delight Lake stayed in freshwater 1 year as 91% of spawners ($N = 11$) were in the age categories 1:2 and 1:3; in Desire Lake 68% of spawners ($N = 81$) were in this age category; the remainder stayed 2 years in freshwater (age categories 2:2 and 2:3). In contrast, 93% of the sockeye spawners ($N = 41$) from upper Delusion Lake in 1992 were in the 2:2 and 2:3 age categories, indicating that in this system the majority of juvenile sockeye salmon remained 2 years in freshwater. Water samples from the two Delusion lakes in June 1991 showed extremely low chlorophyll *a* levels of <0.1 $\mu g\,L^{-1}$ (Table 12.6) and vertical zooplankton hauls indicated the virtual absence of zooplankton (Table 12.7). Consequently, like the Berg Bay lakes, juvenile sockeye fry may be using other food sources than zooplankton. TN levels were low, although as in Lawrence Lake, TP values were higher in upper Delusion Lake, which receives meltwater from remnant ice (Table 12.6).

Table 12.7. Zooplankton Density and Biomass in Delusion Lakes in 1991

Taxa	Numbers (m^{-3}) 6/15	Biomass ($mg\,m^{-3}$)
	Upper Lake	
Bosmina sp.	1	0.002
	Lower Lake	
Daphnia	6/14 2	0.01

Discussion

Invertebrates

Four main sources of invertebrate colonization were identified by Williams and Hynes (1976) namely: downstream migration or drift, upstream migration within the water, vertical upward migration from within the substrate, and aerial oviposition. In the absence of the first three of these sources, aerial oviposition is the initial mechanism of colonization in these new streams formed after deglaciation and hence most insects have a distinct advantage over non-insect forms in possessing a winged terrestrial stage. Drift may become more important as the community becomes established.

The invertebrate colonization and successional patterns observed in Wolf Point Creek from 1978 to 1990 have been shown to result from either more favorable conditions due to increased water temperature or longer dispersal times required by certain taxa to reach the drainage (Milner, 1994). However such taxa as *Pagastia sp.* A, *Baetis sp.*, and *Neaviperla forcipata*, which appear later in the succession, were collected in Nunatak Creek during 1977 and 1978 (Milner, 1987). This clear water creek is located 1.5 km across the fjord (see Fig. 12.3) and it would seem unlikely that this distance would constitute a sufficient barrier to dispersal and establishment of these taxa if stream conditions were suitable. Brundin (1967) considered small-winged and relatively light chironomids to have a high dispersal capacity, even over wide oceanic expanses.

Water temperature increases from 1978 to 1990, as Lawrence Lake increased in area on ablation of the remnant ice mass, would therefore appear to be the principal variable influencing the colonization of additional invertebrate taxa and increased community diversity in Wolf Point Creek.

Are the early successional *Diamesa* chironomids in Wolf Point Creek cold stenothermal species adversely affected by increased water temperature with stream development? The 1989 data from Delusion Creek in Kenai Fjords National Park would seem to indicate otherwise. There *Diamesa* chironomids appear able to maintain populations at 9°C in the presence of low numbers of other taxa due to their adaptations for surviving in unstable channels. The larvae are characterized by long posterior prolegs that allow them to firmly grip substrate in strong currents (Milner and Petts, 1994). The frequent spates in Delusion Creek appear to reduce numbers of other taxa and hold the 1989 community in an early successional stage even though water temperature has attained levels at which certain *Diamesa* species in Wolf Point Creek have become extinct or significantly reduced in number. Hence if colonization by other taxa had been restricted in Wolf Point Creek due to an unstable channel condition, it is suggested that *Diamesa* chironomids could still dominate the community at temperatures of 9°C. The disappearance of *Diamesa* sp. B and other **Diamesa** species, and the significant reduction in numbers of the *Diamesa davisi* group from summer samples occurred before the arrival of the predatory stonefly *Neaviperla forcipata* and salmonids, suggesting that interspecific competition (and not predation) may be an important structuring factor in the Wolf Point Creek successional sequence (Milner, 1994).

It is interesting to compare the time scale of the colonization sequence in Wolf Point Creek to other studies of colonization within a primary successional framework (Table 12.8). Colonization and community development in new channels following relocation and reconstruction projects occurs via primary succession. Gore (1979, 1982) examined source distance effects on community development in a reclaimed river channel after strip mining in the Tongue River in Wyoming, and found that maximum densities of macroinvertebrates were obtained in <90 days with species equilibrium (immigration = extinction) reached in about 200 days when source areas were <200 m upstream. Increased distance from the colonization source resulted in slower recovery rates. For a new stream channel associated with the Ponesk Burn in the uplands of Scotland, Doughty and Turner (1991) reported maximum diversity of macroinvertebrates and similarity to a reference site after 80 to 100 days of colonization and taxa equilibrium attained at one site after 140 to 150 days. In both of these studies, upstream colonizers were present as a source of drift. Malmqvist et al. (1991), examining colonization of a manmade channel (approximately 1.5 km long) running between two lakes in Sweden, found community structure (including noninsect taxa) similar to reference lake-outlet streams within 1 year. Following major flood events, Minshall et al. (1983) reported 600 to 650 days to reach species equilibrium and Giller et al. (1991) over 2 years for a small Irish river.

A number of streams were denuded of fauna with no upstream sources of colonizers by the Mount St. Helens eruption in western Washington during May 1980 and probably represent the closest situation to the colonization of new streams following glacial recession. During 1980, 18 taxa (Ephemeroptera, Plecoptera, and Trichoptera taxa [EPT] = 3) colonized Clearwater Creek increasing to 80 taxa (EPT = 50) by 1989 (Meyerhoff, 1991). Chironomid dominance decreased significantly from 75% (of nu-

Table 12.8. Time Scales for Invertebrates to Attain Species Equilibrium From Other Disturbances

Disturbance	Upstream Colonizers?	Time	Reference
New channel	Yes	140–150 days	Doughty and Turner (1991)
New channel	Yes	200 days	Gore (1979, 1982)
New channel	Yes	<365 days	Malmqvist et al. (1991)
Flood	Yes	600–650 days	Minshall et al. (1983)
Flood	Yes	>2 years	Giller et al. (1991)
Volcanic ash	No	>5 years	Meyerhoff (1991)

merical abundance) in 1980 to 30% in 1989; in Wolf Point Creek dominance decreased from 100% in 1978 to only 84% in 1990. In Clearwater Creek, the Diptera: EPT taxa ratio was 1.5:1 in 1982 falling to 0.3:1 in 1988 compared to 3.5:1 in Wolf Point Creek in both 1986 and 1990. Whereas overall densities decreased with community development in Wolf Point Creek, densities in Clearwater Creek increased three- to fourfold to 45 to 80,000 m^{-2} after 1985 as a result of stabilization of the streambed, arrival of additional colonizing species, and some increase in the availability of allochthonous organic matter associated with the growth of riparian vegetation (Anderson, 1992). Anderson and Wisseman (1987) considered shifting substrates to be the primary factor limiting caddis fly populations in Clearwater Creek. Although 29 species of caddis fly larvae or pupae were present at the most impacted site by 1985, many species were represented by only a few individuals. However large numbers and abundance of species were found in aerial collections. The unstable channel may explain the absence of caddis flies in Delusion Creek in 1989, but the paucity of caddis flies in Wolf Point Creek is more likely attributable to cold water temperature and the relative unavailability of allochthonous organic matter. New taxa were also still colonizing Clearwater Creek after 10 years (Meyerhoff, 1991).

It is evident that biotic community development within a primary successional framework without upstream or downstream sources of colonizers takes a significantly longer period of time than development where these sources are present and time to species equilibrium can be extremely long.

The time scale of colonization in Wolf Point Creek (13 years of study, commencing about 17 years after site creation) is longer than any sequence previously reported in the literature. No noninsect macroinvertebrates became established by 1990, presumably because of the difficulty of crossing mountain and oceanic barriers. A marked temporal difference between the attainment of maximum densities and maximum species diversity existed. Maximum densities were reached in 1978 (or earlier) but maximum diversity was not attained until 1988, at least 10 years apart. Indeed maximum species diversity may not yet be reached in Wolf Point Creek as other EPT groups are likely to be added to the community. However the inhibition of colonization by taxa other than Chironomidae as a result of low water temperatures is a significant factor in the large temporal difference in these end points. Yasuno et al. (1982) notes that the reduction or absence of predators or competitors creates ideal conditions for the rapid expansion of chironomid populations following major disturbances and these factors would account for the larger densities and biomass observed in 1978. It would appear that trophic equilibrium (representative diversity of the functional feeding groups with reference to mature streams) has not yet been attained in Wolf Point Creek or Delusion Creek.

These findings from Wolf Point Creek and Delusion Creek infer deterministic patterns to the sequence of invertebrate taxa colonization and succession of new streams following deglaciation. Burroughs River (see

Fig. 12.3), a turbid meltwater system in Glacier Bay, supported similar chironomids to Wolf Point Creek in 1978 when water temperature was 2°C (*Diamesa davisi* group, *Diamesa* sp. B, and *Orthocladius* sp. A) (Milner, 1983). In 1991 (temperatures 5°C), *Baetis* and *Neaviperla forcipata* were found to have colonized and *Diamesa* sp. B was absent (A. Milner, unpublished data). Longitudinal zonation of taxa in glacier-fed rivers in Alaska, Scandinavia, and the European Alps appears to follow similar trends (Milner and Petts, 1994). Invertebrate communities in glacier-fed rivers are dependent upon distance downstream from the glacier margins or time since deglaciation that results in potential deterministic patterns to community structure. Deterministic patterns appear to be influential at the genera and family levels in this type of lotic biotope. Milner and Petts (1994) propose a model for glacier-fed rivers where the commmunity structure is initially dominated by *Diamesa* chironomids, followed by Orthocladiinae, other Diamesinae, Simuliidae, Baetidae, and Chloroperlidae, as temperatures and channel stability increase. An important modifier to this model is whether lakes are formed within the glacial river system. As with Wolf Point Creek, a lake can enhance stream channel stability and, depending upon lake surface area, increase stream temperature. This accelerates invertebrate community development and increases invertebrate diversity.

Salmonids

The recent colonization of Wolf Point Creek and the initial findings from Delusion Creek again confirm the role of Dolly Varden char as the typical and most well-adapted early fish resident of new streams formed on deglaciation, as first suggested in Milner and Bailey (1989). Coho salmon are also relatively rapid colonizers but juveniles of this species are at a disadvantage compared to Dolly Varden in being predominantly surface sight feeders in streams that are largely devoid of pool habitat and frequently turbid. In systems where lakes are present, sockeye salmon also colonize rapidly and, as illustrated by the Desire Lake and Delight Lake systems in Kenai Fjords, are able to support runs of commercial value within 30 years of being created. It is amazing that upper Delusion Lake, which only began to form in the mid-1970s, now supports a run of almost 1000 sockeye salmon. However the majority of the fry remain 2 years in the lakes to attain smolting size, which is not the optimum situation for enhanced production. Young-of-year fish are competing against year-old fish for limited food resources potentially resulting in their increased mortality.

In Glacier Bay none of the recently formed streams in Muir Inlet have accessible lakes of any magnitude and it is impossible to calculate the time frame of colonization by sockeye salmon in the older streams further south. In addition, reporting districts of the ADFG (Schroeder and Morrison, 1990) are not specific to streams within Glacier Bay, and thus it is very difficult to determine the contribution of these new stocks to the commercial fishery. The phenomenon of glacial recession and the creation of new stream habitat that can support salmon runs is not restricted to Glacier Bay and Kenai Fjords National Parks, although these areas are the sites of some of the more dramatic recessions.

The amount of potential new habitat for salmonids that has been formed in Glacier Bay since the Neoglacial ice recession is extensive. Soseith and Milner (1995) used a simple model based on gradient and water clarity/channel stability to estimate that 310 streams had been formed within the last 200 years of which 60% were predicted to contain salmonids. Additional stream systems have been created in coastal Alaska between Glacier Bay and Kenai Fjords and thus glacial recession within the last century has provided a significant amount of new habitat for salmonid colonization, production, and the development of new stocks. This development should be viewed as a positive attribute of glacial recession, particularly when stocks are under threat of extinction in other areas (Nehlsen et al., 1991).

Is this recession from Neoglacial ice maxima in coastal Alaska related to global climate change? Although the initial phase of the recession of tidewater glaciers in Glacier Bay may have been due to climatic factors, the recent history of retreat is related to water depth of the fjord and glacier terminus dynamics (Powell, 1990). The same is probably true for the tidewater glaciers in Kenai Fjords (R. Powell personal communication). Powell (1990) also suggests that tidewater glaciers can advance independent of climatic factors. The probable prognosis for glaciers in Alaska with their termini on land under warmer conditions of global climate change may be a reduction in mass (Melack et al., in press). Although measurements of the Wolverine Glacier in southcentral Alaska and the Gulkana Glacier in the Alaska Range indicated that increased air temperatures since the early 1970s resulted in an increase in glacial mass and glacial thickness (Mayo and Trabant, 1986; Mayo and March, 1990) evidence indicates a loss in mass balance since 1989 and also overall for the period 1965 to 1995 (Hodge et al. unpublished). The decrease in glacial mass translates into glacial recession at the terminus after a time lag that may initially increase glacial runoff. Increased glacial runoff may have a significant effect on flow, temperature, and sediment regimes in downstream reaches (Oswood et al. 1992), particularly on clear water streams that have been formed since the Neoglacial retreat. Reduced water temperature may cause decreased invertebrate diversity and the increase in the dominance of *Diamesa* chironomids in the benthic community (Melack et al., in press).

Summary

New streams formed after glacial recession provide unique natural sites to study colonization and succession of biotic communities at the scale of entire drainage systems. Site-specific temporal succession has been demonstrated in Wolf Point Creek as changes in dominance of the invertebrate community have been documented and a number of the early *Diamesa* colonizers are now extinct. Interspecific competition may be an important factor in the elimination of early colonizers in Wolf Point Creek; this inference is presently being tested through field experimentation and manipulation. This study has demonstrated the long time scales for macroinvertebrate colonization where source areas of drift are absent and the difficulty for noninsect taxa to cross mountain and oceanic barriers. An extensive amount of potential new salmonid habitat has been created in coastal Alaska during the last 100 years that has been rapidly colonized by salmonids. Within 30 years, salmonid runs in these new streams may potentially be of sufficient magnitude to contribute to commercial fisheries.

Acknowledgments. Grateful appreciation to the many people who have assisted with the fieldwork over the years is expressed to Sally Tanner, Barbara Blackie, Greg Dudgeon, Chris Kondzela, Calum MacNeil, Mark Porter, Michael Dauber, Steve Mackinson, Elizabeth Adamson, Colin Bull, Geoff York, and David Rundio. Special thanks to Geoff York for collecting otoliths from sockeye salmon in the Delusion Lakes and for locating the Harriman photograph. Grateful thanks to Jim Luthy and Bill Stevens, captains of the m.v. *Nunatak* and m.v. *Serac*, respectively, who logistically supported the field camps in Glacier Bay and Kenai Fjords National Parks. I am particularly grateful to Endre Willassen and Peter Cranston for identifying adult chironomids and to Ken Stewart for the identification of the stoneflies. I am indebted to the Limnology laboratory, ADFG, in Soldotra for analysis of lake water and zooplankton samples. Many at the U.S. National Park Service have kindly facilitated and assisted field operations including Greg Streveler, Gary Vequist, Mark Schroeder, Mary Beth Moss, Chad Soseith, Bud Rice, and Jeff Troutman. Maureen Milner supplied the figures. Thanks to Jackie LaPerrierre, James Ward, Gary Kyle, and Geoff York for constructive comments on an earlier version of this chapter. This work has been supported by grants from the U.S. National Park Service, the Central Research Fund of the University of London, The Royal Society, and the U.S. Environmental Protection Agency.

References

Alaska Department of Fish and Game [ADFG] (1992) Lower Cook Inlet area finfish management report. Regional Information Report 2A93–11.

Anderson NH (1992) Influence of disturbance on insect communities in Pacific Northwest streams. Hydrobiologia 248:79–92.

Anderson NH, Wisseman RW (1987) Recovery of the Trichoptera fauna near Mt. St. Helens five years after the 1980 eruption. In Bownaud M, Tachet H (Eds) Proceedings of the 5th International Symposium on Trichoptera. Dr Junk, The Netherlands.

Benson C, Harrison W, Gosnik J, Bowling S, Mayo L, Trabant, D (1986) Workshop on Alaskan hydrology: Problems related to glacierized basins. Geophysical Institute Report UAG-R (306), University of Alaska, Fairbanks.

Brundin L (1967) Insects and the problem of austral disjunctive distribution. Ann Rev Entomol 12:149–168.

Calkin PE (1988) Holocene glaciation of Alaska (and adjoining Yukon Territory, Canada). Quaternary Sci Rev 7:159–188.

Cushing CE, Gaines WL (1989) Thoughts on recolonisation of endorheic cold desert spring-streams. J North Am Benthol Soc 8:277–287.

Doughty CR, Turner MJ (1991) The Ponesk burn diversion: Colonization by benthic invertebrates and fish. Technical Report No. 98. Clyde River Purification Board, East Kilbride, UK.

Downes BJ, Lake PS (1991) Different colonization patterns of two closely related stream insects (*Austrosimulium* spp.) following disturbance. Freshwater Biol 26:295–306.

Field WO (1979) Observations of glacier fluctuations in Glacier Bay National Monument. Proceedings of the first conference on scientific research in the national parks. Natl Park Serv Trans Proc 5:803–808.

Fisher SG (1990) Recovery processes in lotic ecosystems: Limits of successional theory. Environ Manage 14:725–736.

Gabriel WL (1978) Statistics of selectivity. In Lipovski SJ, Simenstad CA (Eds) Fish food habitat studies. Sea Grant Publication, University of Washington, 62–66.

Giller PS, Sangpradub N, Twomey H (1991) Catastrophic flooding and macroinvertebrate community structure. Verh Int Verein Theor Angew Limnol 24:1724–1729.

Goldthwait RP (1966) Glacial history. In Goldthwait RP (Ed) Soil development and ecological succession in a deglaciated area of Muir Inlet, southeast Alaska. Institute of Polar Studies, Ohio State University, Columbus, OH, 1–18.

Gore JA (1979) Patterns of initial benthic recolonization of a reclaimed coal strip-mined river channel. Can J Zool 57:2429–2439.

Gore JA (1982) Benthic invertebrate colonization: Source distance effects on community composition. Hydrobiologia 94:183–193.

Gore JA, Milner AM (1990) Island biogeographical theory: Can it be used to predict lotic recovery rates. Environ Manage 14:737–753.

Hemphill N, Cooper SD (1983) The effect of physical disturbance on the relative abundances of two filter-feeding insects in a small stream. Oecologia 58:378–382.

Lake PS, Doeg TJ (1985) Macroinvertebrate colonization of stones in two upland southern Australian streams. Hydrobiologia 126:199–211.

Malmqvist B, Rundle S, Bronmark C, Erlandsson A (1991) Invertebrate colonization of a new, man-made stream in southern Sweden. Freshwater Biol 26: 307–324.

Mason CF (1991) Biology of freshwater pollution. Longman Scientific Harlow Essex, UK.

Mayo LR, March RS (1990) Air temperature and precipitation at Wolverine Glacier, Alaska; Glacier growth in a warmer, wetter climate. Ann Glaciol 14: 191–194.

Mayo LR, Trabant DC (1986) Recent growth of Gulkana Glacier, Alaska Range, and its relation to glacier-fed river runoff. In Subitzky S (Ed) Selected papers in the hydrological sciences. U.S. Geological Survey Water Supply Paper 2290, 91–99. Washington, DC.

Melack JM, Dozier J, Goldman CR, Greenland D, Milner AM, Naiman RJ (in press) Effects of climate change on inland waters of the Pacific Central Mountains and Western Great Basin of North America. Hydrological Sciences.

Meyerhoff RD (1991) Post-eruption recovery and secondary production of grazing insects in two streams near Mt. St. Helens. Ph.D. thesis, Oregon State University, Corvallis, OR.

Miller MM (1964) Inventory of terminal positions in Alaskan coastal glaciers since the 1750's. Proc Am Philos Soc 108:257–273.

Milner AM (1983) The ecology of post-glacial streams in Glacier Bay. Ph.D. thesis, University of London, London.

Milner AM (1987) Colonization and ecological development of new streams in Glacier Bay National Park, Alaska. Freshwater Biol 18:53–70.

Milner AM (1994) Invertebrate colonization and succession in a new stream in Glacier Bay National Park, Alaska. Freshwater Biol 32:387–400.

Milner AM, Bailey RG (1989) Salmonid colonization of new streams in Glacier Bay National Park, Alaska. Aquacul Fish Manage 20:179–192.

Milner AM, Petts GE (1994) Glacial rivers: Physical habitat and ecology. Freshwater Biol 32:295–307.

Minshall GW, Andrews DA, Manuel–Faler CY (1983) Application of island biogeographical theory to streams: Macroinvertebrate recolonization of the Teton River, Idaho. In Barnes JR, Minshall GW (Eds) Stream ecology: Application and testing of general ecological theory. Plenum Press, New York, 279–297.

Nehlsen W, Williams JE, Lichatowich JA (1991) Pacific salmon at the crossroads: Stocks at risk from California, Oregon, Idaho and Washington. Fisheries 16:4–21.

Oswood MW, Milner AM, Irons JG (1992) Climate change and Alaskan rivers and streams. In Fifth P, Fisher S (Eds) Global warming and freshwater ecosystems. Springer–Velag, New York, 192–210.

Oliver DR, Roussel ME (1982) The larvae of *Pagastia* Oliver (Diptera: Chironomidae) with descriptions of three Nearctic species. Can Ent 114:849–854.

Peckarsky BL (1986) Colonization of natural substrates by stream benthos. Can J Fish Aquat Sci 43:700–709.

Pinder LCV (1978) A key to the adult males of the British Chironomidae (Diptera). Freshwater Biological Association (UK), Special Publication No. 37. Ambleside, Cumbria, UK.

Post A (1980) Preliminary bathymetry of McCarty Fjord and Neoglacial changes of McCarty Glacier, Alaska. U.S. Geological Survey Open File Report 80–424. Tacoma, WA.

Post A, Meier NF (1980) World Glacier Inventory Proceedings of the Riederalp Workshop, September 1978 IAHS-AISH Publication no. 126, 45–47.

Powell RD (1990) Advance of glacial tidewater fronts in Glacier Bay, Alaska. In Milner AM, Wood JD (Eds) Proc Second Glacier Bay Sci Symp. U.S. Department of the Interior, Anchorage, 67–73.

Schroeder TR, Morrison TR (1990) Lower Cook Inlet area: Annual finfish management report. Alaska Department of Fish and Game Regional Information Report No. 2490–03, Anchorage.

Sheldon A (1984) Colonization dynamics of aquatic insects. In Resh VH, Rosenberg DM (Eds) The ecology of aquatic insects. Praeger, New York, 401–429.

Sidle RC, Milner AM (1989) Stream development in Glacier Bay National Park, Alaska, U.S.A. Arct Alpine Res 21:350–363.

Soseith CR, Milner AM (1995) Predicting streams containing salmonids using physical characteristics. In Engstrom DR (Ed) Proc Third Glacier Bay Sci Symp. U.S. Department of the Interior, Anchorage 174–183.

Willassen E (1985) A review of **Diamesa davisi** Edwards and the davisi group. Spixiana Suppl 11:109–137.

Williams DD, Hynes HBN (1976) The recolonization mechanisms of stream benthos. Oikos 27:265–272.

Yasuno M, Fukushima S, Hasegawa J, Shioyama F, Hatakeyama S (1982) Changes in the benthic fauna and flora after application of temephos to a stream on Mt. Tsukuba. Hydrobiologia 89:205–214.

13. Streams and Rivers of Alaska: A High Latitude Perspective on Running Waters

Mark W. Oswood

The Diversity and Unity of Alaskan Running Waters

Alaska is a frontier, both geographically and scientifically. The yearly dearth and seasonal extremes of sunlight create a cold-dominated landscape of short summers and long cold winters (Fig. 13.1), limiting human population densities and molding ecosystem properties. However, Alaska encompasses tremendous physiographic diversity (Milner et al., 1996, Chapter 1). Ecosystems range over the latitudinal gradient from the cool rainforest of southeast Alaska with maritime climate and dense riparian vegetation (Fig. 13.2), to the taiga forest of Interior Alaska, with a continental climate and modest riparian vegetation (Fig. 13.3), to the Arctic tundra of the North Slope, with virtually no riparian vegetation (Fig. 13.4). This latitudinal range makes it impossible to characterize typical "Alaskan" streams, although all Alaskan streams share a suite of high latitude characteristics that distinguish them from the lower latitude streams of temperate climates.

This chapter first discusses the special importance of high latitudes in general, and Alaska in particular, to the ecology and management of running waters. In the second part, I propose a model (framework of hypoheses), tracing the interacting effects of high latitude climate on running waters.

Figure 13.1.
Extensive ice cover
on the Nenana River,
near Denali National
Park, Alaska. For
scale, note highway in
upper right of photo.

a

Figure 13.2. Streams and environs on Prince of Wales Island, southeast Alaska. Note large trees and dense understory in old-growth forest (a), well-developed riparian vegetation (b), and woody debris in stream (c).

a

b

Figure 13.3. A stream in taiga forest of interior Alaska (Monument Creek, at Chena Hot Springs, near Fairbanks).

334

Figure 13.4. A beaded stream in arctic tundra (Imnavait Creek, near Toolik Lake).

The Special Importance of Alaskan Fresh Waters

Study of Systems at Climatic Extremes Broadens the Knowledge Base and Expands the Domain of Theories in Running Water Ecology

Minshall (1988) traces the history of research and ideas in stream ecology, from beginnings in descriptive ecology to later emphasis on quantification and experimentation. Minshall provides an extensive list of exemplary references to illustrate the major stages in the intellectual "succession" of stream ecology. Even a cursory review of these references reveals the almost complete dominance of stream ecology (both empirical studies and coevolving theory) by research in temperate regions, especially North America and northern Europe (Harper, 1981). Contributions to stream ecology from high latitude environments have been modest for several reasons: the very low density of roads (and hence costly access to remote sites via fixed-wing aircraft or helicopters); the small number of stream ecologists working in high latitude regions; and, the logistical difficulties of conducting research in winter, the dominant season at high latitudes.

Facts ("confirmable records of phenomena") are the empirical basis for theory and, in turn, facts take their meaning from the theory in which they are embedded (Pickett et al., 1994). An example from stream ecology illustrates this point. The River Continuum Concept (RRC) (Vannote et al., 1980) is a model, positing several relationships between a stream and its valley. One aspect of these relationships is a hypothesized correspondence between the relative abundance of different kinds of food materials and the relative abundance of stream organisms specialized to consume each kind of food material. Thus, the relative abundance of leaf litter from riparian vegetation, as a proportion of total food resources, should be positively related to the relative abundance of shredder macroinvertebrates, as a proportion of total macroinvertebrates. Measurements of leaf litter inputs to streams in various biomes (e.g., deserts, deciduous forests, grasslands, taiga) and over the river continuum from headwater streams to large rivers constitute a factual data base and one importance of these facts lies in testing of the hypotheses embedded the RCC. The small number of stream ecologists conducting research at high latitudes and the logistical difficulties of work at remote sites (especially in winter) has led to a surprising lack at high latitudes of the most fundamental commodity of science—facts.

Theories have an explicit or implicit domain (or scope). The components of domain are: (1) the concepts or phenomena included; (2) the level of organization (e.g., population or ecosystem); and (3) the spatial and temporal boundaries (Pickett et al., 1994). Using the RCC as an example again, the RCC was developed from facts and concepts indigenous to mid-latitude river systems. Work subsequent to the RCC has linked the RCC to other models (e.g., serial discontinuity; Ward and Stanford, 1983) and an expansion of the domain of the RCC to other biomes has begun (Cushing et al., 1995).

Expanding the domain of theories or models in running water ecology requires the gathering of basic facts from the latitudinal extremes of polar and tropical regions. For example, estimates of secondary production provide a composite measure of the ecological "success" of a consumer population or functional group and are essential to understanding the contributions of various consumers to ecosystem energy flows (Benke, 1993). Consequently, the number of estimates of secondary production of invertebrates in running waters has greatly increased over the past two decades, but the striking feature of these data is the shortage of estimates at the latitudinal extremes of very cold and very warm running waters (Benke, 1993).

High Latitude Landscapes May Serve as Sensitive Indicators of Climate Change

Although there is great uncertainty in predicting climate changes resulting from increasing carbon dioxide concentrations in the atmosphere, climate warming may be amplified at high latitudes, with concordant changes in the amount and seasonal timing of precipitation (Manabe and Wetherald 1980; Mitchell 1989). The consequences for high latitude landscapes may be profound, with complex interactions among permafrost depth and distribution, growing season, and fire frequency (as well as temperature and moisture regimes) producing changes in the distribution and productivity of plants and consumers (McBeath et al., 1984). Changes in climate might drive several kinds of systemic changes to high latitude running water systems (Oswood et al., 1992a). For example, changes in the mass balance of glaciers could cause glacial advance or recession (daming valleys or creating new streams respectively) and influence the contribution of glacial runoff to Alaskan running waters (currently 35%) (Mayo, 1986). Climate changes at high latitudes might affect watershed processes in two ways: (1) changes in the balance of production and decomposition and, hence, the storage of carbon in soils, aquatic sediments, and biota, and; (2) changes in depth and distribution of permafrost. Soils at high latitudes (boreal forest and tundra) contain an estimated 13% of soil carbon stores of the earth (Post et al., 1982). Soils are cold and often waterlogged (especially in regions of permafrost), conditions that slow decomposition and favor storage of carbon. Warming of soils would likely cause interacting changes in the factors controlling storage of soil carbon and controlling soil hydrology. For example, drainage of wet organic soils will likely lead to oxidation of soil carbon (Armentano, 1980) and possibly changes in delivery of dissolved organic carbon to surface waters (Oswood et al., 1992a). Melting of permafrost (Goodwin et al., 1984; Ostercamp, 1984) would be a likely response to climate warming with subsequent changes in hydrology and watershed biogeochemistry (Oswood et al., 1992a).

Organisms may serve as excellent integrators of the complex physical effects of climate change at high latitudes because many organisms are near adaptational limits. As a result, small changes in climate may produce disproportionate biotic changes. Danks (1992) suggests several means of using arctic insects as indicators of environmental changes. Over the latitudinal gradient from temperate regions to the high arctic, terrestrial and aquatic insects show well-described changes in taxonomic composition (Danks 1992a; McLean 1975; Oswood 1989). For example, at the ordinal level, Diptera constitute an increasing proportion of the insect fauna with increasing latitude. Similar gradients in faunal composition with latitude occur at lower taxonomic levels (families within orders; genera within families; Danks, 1992; Oswood, 1989). Long-term monitoring of the composition of the insect fauna at representative sites might reveal climate changes as shifts in faunal composition (e.g., climate amelioration might cause a decrease in the proportion of Diptera in the insect fauna). Similarly, many insect taxa reach their northern limits of distribution along the transect from temperate to arctic latitudes and so biogeographic changes (e.g., northward extension of ranges) might indicate climate changes. Finally, generation times of many taxa of aquatic insects are longer at high latitudes than at temperate latitudes (Danks, 1992b), likely because growth rates are limited by low food availability and cold water temperatures.

Although changes in life histories (e.g., a shift from semivoltine to univoltine life cycle) might nicely integrate watershed changes in primary production and thermal regime, monitoring life histories would require greater sampling effort than monitoring community composition or geographic ranges.

Alaska Is a Warehouse of Pristine Running Water Systems

The same factors that have limited research on Alaska's fresh water systems (poorly developed road system, low population density, long and severe winters) have also prevented most large scale resource development and consequent impacts on fresh waters. Much of Alaska's economy is based on extraction and export of natural resources, such as timber, minerals, and oil. Timber harvest in the productive coastal forests of southeast Alaska affects resident and anadromous fishes (reviewed in Murphy and Milner, 1996, Chapter 9). Mining, for example, placer mining for gold (LaPerriere and Reynolds, 1996, Chapter 10), has impacted running waters by increasing sediment loads and concentrations of metals. Although sediments and metals may have severe effects on stream biota and degrade water quality, streams affected by mining are a very small proportion of Alaska's running waters. Export of oil is the engine of Alaska's economy. Construction and operation of the trans-Alaska pipeline, connecting the oilfields on the North Slope with shipping facilities at Valdez on Prince William Sound, has occurred with remarkably little impact on terrestrial or freshwater ecosystems (Alexander and VanCleve, 1983). However, oil development is not always benign; the Exxon Valdez oil spill catastrophically impacted coastal marine ecosystems in Alaska.

Low population densities have also prevented extensive effects of urbanization on Alaskan rivers and streams. Anchorage is by far the most populous urban area of Alaska. Streams originate on the slopes of the Chugach mountains surrounding Anchorage and flow into Cook Inlet, passing through Anchorage. Some streams show no discernible impairment but others show downstream degradation (as measured by changes in macroinvertebrate communities) from both nonpoint and point sources of pollutants (Milner et al., unpublished data).

In spite of some impacts from resource extraction and urbanization, Alaska is a museum of free-flowing, unregulated rivers. Alaska has enormous potential for hydropower development. In 1980, 695 potential hydropower sites were identified (with 33 million KW capacity), but only 40 hydrolelectric plants were in operation (mostly small) (Whitehead, 1983). The Yukon River is the largest river system in the northern latitudes (North America, northern Europe, former Soviet Union) unimpacted by impoundment or water diversion and no Alaskan rivers ($>350\,m^3sec^{-1}$ discharge) are considered impacted (Dynesius and Nilsson, 1994). In the contiguous 48 states of the United States, there are only 42 free-flowing rivers greater than

200 km in length, and only one, the Yellowstone River in Montana, is greater than 1000 km (Benke, 1990). Alaska's rivers are analogous to reserves of remnant areas of native terrestrial landscapes and provide increasingly rare opportunities to study river corridors uninterrupted over the whole of the river continuum, from headwaters to mouth.

A Model of Key Environmental Factors and Characteristics of Consumers in Alaskan Running Waters

Starfield and Bleloch (1986) discuss the various uses of models in science, using Holling's 1978 classification of modeling problems. In brief, Holling suggests classes of models based on the quality/quantity of data available and on our understanding of the structure of the problem. As discussed above, the small number of stream ecologists working in high latitudes and the logistical difficulties of work at remote sites (especially in winter) means that we have relatively poor data for constructing models of Alaskan running waters. However, the rich literature on running water ecology in general can supply ideas and hypotheses applicable, often with modification, to high latitude running waters. Therefore, our understanding of the problem is solid. As Starfield and Bleloch (1986) point out, such models are bound to be speculative, but allow us to ". . . explore the consequences of what we believe to be true . . ." by ". . . living with our models, by exercising them, manipulating them, questioning their relevance, and comparing their behavior with what we know (or think we know) about the real world" and that even these initially simple and flawed models are preferable to blind collection of data, because of the synergy of interplay between data collection and model development.

Figure 13.5 is just such a model, generally based upon sparse data but plausible ideas. The model examines the consequences of the annual light regime at high latitudes: low annual solar radiation and seasonally eccentric solar radiation. Cold permeates the ecology of Alaskan fresh waters, especially at high latitudes. The model suggests that the direct consequences of high latitude climate on running waters can be grouped into three broad categories: creation of ice in both aquatic and terrestrial systems, limited inputs and retention of inorganic nutrients and carbon, and thermal effects on rates of biological processes of aquatic organisms. Finally, the model posits interactions among these direct effects that codetermine some characteristic features of consumers in Alaskan running waters: low biotic diversity and low growth and production.

Other authors have similarly constructed frameworks of hypotheses to promote reciprocal development of theory and data gathering. For example, Glazier (1991) examined (mainly via literature review) the fauna of North American freshwater springs and noted the prevalence of noninsect taxa in hard-water limestone springs, in contrast to insect dominance in

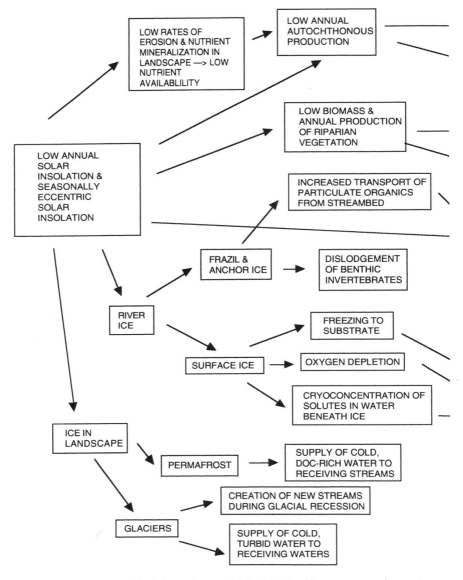

Figure 13.5. A model of the effects of high latitude climate on running waters.

most temperate running waters. Glazier constructed a hypothetical framework, showing how the observed characteristics of cold, hard-water springs
(flow and thermal constancy, small isolated habitats, paucity of large predators) might interact to promote high population densities and selection for
organisms with nonemergent lifestyles, conditions that could account for

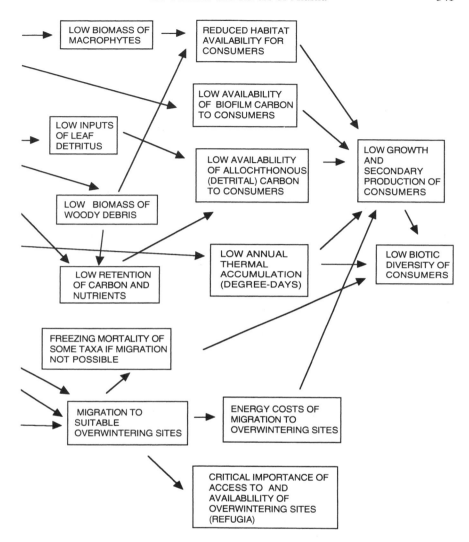

Figure 13.5. *Continued*

the relative dominance of noninsect macroinvertebrates. Glazier's framework, and the conceptually similar framework of Alaskan running waters (Fig. 13.5), provide families of nested hypotheses. I hope that the model will sufficiently irritate workers who recognize omissions to add new components and will encourage systematic collection of data, disproving some tenets of the model but revealing new patterns. The highest goal for all such models is mutation beyond recognition. The remainder of this chapter is a brief exposition of this model of Alaskan running waters.

Low Annual Solar Insolation and Seasonally Eccentric Solar Insolation: A Landscape Dominated by Cold

The climate of Alaska is driven by the low annual solar insolation and seasonally eccentric solar insolation of high latitudes. In consequence, air temperatures are low compared to temperate regions (see Chapter 1), with average annual air temperatures ranging from $+2°C–+4°C$ in the maritime region of southeast Alaska to $-2°C--4°C$ in interior Alaska to $-10°C--12°$ on the Arctic Slope (Hultén, 1968). The effects of cold on river and stream ecosystems ranges from the creation of ice in aquatic habitats and terrestrial landscapes to limitations on inputs of inorganic nutrients and carbon to running water food webs to thermal limitations on rates of biological processes. These three topics are taken up in turn below.

Effects of Cold: Creation of Ice in Both Aquatic and Terrestrial Systems

River ice has been recently reviewed in Prowse and Gridley (1993) which, along with Ashton (1979), (Prowse, 1994), (Oswood et al., 1991) and (Scrimgeour et al., 1994), provides an overview of the physical and biological effects of ice in running waters. The notes below are largely based on these summaries.

Frazil ice typically forms in turbulent flows when water temperatures drop slightly below 0°C and the water becomes supercooled. Frazil ice particles grow into small discoids which agglomerate into larger slushy accumulations known as frazil flocs (Fig. 13.6). In slowly flowing water, frazil ice floats up to the surface, contributing to the permanent ice cover for the winter season. In more rapid flows, turbulence may be sufficient to allow continued generation of frazil ice.

Frazil ice accumulations can build up on any objects within the stream, such as boulders, woody debris, and vegetation, to form deposits of anchor ice. Formation of anchor ice over large areas of streambed can raise water height and seal the streambed (and benthic organisms) from surface flow (Beltaos et al., 1993). Anchor ice may also be sufficiently buoyant to float to the surface, carrying with it adhering substrate materials.

The biological effects of frazil and anchor ice have been little studied. Frazil ice is near 0°C and so poses no threat to lotic organisms in terms of freezing mortality. However, a river full of frazil ice is about the consistency of a Margarita (Fig. 13.6) and frazil ice formation has the potential to be the lotic equivalent of an industrial-grade enema, scouring surfaces and moving organisms and stored organic matter downstream. I have not located any studies of the effects of frazil or anchor ice on periphyton or biofilm. Winter loss of biofilm due to underwater ice (or the hydraulic effects of ice dams or spring break up) could remove senescent or moribund algae, and thereby

Figure 13.6. Frazil ice in Nenana River, near Denali Park, Alaska.

contribute to very high P/B ratios for benthic primary producers in summer. The few studies investigating the effects of frazil and anchor ice on benthic invertebrates have been equivocal (Oswood et al., 1991; Power et al., 1993), with only limited indirect evidence of detrimental effects (fewer organisms in areas subject to repeated anchor ice formation). The potential effects of underwater ice on stream fishes seem to be largely hydraulic. Fishes can be stranded in temporary channels following cyclic formation (night) and loss (day) of frazil and anchor ice dams. Anchor ice also may seal the substrate, changing the flow of water through streambed materials and so possibly affect developing eggs, fry or overwintering juveniles.

Frazil and anchor ice events are episodic and occur under conditions very poor for direct observation or experimental study. Artificial streams or natural streams small enough to be manipulated (e.g., by maintaining water temperatures at 0°C, preventing formation of frazil or anchor ice) may offer a means of assessing the impact of underwater ice on organisms and upon transport of streambed organic matter.

Like an unwelcome guest, surface ice at high latitudes arrives early and stays late. The consequences of surface ice must be one of the key factors determining organismal survival in small streams. The deep winter ice cover, with surface ice reaching the stream bottom over long reaches, forces some organisms, especially fish, to migrate to suitable habitat (Craig, 1989). Chemical conditions in the water beneath deep ice cover may be very inhospitable to biota, because of exclusion of dissolved material from the ice (increasing solute concentrations of the water) and because of low oxygen concentrations (Craig, 1989; Oswood et al., 1991; Prowse, 1994).

In winter, thickening ice cover contacts the stream bed at the margins of streams, allowing freezing to substrate. Benthic habitats of small streams at high latitudes may be completely frozen. Freshwater organisms facing freezing have two choices: move to suitable habitat with free-flowing water or remain in place, to be surrounded by frozen substrate. Many invertebrates can survive in frozen habitats by either avoiding freezing by supercooling or by tolerating freezing of tissues (often associated with production of cryoprotectant substances; Lee, 1991). Fishes cannot tolerate freezing of tissues and so must seasonally migrate to suitable overwintering habitat (Reynolds, Chapter 11). For fishes in arctic Alaska, spring-fed streams (Fig. 13.7), deep pools of large rivers, and deeper lakes connected to rivers are overwintering refugia in inland waters (Craig, 1989). The scarcity and patchiness of overwintering refugia are likely important in shaping not only life histories and migratory behavior of arctic fishes, but are likely also important in limiting production of arctic fishes because of the energy costs incurred in migration to and from winter refugia and winter mortality of fishes failing to find an overwintering site (Craig, 1989).

One might think that freezing to the substrate would be such a routine event in the lives of high latitude stream invertebrates that all taxa would be freezing tolerant or be able to avoid freezing via supercooling. However,

Figure 13.7. Spring and spring-fed stream, North Slope, Alaska (photo courtesy of David F. Murray, University of Alaska Museum, Fairbanks).

the limited evidence from stream invertebrates of some Alaskan subarctic streams suggests that, with few exceptions, stream organisms are susceptible to freezing, show little supercooling capability, and that nearly all stream invertebrates actively move away from the ice (Irons et al., 1993; Oswood et al., 1991). However, at least some individuals or taxa of two families of Diptera, Chironomidae and Empididae, show substantial survival upon thawing of frozen streambed substrates and also show substantial supercooling capabilities (as low as $-26°C$ for Empididae) (Irons et al., 1993; Oswood et al., 1991). These physiological adaptations to freezing might contribute to the dominance of Diptera in the benthos of Alaskan running waters.

The two major landscape consequences of cold climate that effect running waters are permafrost and glaciers. The contribution of glaciers to runoff in Alaska is substantial (35%) (Mayo, 1986). Even modest contributions of glacial runoff to streamflow markedly affect the channel dynamics and water temperature of streams and rivers (Milner and Petts, 1994). Glacially influenced running waters generally show peak flow in mid-summer, when higher areas of glaciers contribute to glacial melt. Glacial meltwater is, of course, cold, and contains high concentrations of suspended sediments ("glacial flour"). Presumably, light limitations from high turbidity, cold water temperatures, and the short growing season combine to severely limit annual primary production, although there appear to be no estimates of autochthonous production in streams or rivers dominated by glacial meltwater. Similarly, inputs of leaf litter from riparian vegetation would likely be limited in areas of unstable substrates close to the glacial margin. Finally, benthic invertebrate communities appear to develop a characteristic longitudinal progression, with chironomids of the genus *Diamesa* dominating near the glacial margin and increasing taxonomic richness downstream as water temperature and channel stability increase (Milner and Petts, 1994).

Permafrost is soil or other matter that has remained frozen for two or more years. In Alaska, permafrost is continuous (underlies the entire region except beneath large water bodies) north of the Books Range. Discontinuous permafrost (mostly on north-facing slopes and in poorly drained valleys) is largely found south of the Brooks Range and north of the Alaska Range. South of the Alaska Range, areas of maritime influence are generally permafrost free. Permafrost has several consequences for freshwater systems (see Milner et al., Chapter 1). Local melting of permafrost creates thermokarst lakes and beaded streams. Streams draining landscapes dominated by permafrost differ biogeochemically and thermally from landscapes without permafrost. Permafrost is largely impervious to percolation of water and so summer precipitation falling on soils underlain by permafrost moves quickly through the unfrozen active layer and flows over permafrost to the stream, short-circuiting the groundwater portion of the hydrological cycle. The consequences for stream organisms are a "flashy" hydrograph in response to storms, very cold water temperatures (Irons and Oswood, 1992), and high concentrations of dissolved organic carbon (R. MacLean and J.G. Irons III, University of Alaska, Fairbanks, unpublished data).

Effects of Cold: Limiting Inputs and Retention of Inorganic Nutrients and Carbon

Litter inputs for undisturbed Alaskan streams (Oswood et al., 1995) appear to range from negligible in Arctic tundra streams to near 90 g $AFDW m^{-2} yr^{-1}$ in a heavily canopied (old-growth) coastal forest stream in Southeast Alaska. Two headwater streams in the taiga forest of Interior Alaska, Monument Creek and Little Poker Creek, had inputs of 63 and 37 $AFDW m^{-2} yr^{-1}$ respectively. For comparison, Webster et al. (1995), summarized extensive data on litterfall for upland and for flood–plain/wetland streams in the eastern United States. Leaf inputs to upland streams averaged 345 g $DW \cdot m^{-2} \cdot yr^{-1}$ (range 202–538 g $DW m^{-2} yr^{-1}$) while inputs to floodplain/wetland streams averaged 588 g $DW m^{-2} yr^{-1}$ (range 491–839 g $DW m^{-2} yr^{-1}$). For streams of western United States (Fisher, 1995), inputs ranged from <6 to 350 g $C m^{-2} yr^{-1}$, approximately <12–700 g $DW m^{-2} yr^{-1}$ (median = 330 g $DW m^{-2} yr^{-1}$, excluding one value for wood input only).

Thus, based upon scanty information for Alaska, it appears that allochthonous inputs of riparian leaf litter are low in Alaskan waters compared to temperate regions of the United States, with Alaskan streams ranging from <1/3 of average inputs to temperate streams (in old-growth forest of southeast Alaska) to negligible inputs (in arctic streams). Meentmeyer et al. (1982) analyzed world-wide patterns in production of plant litter (both total litter and leaf litter) in relation to various predictive variables. Annual leaf litterfall was highly correlated ($r = 0.89$) with annual actual evapotranspiration and was nearly as highly correlated (negatively; $r = 0.83$) with latitude. Both actual evapotranspiration and latitude apparently serve as a useful composite variables amalgamating complex interactions of energy and moisture in controlling terrestrial plant productivity. We therefore expect decreasing production of riparian plant communities resulting in decreasing inputs of leaf litter and woody debris to running waters, over the temperate to tundra latitudinal gradient.

Regional syntheses of stream ecology in Cushing et al. (1995) provide for comparison of primary production in streams in the western United States, and streams in the eastern United States to streams in Alaska. As was the case for estimates of allochthonous inputs (discussed above), there are relatively few data for Alaskan streams. In-stream primary production (GPP) during the (ice-free) growing season in Alaska ranges from approximately 0.25 to 0.45 g $C m^{-2} d^{-1}$ (Oswood et al., 1995), which is similar to values reported for forested streams in the western U.S., but substantially lower than for streams in arid regions, which range very approximately from about 1.5 to 7.5 g $C m^{-2} d^{-1}$ (Fisher 1995), about an order of magnitude greater than Alaskan streams. Webster et al. (1995) analyzed data on primary production in streams in the eastern U.S. For studies that estimated primary production by open-system dissolved oxygen change ($n = 28$), NPP averaged $0.76 \, g \, C m^{-2} d^{-1}$, $\approx 1.37 \, g \, C m^{-2} d^{-1}$ GPP, a value about 3 to 6 times greater than the range of values for undisturbed Alaskan streams.

Therefore, although surprisingly few data are available for Alaskan systems, carbon inputs at the base of food webs, namely primary production by stream primary producers and inputs of leaf litter by riparian vegetation, appear to be low compared to temperate regions. However, availability of particulate carbon to consumers (and dissolved nutrients to biofilm organisms) is controlled not only by inputs but also by ecosystem retention of carbon and nutrients. Woody debris and macrophytes act as a retention structures for carbon and nutrients (e.g., Aumen et al., 1990; Bilby, 1981). I could locate no systematic surveys of biomass of woody debris in Alaskan streams and rivers—yet another sorely needed component of a general synoptic database for Alaskan running waters. However, abundance of woody debris is negligible in tundra streams and appears to be sparse in most taiga streams (see Figs. 13.2–13.4), reflecting latitudinal trends in production of plant detritus (Meenteyer et al., 1982). Low abundance of woody debris, coupled with hydrologic regimes often characterized by ice-rich flows at winter freeze-up and at spring breakup or by spates from summer rainstorms, might well combine to produce low ecosystem retention of carbon and nutrients.

Effects of Cold: Thermal Effects on Biological Processes

At high latitudes, low annual solar radiation and consequent low air temperatures not only promote extensive formation of ice (discussed above) but also result in several aspects of the annual water temperature regime that are likely to codetermine organismal adaptations and ecosystem productivity. Poikilothermic freshwater organisms (from microbes to fish) might respond to many aspects of the thermal regime, over time scales ranging from hours to years. The thermal ecology of aquatic insects has received considerable attention. Growth, maturation, emergence (as adults), voltinism, and relative allocation of resources to bioenergetic pathways are codetermined by one or more aspects of the thermal regime, including thermal sum (degree-days), maximum and minimum temperature, and the duration of periods of suitable temperature (Sweeney, 1984; Vannote and Sweeney, 1980; Ward and Stanford, 1982).

The thermal sum (degree-days above 0°C) is a useful amalgamative measure of the thermal regime of a given location and serves as a useful predictor of life history characteristics for at least some taxa of aquatic insects. As was the case for other kinds of data, there are relatively few data on annual degree-days for Alaskan running waters but these can serve as a basis for a rough comparison to lower latitudes. As would be expected, annual degree-days for Alaskan streams loosely follow a latitudinal gradient (modified by maritime climatic influence) as follows (Oswood et al., 1995): 2300 degree-days in a southeast Alaska stream, 1780 in southcentral Alaska stream, 950 and 409 for two streams in Interior Alaska, and 1403 for a tundra stream on the North Slope. Vannote and Sweeney (1980) showed a strong correlation between latitude and degree-days for streams in the eastern U.S. and in the central U.S. For comparison to Alaskan running waters, degree-days ranged from about 7000 degree-days at 32° latitude (Georgia, Alabama) to about 3000 degree-days at 46° (New York, Michigan), compared to a range of approximately 400 to 2300 degree-days for Alaska. Clearly, accumulation of thermal resources (i.e., degree-days) is low in Alaskan running waters. Some likely consequences of this thermal impoverishment to secondary production and to biotic diversity are discussed below.

How Key Environmental Factors Might Produce Characteristic Features of Consumers in High Latitude Running Waters

Growth and Production. There are very few estimates of animal production in Alaskan (or other northern high latitude) running waters. Hershey et al. (1996, Chapter 4) have estimated secondary production of the major taxa of benthic macroinvertebrates in the Kuparuk River, a tundra stream on Alaska's North Slope. Estimates of total production over the ice-free season ranged from 5.72 to $14.38 \, \mathrm{g \, DW \, m^{-2} \, yr^{-1}}$ with black flies (Diptera: Simuliidae) dominating secondary production. Production of benthic macroinvertebrates in Bridge Creek, Amchitka Island (Aleutian Islands) was estimated at $32.6 \, \mathrm{g \, DW \, m^{-2} \, yr^{-1}}$ for insects (dominated by chironomids) and at $9.1 \, \mathrm{g \, DW \, m^{-2} \, yr^{-1}}$ for non-insect invertebrates, for a total of $41.7 \, \mathrm{g \, DW \, m^{-2} \, yr^{-1}}$ (Neuhold, 1971; Valdez et al., 1977). Bridge Creek receives substantial subsidies of marine-derived nutrients from spawning of anadromous Dolly Varden charr and pink salmon. In southeast Alaska, Duncan et al. (1989) estimated production of benthic macroinvertebrates by multiplying mean biomass (June–October) by three P/B ratios (5, 10, and 20), yielding estimates from 0.68 to $2.74 \, \mathrm{g \, AFDW \, m^{-2} \, yr^{-1}}$.

As with comparisons of allochthonous inputs and primary production above, Webster et al. (1995) and Fisher (1995) provide summaries of secondary production estimates for eastern and western U.S. streams respectively. Estimates of whole community secondary production in western streams range from 13.2 to $23 \, \mathrm{g \, DW \, m^{-2} \, yr^{-1}}$, but with much higher production $(135 \, \mathrm{g \, DW \, m^{-2} \, yr^{-1}})$ for a desert stream in Arizona (Fisher, 1995). Estimates from eastern streams (Webster et al., 1995) average $9.03 \, \mathrm{g \, AFDW \, m^{-2} \, yr^{-1}}$ (range 1.77–$30.41 \, \mathrm{g \, AFDW \, m^{-2} \, yr^{-1}}$).

Data for growth and production of running water fishes in Alaska are likewise sparse. Oswood et al. (1992b) compare growth rates of four species of fishes in the Chena River (Interior Alaska, near Fairbanks) with the same species in temperate regions. Growth rates of these fishes in the Chena River are lower than growth rates of fishes in the contiguous United States. For example, Arctic grayling (*Thymallus arcticus*) in the Chena River take 5 to 7 years to reach the same length obtained in only 2 to 3 years in Montana streams. Sonnichsen (1981) estimated the production of slimy sculpin (*Cottus cognatus*) at $0.85\,g\,m^{-2}\,yr^{-1}$ (possibly an overestimate because of sampling bias) compared to $5.94\,g\,m^{-2}\,yr^{-1}$ for sculpin in a Minnesota stream (Petrosky and Waters, 1975).

Growth, and possibly secondary production, of benthic invertebrates and fishes in Alaskan running waters are apparently low compared to running waters in temperate regions, although the estimates of secondary production of benthic macroinvertebrates in an Alaskan tundra stream (Hershey et al., Chapter 4) and in an Amchitka island stream (Neuhold, 1971; Valdez et al., 1977) are well within the range for values from streams in temperate regions. Butler's (1982) 7 year life cycle for *Chironomous* in an arctic tundra pond provided the exemplar of slow growth and an extended life cycle in high latitude freshwater insects. However, secondary production of this population is comparable to production of *Chironomous* larvae elsewhere, including some temperate locations, because seven overlapping cohorts maintain high biomass (Butler, 1982). The warning is that slow growth rates and long life cycles of consumers do not necessarily give rise to low production. However, there are so few estimates of secondary production for Alaskan running waters that even this very tentative discussion seems premature. Secondary production integrates most of the components (e.g., density, growth, survivorship) of success in a population or set of populations (e.g., functional group) and is therefore an ideal response variable for many of the questions of stream ecology (Benke, 1993). The paucity of research on high latitude streams in general guarantees that "more research is needed," at every level from the autecology of individual organisms to regional biogecochemistry. However, constraints of funding and time make it necessary to distinguish the essential from the merely interesting. Determinations of secondary production across the range of stream types in Alaska (from coastal rainforest to tundra; from clearwater to brown water to glacier-fed streams; over the river continuum from headwater streams to large rivers) would provide a means of answering some fundamental questions in basic and applied ecology.

Low Biotic Diversity. The taxonomic composition of benthic macroinvertebrates in Alaskan running waters has been reviewed in Oswood (1989) and in Oswood et al. (1995), with some additional information in Bruns et al. (1992) for northwest Alaska. In brief, Diptera dominate the lotic macroinvertebrates, similar to the situation for terrestrial insects (Danks, 1981; McLean, 1975). For example, in the Kuparuk River (Hershey et al., 1996) there are four species of Ephemeroptera, two of Plecoptera, and two of Trichoptera but at least 39 dipteran species, including 30 species of Chironomidae. Compared to streams in north temperate regions, Diptera and Plecoptera are relatively (% composition based upon numerical abundance) more abundant, Ephemeroptera and Trichoptera are less abundant, and some insect orders are apparently absent (or very rare) in Alaskan running waters, such as Megaloptera, Neuroptera, Odonata, and Coleoptera. There are similar patterns at lower taxonomic levels. For example, over the North American latitudinal gradient from subtropical to arctic, the trichopteran suborders Hydropsychoidea and Rhyacophiloidea decrease in relative abundance and the Limnephiloidea increase (Irons, 1985). At the family level, the Nemouridae, Capniidae, and Chloroperlidae dominate the Plecoptera while other families are rare (e.g., Pteronarcyidae, Perlidae, Peltoperlidae). Clearly, at all taxonomic levels, there is a selective winnowing of taxa with increasing latitude, culminating in a fauna dominated by cold-adapted Diptera. The physical conditions of high latitude running waters are loosely analogous to an electrophoresis gel, with different kinds of molecules (taxa) progressing varying distances along the gel (latitudinal gradient). This analogy raises two questions: What properties (environmental factors) of the latitudinal gel retard northward movements of taxa? What biological properties of various taxa determine their relative mobility in the latitudinal gel?

Growth, Production, and Biotic Diversity: Energy as a Unifying Factor. I have argued above that a syndrome of conditions at high latitudes produces characteristic limitations of consumers at the level of the individual (growth), the population (production), and the community (taxonomic diversity). Investigation of these assertions would traditionally fall within the domain of separate subdisciplines of ecology—the physiological ecologist, the population ecologist, and the community ecologist respectively. Finding the "One Ring to Rule Them All" (J.R.R. Tolkien) has been difficult because the separate subdisciplines traditionally measure different properties of biological systems. For population and evolutionary ecologists, the variable of interest is often some variation of change in population size (dN/dt) while for physiological ecologists (and for community ecologists studying trophic ecology) rates of energy exchange (dE/dt) are measured (Brown, 1995a). Brown (1995a; 1995b) suggests that dN/dt and dE/dt are fundamentally the same and that ". . . both survival and reproduction are thermodynamic phenomena." Hall et al. (1992) make essentially the same argument, that species' responses to the N-dimensional hyperspace of environmental and resource gradients (i.e., the Hutchinsonian niche) can be interpreted (and assessed) as energy costs and gains.

The model of high latitude running waters presented above (Fig. 13.5) is based upon this view of energy as an integrating ecological currency. I suggest that growth (of individuals) and production (of populations) is fundamentally controlled by availability of food resources (chemical energy) and is constrained by the energetic costs of adaptation to the high latitude thermal environment. This might, at first glance, seem like a unpardonable oversimplification of complex ecosystem dyanmics. However, I am only suggesting (following Hall et al., 1992) that the many complexly interacting processes of ecosystems (e.g., Fig. 13.5) ultimately manifest themselves to consumers in bioenergetic terms. Joint control of growth and production of stream consumers by food and temperature is not a new idea. For example, Sweeney and colleagues (Sweeney, 1986a; Sweeney, 1986b; Sweeney and Vannote, 1984; Sweeney and Vannote, 1986) used laboratory experiments and manipulations of litter inputs at field sites to examine the interaction of temperature and food type (quality) on growth and production of a nemourid stonefly and of baetid and leptophlebiid mayflies.

Bioenergetics also offers a tool for investigating biogeographic questions in high latitude running waters. Vannote and Sweeney (1980) proposed a model, linking geographic ranges of aquatic insects to the thermal regime of the habitat. Water temperature affects the bioenergetics of larval insects in several important ways: rate of assimilation of food, relative allocation of assimilated food energy to growth vs. respiration, and relative development of adult and larval tissues. These variables control adult body size and the proportion of adult mass allocated to eggs such that organisms in an optimum thermal regime have greater egg production (and hence population sizes) compared to organisms in suboptimal (either too warm or too cool) temperature regimes. Vannote and Sweeney's model (Vannote and Sweeney, 1980) suggests that organisms at the "boreal" edge of their distribution (i.e., at suboptimally cool water temperatures) will show (compared to the optimum temperature regime): reduced assimilation rates of food; slower maturation of adult tissues, thus prolonging the duration of the larval stage; and, reduced adult size and incomplete conversion of stored materials into eggs. Food availability (allochthonous inputs and autochthonous production) very likely decreases over the temperate to polar latitudinal gradient and so both food availability and rate of assimilation of available food may decrease in the colder water temperatures of high latitude environments. Over a latitudinal gradient from temperate regions to the polar tundra, there is a selective winnowing of taxa, culminating in dominance of the benthic macroinvertebrates by Diptera and fishes by salmonids. The northern edge of the distribution of a species is likely the point where the joint limitations of food and temperature produce too few offspring to sustain a population. Hall et al. (1992) make the same argument, conched in economic terms: "Persistence of a population and, ultimately, the species, in a given locality, will occur only where energy return on investment allows a significant energy profit in the form of propagules."

This hypothesis—joint control by food and temperature of growth, production, and geographic distribution of consumers—offers a framework for investigating the ecology of consumers in high-latitude running waters. An integrated descriptive and manipulative approach would be especially useful. Field measurements of consumer growth and production, with associated data on food availability and temperature regime, could be obtained at sites over latitudinal and elevational gradients, and at sites in distinct habitats (e.g., clearwater vs. glacier-fed streams or non-salmon vs. salmon-enriched streams). Temperature regime and food availability could be independently manipulated in stream mesocosms (e.g., leaf litter inputs characteristic of temperate climates with a water temperature regime characteristic of tundra streams), allowing assessment of the interactive effects of temperature and food availability on consumers. An understanding of the fundamental constraints controlling the productivity of Alaska's streams and rivers is essential information for setting reasoned limits for sport, subsistence, and commercial fisheries.

This bioenergetic approach is, of course, only one of many approaches likely to be useful in understanding the ecology of high latitude running waters. For example, winter mortality caused by largely density-independent factors such as severity of streamed freezing or icing events, may be a major determinant of year-to-year changes in population sizes of organisms, ranging from biofilm microbes to fishes (Murphy and Milner, Chapter 9). The rivers and streams of Alaska provide both opportunity and obligation for scientists—opportunity to study one of the earth's last great preserves of pristine lotic systems and an obligation to provide the information necessary for effective stewardship.

Acknowledgments. My thanks to Drs. Alexander Milner, Robert Piorkowski, and Stephen Wagener for comments on the manuscipt. Funding from the National Science Foundation LTER Program (DEB-9211769) has supported much of my recent research on (and thinking about) the ecology of high latitude running waters.

References

Alexander V, VanCleve K (1983) The Alaska Pipeline: a success story. Ann Rev Ecol System 14:443–463.

Armentano TV (1980) Drainage of organic soils as a factor in the world carbon cycle. BioScience 30:825–830.

Ashton GD (1979) River Ice. Am Scientist 67:38–46.

Aumen NG, Hawkins CP, Gregory SV (1990) Influence of woody debris on nutrient retention in catastrophically disturbed streams. Hydrobiologia 190:183–192.

Beltaos S, Calkins DJ, Gatto LW et al. (1993) Physical effects of river ice. In Prowse TD, Gridley NC (Eds) Environmental aspects of river ice. Ministry Supply and Services Canada, Saskatoon, Saskatchewan, Canada, pp 3–74.

Benke AC (1990) A perspective on America's vanishing streams. Journal North Am Benthol Soc 9:77–88.

Benke AC (1993) Concepts and patterns of invertebrate production in running waters. Verh Intern Verein Limnol 25:15–38.

Bilby RE (1981) Role of organic debris dams in regulating the export of dissolved and particulate matter from a forested watershed. Ecology 62:1234–1243.

Brown JH (1995a) Macroecology. University of Chicago Press, Chicago.

Brown JH (1995b) Organisms and species as complex adaptive systems: linking the biology of populations with the physics of ecosystems. In Jones CG, Lawton JH (Eds) Linking species and ecosystems. Chapman and Hall, New York, pp 16–24.

Bruns DA, Wiersma GB, Minshall GW (1992) Problems of long-term monitoring of lotic ecosystems. In Firth P, Fisher SG (Eds) Global climate change and freshwater ecosystems. Springer Verlag, New York, pp 285–307.

Butler MG (1982) Production dynamics of some arctic *Chironomus* larvae. Limnol. Oceanogr. 27:728–736.

Craig PC (1989) An introduction to anadromous fishes in the Alaskan arctic. In Norton DW (Ed) Research advances on anadromous fish in arctic Alaska and Canada: nine papers contributing to an ecological synthesis. Institute of Arctic Biology, Fairbanks, Alaska, pp 27–54.

Cushing CE, Cummins KW, Minshall GW (Eds) (1995) River and stream ecosystems. Elsevier, Amsterdam.

Danks HV (1981) Arctic arthropods: A review of systematics and ecology with particular reference to the North American fauna. Entomological Society of Canada, Ottawa.

Danks HV (1992a) Arctic insects as indicators of environmental change. Arctic 45:159–166.

Danks HV (1992b) Long life cycles in insects. Can Ent 124:167–197.

Duncan WFA, Brusven MA, Bjornn TC (1989) Energy-flow response models for evaluation of altered riparian vegetation in three southeast Alaskan streams. Water Res 23:965–974.

Dynesius M, Nilsson C (1994) Fragmentation and flow regulation of river systems in the northern third of the world. Science 266:753–762.

Fisher SG (1995) Stream ecosytems of the western United States. In Cushing CE, Cummins KW, Minshall GW (Eds) River and stream ecosystems. Elsevier, Amsterdam. pp 117–187.

Glazier DS (1991) The fauna of North American temperate cold springs: patterns and hypotheses. Freshwater Biol 26:527–542.

Goodwin CW, Brown J, Outcalt SI (1984) Potential responses of permafrost to climate warming. In McBeath JH, Juday GP, Weller G, Mayo M (Eds) The potential effects of carbon dioxide-induced climatic changes in Alaska. School of Agriculture and Land Resources Management, University Alaska Fairbanks, Fairbanks, Alaska, pp 106–113.

Hall CAS, Stanford JA, Hauer FR (1992) The distribution and abundance of organisms as a consequence of energy balances along multiple environmental gradients. Oikos 65:377–390.

Harper PP (1981) Ecology of streams at high latitudes. In Lock MA, Williams DD (Eds) Perspectives in running water ecology. Plenum Press, New York, pp 313–337.

Hershey AE, Bowden WB, Deegen LA et al. (1996) The Kuparuk River: a long-term study of biological and chemical processes in an arctic river. In Milner AM, Oswood MW (Eds) Fresh waters of Alaska: Ecological syntheses. Springer-Verlag, New York.

Hultén E (1968) Flora of Alaska and neighboring territories. Stanford University Press, Stanford, California.

Irons JG, III (1985) Life histories and community structure of the caddisflies (Trichoptera) of two Alaskan subarctic streams, MS Thesis, University of Alaska, Fairbanks.

Irons JG III, Miller LK, Oswood MW (1993) Ecological adaptations of aquatic macroinvertebrates to overwintering in interior Alaska (U.S.A.) subarctic streams. Can J Zool 71:98–108.

Irons JG III, Oswood MW (1992) Seasonal temperature patterns in an arctic and two subarctic Alaskan (U.S.A.) headwater streams. Hydrobiologia 237:147–157.

LaPerriere JD, Reynolds JB (1996) Gold placer mining and stream ecosystems of interior Alaska. In Milner AM, Oswood MW (Eds) Fresh waters of Alaska: Ecological syntheses. Springer-Verlag, New York.

Lee RE Jr. (1991) Principles of insect low temperature tolerance. In Lee RE, Jr., Denlinger DL (Eds) Insects at low temperature. Chapman and Hall, New York, pp 17–46.

Manabe S, Wetherald RT (1980) On the distribution of climate change resulting from an increase in CO2 content of the atmosphere. J Atmos Sci 37: 99–118.

Mayo LR (1986) Annual runof rate from glaciers in Alaska: A model using the altitude of glacier mass equilibrium. Cold Regions Hydrology Symposium, American Water Resources Association, Technical Publication Series TPS-86-1:509–517.

McBeath JH, Juday GP, Weller G, Mayo M (Eds) (1984) The potential effects of carbon dioxide-induced climatic changes in Alaska. School of Agriculture and Land Resources Management, University Alaska Fairbanks, Fairbanks.

McLean SFJ (1975) Ecological adaptations of tundra invertebrates. In Verburg FJ (Ed) Physiological adaptation to the environment. Intext Educational Publishers, New York, pp 269–299.

Meenteyer V, Box EO, Thompson R (1982) World patterns and amounts of terrestrial plant litter production. BioScience 32:125–128.

Milner AM, Irons JG III, Oswood MW (1996) The Alaskan landscape: an introduction for limnologists. In Milner AM, Oswood MW (Eds) Fresh waters of Alaska: Ecological syntheses. Springer-Verlag, New York.

Milner AM, Petts GE (1994) Glacial rivers: physical habitat and ecology. Freshwater Biol 32:295–307.

Minshall GW (1988) Stream ecosystem theory: A global perspective. J North Am Benthol Soc 7:263–288.

Mitchell JFB (1989) The "greenhouse" effect and climate change. Rev Geophys 27:115–139.

Murphy ML, Milner AM (1996) Alaska timber harvest and fish habitat. In Milner AM, Oswood MW (Eds) Fresh waters of Alaska: Ecological syntheses. Springer-Verlag, New York.

Neuhold JM (1971) The lowland stream ecosystem of Amchitka Island, Alaska. BioScience 21:683–686.

Ostercamp TE (1984) Potential impact of a warmer climate on permafrost in Alaska. School of Agriculture and Land Resources Management, University of Alaska, Fairbnaks.

Oswood MW (1989) Community structure of benthic invertebrates in interior Alaskan (USA) streams and rivers. Hydrobiologia 172:97–110.

Oswood MW, Irons JG III, Milner AM (1995) River and stream ecosystems of Alaska. In Cushing CE, Cummins KW, Minshall GW (Eds) River and stream ecosystems. Elsevier, Amsterdam.

Oswood MW, Miller LK, Irons JG III (1991) Overwintering of freshwater benthic invertebrates. In Lee REJ, Denlinger DL (Eds) Insects at low temperatures. Chapman and Hall, New York, pp 360–375.

Oswood MW, Milner AM, Irons JG III (1992a) Climate change and Alaskan rivers and streams. In Firth P, Fisher SG (Eds) Global climate change and freshwater ecosystems. Springer-Verlag, New York, pp 192–210.

Oswood MW, Reynolds JB, LaPerriere JD et al. (1992b) Water quality and ecology of the Chena River, Alaska. In Becker CD, Neitzel DA (Eds) Water quality in North American river systems. Battelle Press, Columbus, OH, pp 5–27.

Petrosky CE, Waters TF (1975) Annual production by the slimy sculpin population in a small Minnesota watershed. Trans Am Fish Soc 104:237–244.

Pickett STA, Kolasa J, Jone CG (1994) Ecological understanding. Academic Press, San Diego, CA.

Post WM, Emmanuel WR, Zinke PJ, Stangenberger AG (1982) Soil carbon pools and world life zones. Nature 298:156–159.

Power G, Cunjak R, Flannagan J, Katopodis C (1993) Biological effects of river ice. In Prowse TD, Gridley NC (Eds) Environmental aspects of river ice. Ministry Supply and Services Canada, Saskatoon, Saskatchewan, Canada, pp 97–119.

Prowse TD (1994) Environmental significance of ice to streamflow in cold regions. Freshwater Biol 32:241–259.

Prowse TD, Gridley NC (Eds) (1993) Environmental aspects of river ice. Minister of Supply and Services Canada, Saskatoon, Saskatchewan, Canada.

Reynolds JB (1996) Ecology of overwintering fishes in Alaskan fresh waters. In Milner AM, Oswood MW (Eds) Fresh waters of Alaska: Ecological syntheses. Springer-Verlag, New York.

Scrimgeour GJ, Prowse TD, Culp JM, Chambers PA (1994) Ecological effects of river ice break-up: a review and perspective. Freshwater Biol 32:261–275.

Sonnichsen SK (1981) Ecology of slimy sculpin (*Cottus cognatus*) in the Chena River, Alaska. M.S. Thesis, University of Alaska, Fairbanks.

Starfield AM, Bleloch AL (1986) Building models for conservation and wildlife management. Macmillan, New York.

Sweeney BW (1984) Factors influencing life-history patterns of aquatic insects. In Resh VH, Rosenberg DM (Eds) The ecology of aquatic insects. Praeger Scientific, New York, pp 56–100.

Sweeney BW (1986a) Effects of temperature and food quality on growth and development of a mayfly, *Leptophlebia intermedia*. Can J Fish Aquat Sci 43:12–18.

Sweeney BW (1986b) The relative importance of temperature and diet to larval development and adult size of the winter stonefly, *Soyedina carolinensis* (Plecoptera: Nemouridae). Freshwater Biol 16:39–48.

Sweeney BW, Vannote RL (1984) Influence of food quality and temperature on life history characteristics of the parthenogenetic mayfly, *Cleon triangulifer*. Freshwater Biol 14:621–630.

Sweeney BW, Vannote RL (1986) Growth and production of a stream stonefly: influences of diet and temperature. Ecology 67:1396–1410.

Valdez RA, Helm WT, Neuhold JM (1977) Aquatic ecology. In Merritt ML, Fuller RG (Eds) The environment of Amchitka Island, Alaska. Technical Information Center, Energy, Research, and Development Administration, Springfield, VA, pp 287–313.

Vannote RL, Minshall GW, Cummins KW et al. (1980) The river continuum concept. Can J Fish Aquat Sci 37:130–137.

Vannote RL, Sweeney BW (1980) Geographic analysis of thermal equilibria: A conceptual model for evaluating the effect of natural and modified thermal regimes on aquatic insect communities. Am Natur 115:667–695.

Ward JV, Stanford JA (1982) Thermal responses in the evolutionary ecology of aquatic insects. Ann Rev Entomol 27:97–117.

Ward JV, Stanford JA (1983) The serial discontinuity concept of lotic ecosystems. In Fontaine TD, III, Bartell SM (Eds) Dynamics of lotic ecoystems. Ann Arbor Science Publishers, Ann Arbor, Michigan, pp 29–42.

Webster JR, Wallace JB, Benfield EF (1995) Organic processes in streams of the eastern United States. In Cushing CE, Cummins KW, Minshall GW (Eds) River and stream ecosytems. Elsevier, Amsterdam, pp 117–187.

Whitehead J (1983) Hydropower in twentieth century Alaska: Anchorage, Juneau, Ketchikan, and Sitka. Bulletin of the Institute of Water Resources, University Alaska Fairbanks, Fairbanks.

Subject Index

A

Algae
  blooms, 40, 132, 136
  blue-green, 95, 145
  epilithic, 79, 95, 100, 115, 144, 246,
    274–275
  periphytic, 186
  and water clarity, 206
Alkalinity, 73, 133
Allochthonous inputs, 77, 112, 138,
    245–246, 325, 336, 345–346
Amino acids, 110–111, 114
Anadromous fish
  abundance, 185, 206, 207–208, 211–
    212, 213, 215, 224, 313–314, 321
  habitat degradation, 23, 40, 232–236,
    239, 240
  harvesting, 3, 40, 47, 198
  migration, 185, 222–223, 288–289
  overwintering strategies, 249–251,
    286, 288–291, 344
  recruitment, 210–212
  salmon carcasses, 19, 23, 179–180,
    183–201, 205

salmonid production, 181–183, 200,
    219–224, 248
  spawning, 23, 40, 200, 236, 239
Arctic coastal tundra. *See* Wetlands
Arsenic, 276
Atomic testing, 50–54
Aufwuchs, 144

B

Bacterial productivity, 87–92
Bering, Vitus, 3
Best Management Practices (BMPs),
    255–256, 277–278
Biodiversity
  and climate change, 327, 337
  in rivers, 349–350
Bioenergetics, 350–352
Biogeography, 351
Biomass, 82–83, 215
Blowdown, 251
Boreal Forest, 13, 230. *See also*
    Wetlands
Brownwater, 22, 38

Taxonomic Index

Algae and Phytoplankton

Vascular Plants

Place Index

Ecological Studies

Ecological Studies

Ecological Studies

DATE DUE

APR 1 1 2001			
			Printed in USA

HIGHSMITH #45230